AF361763

High-Temperature Behavior of Metals

High-Temperature Behavior of Metals

Editors

Elisabetta Gariboldi
Stefano Spigarelli

MDPI • Basel • Beijing • Wuhan • Barcelona • Belgrade • Manchester • Tokyo • Cluj • Tianjin

Editors
Elisabetta Gariboldi
Politecnico di
Milano—Mechanical
Engineering Department
Italy

Stefano Spigarelli
DIISM—Università Politecnica
delle Marche
Italy

Editorial Office
MDPI
St. Alban-Anlage 66
4052 Basel, Switzerland

This is a reprint of articles from the Special Issue published online in the open access journal *Metals* (ISSN 2075-4701) (available at: https://www.mdpi.com/journal/metals/special_issues/high_temperature_behavior_metals).

For citation purposes, cite each article independently as indicated on the article page online and as indicated below:

LastName, A.A.; LastName, B.B.; LastName, C.C. Article Title. *Journal Name* **Year**, *Volume Number*, Page Range.

ISBN 978-3-0365-2199-2 (Hbk)
ISBN 978-3-0365-2200-5 (PDF)

Contents

About the Editors

Professor Elisabetta Gariboldi graduated in Mechanical Engineering in 1990 and obtained her PhD in Metallurgical Engineering in in 1994. She is Full Pofessor of Metallurgy at the Mechanical Engineering Department of Politecnico di Milano. Her main research interest in the high-temperature behaviour of metallic materials has covered a range of topics, from mechanical features and properties to microstructural stability.

Professor Stefano Spigarelli gradutated in Mechanical Engineering in 1990, and obtained his PhD in 1994. He is Full Pofessor of Metallurgy at Università Politecnica delle Marche, Ancona, Italy. His main research interests are the mechanical properties and the deformation response at high temperature of metals and alloys..

Editorial

High-Temperature Behavior of Metals

Elisabetta Gariboldi [1],* and **Stefano Spigarelli** [2],*

1   Department of Mechanical Engineering, Politecnico di Milano, Via La Masa 1, 20156 Milano, Italy
2   Department of Industrial Engineering and Mathematical Sciences, Marche Polytechnic University, Via Brecce Bianche I, 60131 Ancona, Italy
*   Correspondence: elisabetta.gariboldi@polimi.it (E.G.); s.spigarelli@staff.univpm.it (S.S.); Tel.: +39-0223998224 (S.S.)

Citation: Gariboldi, E.; Spigarelli, S. High-Temperature Behavior of Metals. *Metals* **2021**, *11*, 1128. https://doi.org/10.3390/met11071128

Received: 12 July 2021
Accepted: 15 July 2021
Published: 16 July 2021

Publisher's Note: MDPI stays neutral with regard to jurisdictional claims in published maps and institutional affiliations.

1. Introduction and Scope

The design of new alloys as well as the optimization of processes involving whichever form of high-temperature deformation cannot disregard the characterization and/or modelling of the high-temperature structural response of the material.

If this has been quite extensively investigated for conventional hot working processes, there is still a lot to do to accumulate data and models and properly manage innovative deformation processes, including more complex time and temperature combinations, where process-related microstructural changes can severely affect the same processability of the material as well as its final properties (structural or functional).

Similar considerations hold in the case of conventional or innovative metallic materials for which 'high-temperature deformation' occurs during the high-temperature service of the structural components. If extensive scientific literature is available for conventional material and for 'conventional' service conditions reproduced by the constant load creep test condition, there is much to investigate in the field of alloy and processing design. The ultimate task is to widen the temperature and loading ranges of existing or innovative materials, including improving the characterization methods and data for material behavior under complex service conditions (i.e., environmental effects and combination of repeated cycles, stress relaxation, presence of flaws, etc.).

Proper modelling of high-temperature material behavior in all these situations, while considering the need of extending the ranges of applicability of these models, is another important task to be considered in view of optimizing high temperature processing and service of materials. In the latter case, the knowledge of the effects on the initial microstructure as well as the microstructural changes taking place during in-service deformation is of course of paramount importance for the optimization of high-temperature structural alloys.

The present 'High-Temperature Behavior of Metals' Special Issue and book is a collection of contributions presenting the recent advances in the field of high-temperature structural behavior of metallic materials for a wide range of metallic materials. This collection ideally follows the 'Creep and High Temperature Deformation of Metals and Alloys' Special Issue and book [1], as the reader can witness the continuous evolution in the field of alloy optimization, experimental methods, material modelling on the topic of the high temperature behavior of metallic materials.

2. Contributions

Scholars have been invited to submit research papers dealing with innovative research on specific aspects of high-temperature deformation behavior, so that the readers could get at the same time the common basic approach to this topic as well as different experimental methodologies for microstructural and structural high temperature material characterization and establishing of their correlations in view of modelling. Among the submitted manuscripts, 12 papers have been published in the issue. They are here introduced, considering their main aim/s.

2.1. Mmicrostructural Changes Induced by High Temperature Exposure

The strict correlation between microstructure and structural response has always been of paramount importance in the field of high temperature alloys, both for the need to initially optimize the microstructure in view of material performance, and for obtaining a microstructural stability for bulk and surface material. The initial microstructure also plays a role when hot deformation behavior is considered, even if in this case the optimized final microstructure often remains the main target of the process.

A first contribution dealing with microstructure evolution at high temperature was that by Balaško et al. [2]. The tool steel AISI H11 initially soft annealed or hardened and tempered was exposed for up to 100 h in air in a temperature in the range 400–700 °C, in order to investigate the effect of oxide formation by combining experimental analysis and thermodynamic predictions. It was thus noticed that in the highest temperature range (600–700 °C) the oxidation kinetics is slower for the hardened and tempered steel, since the higher amount of dissolved elements in its matrix form a denser inner oxide sublayer, preventing further diffusion of iron and oxygen.

The effects of reheat to 820 or 890 °C of a 9CrMoW steel (92 grade) weld metal produced by GTAW were investigated by Chuang et al [3]. One-minute exposure at the highest reheat temperature was enough to change microstructural features of the steel, which displayed the tendency to early creep failures in creep tests performed at 630 °C.

The paper by Fuyang et al. [4] dealt with a heat resistant casting steel, the HP40Nb grade. The microstructural changes occurring during its exposure at 900 and 950 °C, with or without the concurrent presence of stress, affected the carbides' composition amount and morphology in intra- and interdendritic regions.

The effects of microstructure or microstructural stability, either directly correlated to the manufacturing process, or induced by heat treatment or by prolonged high temperature exposure simulating service, have also been considered in other contributions: for the two Al-SiMg alloys differently produced and heat treated by Gariboldi et al. [5] and Paoletti et al. [6], for the friction stir welded Mg alloy WE54 studied by Álvarez-Leal et al. [7], for the dynamic recrystallization in martensitic steels changes reviewed by Derazkola et al. [8], the Ni-based alloys investigated by Llizzi et al. [9] and Engels [10], and finally the Mo-Si alloys investigated by Krauss et al. [11].

2.2. Hot Deformation

The hot deformation response of metallic materials with given composition is strictly correlated to their initial microstructure as well as to their changes occurring during deformation at high temperature. The high deformation temperature here acts concurrently to the other parameter typically controlled to characterize the high temperature behavior of metals, i.e. strain rate. The analysis of the stress–strain curves obtained at different temperature and strain rate levels can help to interpret what is going on within the material during its deformation.

The contribution by Derazcola et al. [8] considers the above effects in a review on dynamic recrystallization taking place during hot deformation in martensitic stainless steels. The literature analysis led the authors to state that, among the three types of dynamic recrystallization (discontinuous, geometrical and continuous), all taken into account in literature on martensitic stainless, the continuous dynamic recrystallization is the most common for the investigated material class, characterized by high stacking fault energy.

The optimization of processing parameters was the target of the paper by Liang et al. [12]. The investigated material was in this case a Ci-Ni-Al brass, for which hot deformation behavior was characterized by the temperature range 600–800 °C at strain rates ranging from 0.01 to 10 1/s. The analysis of data led to the adoption of a Zener–Hollomon type model for which strain-related parameters were identified. The processing map was constructed based on the dynamic material model and optimized strain-rate/temperature for the alloy processing were identified.

The contribution by Lizzi et al. [9] also analyses the mechanical response/microstructure correlation. The investigated material here is IN718PW, a prototype alloy with chemical composition close to the matrix of classical Inconel®718 alloy but produced by powder metallurgy processes. Compression tests performed at temperatures in the range 900–1025 °C and strain rates in the 0.001–10 1/s range showed that the combination of high temperature and controlled strain caused stress-softening. The phenomenon is partly due to dynamic recrystallization, similar to the reference alloy, but is also partly due to diffusion-controlled pore coarsening. The analyses of the mechanical response of IN718PW alloy led to the estimation of the activation energy and strain rate-stress exponent, both close to those of the reference alloy at a temperature range exceeding solvus temperature, where precipitates do not exist.

The hot deformation behavior was also investigated by Álvarez-Leal et al. [7] for WE54 Mg alloy, where the friction stir welding process produced a fine-grained structure, correlated to process severity. The material characterization was performed from room temperature up to 450 °C, with different testing methods so that peak stress under strain rates from 10^{-4} to 10^{-1} 1/s were obtained. A superplastic behavior for the alloy was possible in the temperature range from about 250 to 400 °C. In cases where this behavior was reached, the analyses of strain rate vs. stress double logarithmic plot show data aligned with a slope close to 2, suggesting grain boundary sliding to be the main deformation mechanism. The microstructural instability of the fine-grained structure was responsible for the change in material behavior above 400 °C.

2.3. Creep and Creep Modeling

Constant load creep tensile or compression tests are a common method to characterize the high temperature behavior of metals, obtaining data later considered for select and fitting proper creep models. In addition to conventional constant-load tests, under specific conditions step-loading could be (carefully) considered as a minor-effort method to obtain at least some material creep data.

In the case of modified Al-7Si-Mg casting alloys investigated by Gariboldi et al. [5], the strengthening role played by nanometric-size precipitate or dispersoids and by their thermal stability at 300 °C for the deformation resistance of the material has been discussed. The amount and morphology of the eutectic Si and of coarse intermetallic phases, both of them at the interdendritic region and responsible for the microstructural damage leading to the final fracture, have also been considered by the above authors.

In the Al10SiMg alloy produced by additive manufacturing by Paoletti et al. [6], the microstructure and its modification during material production, heat treatment and creep tests has been studied. The microstructure role on the creep resistance at 150–225 °C has then been described by physically based models for as-produced and annealed alloy conditions.

In the contribution by Fuyang et al. [4], the heat resistant casting steel HP40Nb was characterized in hot tension and tension creep in the as-cast and aged condition. Similar tests have been performed on fresh-to-aged weld joint. The creep deformation and damage were modelled; high temperature behavior was then used to simulate the high temperature response of fresh-to-aged reformer pipe weld joint in terms of the progression and redistribution of Von Mises stress and material damage.

The paper by Krauss et al. [11] underlined the effect of microstructure and the need for a clear understanding of it in a different way. Clear differences have been demonstrated between the creep properties of Mo, Mo solid solution, Mo5SiB2 and some Mo-Si-B alloys, these latter candidate materials for future gas turbine engines operating at temperatures exceeding those possible for Ni-based alloys. Future modelling of the complex multiphase materials can save experimental efforts and time in the process of alloy optimization. In this process, the creep characterization and modelling of Mo3Si, as one of the phases present in the Mo-Si-B alloys, is a fundamental step. The interesting and careful compression multistep creep characterization performed on small specimens in Ar/H or vacuum environment in the temperature range 1093–1300 °C has been illustrated in the contribution. The identification of the optimal model (and of its parameters) to describe the high-temperature behavior of the Mo3Si has been finally considered.

Creep deformation mechanisms are typically described in terms of secondary creep strain rates. Their analysis has often been correlated to phenomena taking place at a microscopical level and leading to deformation. There is still work to do in this field on understanding and modelling the creep deformation behavior of metals and alloys. The contribution by Kassner [13] covers some of these features. Specifically, he focuses the attention on the so called Harper–Dorn, power-law and power-law-breakdown creep regimes, occurring in a metal at increasing applied stress (or decreasing temperatures) with increasing slope in the double logarithmic plot (temperature normalized) strain rate-stress plot. Results of recent experimental tests or computational works suggest that the presence/absence of Harper–Dorn creep can be correlated to the initial dislocation density in the material. On the other hand, the power law creep is well explained considering recent calculations on dislocation climb-rate. Finally, power-law breakdown can be explained by taking into account the vacancy concentration of plastically deformed materials.

The description of microstructural features of the alloy was also at the base of models describing creep-resistance of Al10SiMg alloy produced by additive manufacturing and differently heat treated, proposed by Paolucci in [6].

2.4. High Temperature Behaviour in the Presence of Loading/Temperature Changes and Environmental Effects

In addition to the classical short-time high-temperature material characterization by means of tension or compression tests, or of its long-term characterization by means of constant load/stress conditions in creep tests, different loading conditions should be taken into account when the expected service conditions significantly differ from those used in conventional material characterization. Among these conditions, isothermal fatigue characterization is one of the most widely performed.

In the present Special Issue and book, the mechanical behavior of the polycrystalline Ni-based superalloy Rene80 has been investigated by means of isothermal strain controlled fatigue tests performed at 850 °C by Engel and co-authors [10]. The alloy was investigated in two different microstructural conditions: with coarse randomly oriented grain structure and fine textured structure. In each case, two sets of isothermal fatigue tests have been performed, one on notched and the other on smooth specimens. The experimental behavior of the alloys was compared to finite element simulations of the specimens reproducing the polycrystalline/texture structure of the material. The presence of the textured structure affected the mechanical behavior of smooth specimens but had a minor effect on that of notched specimens.

3. Conclusions

The Special Issue and Book 'High Temperature Behavior of Metals' includes 12 papers covering in innovative way the relevant topics in the field. All contributions highlight the importance of microstructure and microstructural stability on the high temperature materials' behavior. The quality and variety of materials, approaches and methodologies presented can inspire both scientists and technicians in their approach to structural response of metallic materials at high temperature.

Acknowledgments: The Guest Editors thank all who contributed effectively to the development of this Special Issue. Thanks to the authors who submitted manuscripts to share results of their research activity, and to the reviewers who agreed to read them and gave suggestions to improve their final quality. Thank you to the Editors and to Assistant Editor Toliver Guo and to all the staff at the Metals Editorial Office, for their management and practical support in the publication process of the Issue.

Conflicts of Interest: The authors declare no conflict of interest.

References

1. Spigarelli, S.; Gariboldi, E. *Creep and High Temperature Deformation of Metals and Alloys*; MDPI: Basel, Switzerland, 2019; ISBN 978-3-03921-878-3 (Pbk). ISBN 978-3-03921-879-0 (PDF). [CrossRef]
2. Balaško, T.; Vončina, M.; Burja, J.; Šetina Batič, B.; Medved, J. High-Temperature Oxidation Behaviour of AISI H11 Tool Steel. *Metals* **2021**, *11*, 758. [CrossRef]
3. Chuang, Y.-L.; Wang, C.-C.; Chen, T.-C.; Shiue, R.-K.; Tsay, L.-W. Microstructural Evolution of 9CrMoW Weld Metal in a Multiple-Pass Weld. *Metals* **2021**, *11*, 847. [CrossRef]
4. Fuyang, C.; Zhou, Y.; Shao, B.; Zhang, T.; Guo, X.; Gong, J.; Wang, X. Weldability and Damage Evaluations of Fresh-to-Aged Reformer Furnace Tubes. *Metals* **2021**, *11*, 900. [CrossRef]
5. Gariboldi, E.; Confalonieri, C.; Colombo, M. High Temperature Behavior of Al-7Si-0.4Mg Alloy with Er and Zr Additions. *Metals* **2021**, *11*, 879. [CrossRef]
6. Paoletti, C.; Cerri, E.; Ghio, E.; Santecchia, E.; Cabibbo, M.; Spigarelli, S. Effect of Low-Temperature Annealing on Creep Properties of AlSi10Mg Alloy Produced by Additive Manufacturing: Experiments and Modeling. *Metals* **2021**, *11*, 179. [CrossRef]
7. Álvarez-Leal, M.; Carreño, F.; Orozco-Caballero, A.; Rey, P.; Ruano, O.A. High Strain Rate Superplasticity of WE54 Mg Alloy after Severe Friction Stir Processing. *Metals* **2020**, *10*, 1573. [CrossRef]
8. Derazkola, H.A.; García Gil, E.; Murillo-Marrodán, A.; Méresse, D. Review on Dynamic Recrystallization of Martensitic Stainless Steels during Hot Deformation: Part I—Experimental Study. *Metals* **2021**, *11*, 572. [CrossRef]
9. Lizzi, F.; Pradeep, K.; Stanojevic, A.; Sommadossi, S.; Poletti, M.C. Hot Deformation Behavior of a Ni-Based Superalloy with Suppressed Precipitation. *Metals* **2021**, *11*, 605. [CrossRef]
10. Engel, B.; Chneseit, S.; Mäde, L.; Beck, T. Influence of Grain Orientation Distribution on the High Temperature Fatigue Behaviour of Notched Specimen Made of Polycrystalline Nickel-Base Superalloy. *Metals* **2021**, *11*, 731. [CrossRef]
11. Kauss, O.; Obert, S.; Bogomol, I.; Wablat, T.; Siemensmeyer, N.; Naumenko, K.; Krüger, M. Temperature Resistance of Mo_3Si: Phase Stability, Microhardness, and Creep Properties. *Metals* **2021**, *11*, 564. [CrossRef]
12. Liang, Q.; Liu, X.; Li, P.; Zhang, X. Hot Deformation Behavior and Processing Map of High-Strength Nickel Brass. *Metals* **2020**, *10*, 782. [CrossRef]
13. Kassner, M.E. New Developments in Understanding Harper–Dorn, Five-Power Law Creep and Power-Law Breakdown. *Metals* **2020**, *10*, 1284. [CrossRef]

metals

Article

High-Temperature Oxidation Behaviour of AISI H11 Tool Steel

Tilen Balaško [1,*], Maja Vončina [1], Jaka Burja [2], Barbara Šetina Batič [2] and Jožef Medved [1]

1 Faculty of Natural Sciences and Engineering, University of Ljubljana, Aškerčeva cesta 12, 1000 Ljubljana, Slovenia; maja.voncina@omm.ntf.uni-lj.si (M.V.); jozef.medved@ntf.uni-lj.si (J.M.)
2 Institute of Metals and Technology, Lepi pot 11, 1000 Ljubljana, Slovenia; jaka.burja@imt.si (J.B.); barbara.setina@imt.si (B.Š.B.)
* Correspondence: tilen.balasko@omm.ntf.uni-lj.si; Tel.: +38-612-000-457

Abstract: The high-temperature oxidation of hot-work tool steel AISI H11 was studied. The high-temperature oxidation was investigated in two conditions, the soft annealed condition, and the hardened and tempered condition. First, calculations of the compositions of the oxide layers formed were carried out using the CALPHAD method. The samples were oxidised in a chamber furnace and in a simultaneous thermal analysis instrument, for 100 h in the temperature range between 400 °C and 700 °C. The first samples were used for metallographic (optical microscopy and scanning electron microscopy) and X-ray diffraction analysis of the formed oxide layers, and the second ones for the analysis of the oxidation kinetics by thermogravimetric analysis. Equations describing the high-temperature oxidation kinetics were derived. The kinetics can be described by three mathematical functions, namely: exponential, parabolic, and cubic. However, which function best describes the kinetics depends on the oxidation temperature and the thermal condition of the steel. Hardened and tempered samples have been shown to oxidise less, resulting in a slower oxidation rate. The oxide layers consist of three sublayers, the inner one being spinel-like oxide $(Fe, Cr)_3O_4$, the middle one a mixture of magnetite and hematite and the outer one of hematite. At 700 °C there is also some wüstite in the inner oxide sublayer of the soft annealed sample.

Keywords: high-temperature oxidation; AISI H11 tool steel; thermogravimetry; CALPHAD; kinetics

check for **updates**

Citation: Balaško, T.; Vončina, M.; Burja, J.; Šetina Batič, 3.; Medved, J. High-Temperature Oxidation Behaviour of AISI H11 Tool Steel. *Metals* **2021**, *11*, 758. https://doi.org/10.3390/met11050758

Academic Editor: Stefano Spigarelli

Received: 20 April 2021
Accepted: 30 April 2021
Published: 4 May 2021

Publisher's Note: MDPI stays neutral with regard to jurisdictional claims in published maps and institutional affiliations.

1. Introduction

AISI H11 (1.2343) hot-work tool steel, is a typical representative of the group of chromium hot-work tool steels. It is widely used for the production of tools and dies for high-pressure die casting and forging. The higher concentration of carbide-forming alloying elements improves the high-temperature softening resistance of the steel [1,2]. Since hot-work tool steels usually operate at elevated temperatures, the high-temperature oxidation behaviour of H11 hot-work tool steel is of great importance. Several authors [3–9] have studied the effect of elevated temperatures on the mechanical and physical properties of hot-work tool steels. With this consideration, we investigate the oxidation kinetics of AISI H11 in the temperature range between 400 °C and 700 °C.

Oxidation is one of the most important types of corrosion. Metals and alloys begin to oxidize in air or other atmospheres containing oxygen [10–13]. Assuming that oxygen is in the gaseous state, metals (M) are converted to oxides (M_aO_b), and the reaction proceeds according to the following chemical reaction [12]:

$$\frac{2a}{b}M(s) + O_2(g) \rightleftharpoons \frac{2}{b}M_aO_b(s) \tag{1}$$

Since the chemical reactions of oxidation are well known, one can calculate the Gibbs free energy of the reactions. The Equation (2) for the Gibbs free energy of reaction (1), is given as [12]:

$$G_r = \frac{2}{b}\mu^0_{M_aO_b} + \frac{2}{b}RT\ln\left(a_{M_aO_b}\right)^{2/b} - \frac{2a}{b}\mu^0_M - \frac{2a}{b}RT\ln(a_M)^{2a/b} - \mu^0_{O_2} - RT\ln\frac{p_{O_2}}{p_0} \tag{2}$$

where μ is the chemical potential, a is the activity, R is the gas constant, p is the pressure and T is the temperature. Assuming that $a_{M_aO_b} = 1$; $a_M = 1$ and $p_O = 1$ atm (101 325 Pa), Equation (3) can be derived from Equation (2) [12]:

$$\Delta G_r^0 = \frac{2}{b}\mu^0_{M_aO_b} - \frac{2a}{b}\mu^0_M - \mu^0_{O_2} \tag{3}$$

Then the Gibbs free energy for the chemical reaction (1) can be determined [12]:

$$G_r = \Delta G_r^0 + RT\ln\frac{1}{p_{O_2}} = \Delta G_r^0 + RT\ln K \tag{4}$$

where K is the equilibrium chemical reaction constant and ΔG_r^0 is the Gibbs free energy change per mole of the reaction at standard conditions. Some of the chemical reactions for the oxidation of metals can be found in the well-known Ellingham diagram [14]. Let us first consider the oxidation of pure metals, it depends on the growth kinetics of the oxide layer, on whether it will protect the metal from oxidizing or not. From a thermodynamic point of view, the chemical reaction (1) proceeds spontaneously from left to right when the total change in Gibbs free energy is negative [10–13,15].

The oxidation kinetics of metals and alloys can follow several different laws, the most common of which are linear, parabolic, cubic, logarithmic, and inverse logarithmic. In fact, the kinetics is very complex and can also consist of a combination of several laws (linear and parabolic, etc.). The chemical reaction (1) follows a kinetic law that can be described by the following equation [10–13,15–17]:

$$\frac{d\xi}{dt} = f(t) \rightarrow d\xi = dn_{M_aO_b} = -\frac{dn_M}{a} = -\frac{2dn_{O_2}}{b} \tag{5}$$

where ξ is a measure of the extent of the reaction at time t, and n is the number of moles [10–13,15–17]. However, the most common method for studying the oxidation rate is to measure the weight gain of the sample over time. Gravimetry is a particularly suitable method, where the change in weight over time can be measured continuously or discontinuously [17].

Several authors studied high-temperature oxidation of iron in air or pure oxygen atmosphere [11,13,17–21]. The high-temperature oxidation kinetics of iron in the temperature range 700–1250 °C can be expressed by Tammann-type parabolic equation [21]:

$$dx/dt = k'_x/x \text{ ali } x^2 = k_x t + x_0^2 \tag{6}$$

where, x is the total thickness of the oxide layer, t is the oxidation time, x_0 is the initial thickness of the oxide layer (if some oxides are already present on the surface, otherwise the value is 0) of the parabolic oxidation, and $k_x = 2k'_x$ is a constant of the parabolic rate usually given in cm^2 s^{-1}. However, since oxidation kinetics are usually studied by a change in mass as a function of oxidation time, a parabolic law can be expressed by the Pilling-Bedworth equation [21,22]:

$$dW/dt = k'_p W \text{ ali } W^2 = k_p t + W_0^2 \tag{7}$$

where W is the change in mass per unit area, due to oxidation of iron, and k_p (=$2k'_p$) is the parabolic constant in g^2 cm^{-4} s^{-1}, and W_0 is the initial mass at the time ($t = 0$) of parabolic oxidation. In the original Pilling-Bedworth equation [22], W_0 is equal to 0. In the considered temperature range, the oxidation follows the parabolic law, the resulting oxide layer consists of a very thin outer sublayer of hematite, a thin intermediate sublayer of magnetite and a thick inner sublayer of wüstite [21,23].

On the other hand, the oxidation of iron below 700 °C is much more complex, so that even the published results of different authors [21,24–27] are less consistent. In previous studies, it was found that the kinetics of isothermal oxidation of iron in the temperature range (570–700 °C) follows a parabolic law. However, the thicknesses of the individual oxide layers differ from study to study [24–27].

Regarding the high-temperature oxidation of steels, many studies [21,28–37] in which the high-temperature oxidation of various steels has been investigated have been carried out. Many ways have been proposed to calculate the rate constant of oxidation kinetics. The most suitable method to obtain the equation to calculate the oxidation rate constant of steel under continuous heating or cooling was developed by Kofstad [38]. According to his interpretation, oxidation following a linear, parabolic, or cubic law can be generally expressed as:

$$W^{n-1}[dW/dt] = k_n \tag{8}$$

where W is defined as the change in mass per unit area over time t, n is a constant with values of 1, 2, or 3 for the linear, parabolic, or cubic law, and k_n is a time-independent rate constant expressed as:

$$k_n = B\,exp(-Q/RT) \tag{9}$$

where, B is the constant, T is the absolute temperature, R is the gas constant, and Q is the activation energy. There is also an alternative [39], but since the mass change was measured over time, we used the Kofstad method. Regarding the high-temperature oxidation of hot-work tool steels, there are some studies [40–44] have been carried out. Bidibadi et al. [40] investigated the oxidation behaviour of CrMoV steel, the study sets out the composition of the oxide layers in detail and proves that the inner oxide layer contains spinel-like oxides. Zhang et al. [41] studied the high-temperature oxidation resistance of H13 steel, from the TG curves, there is a mass decrease between 250 °C and 500 °C during the heating, when the oxide layer has already formed, but unfortunately, no further analysis was performed regarding the cause the of mass decrease. On the other hand, Min et al. [42] carried out a prediction and analysis of the oxidation of H13 hot-work tool steel. Using the software Thermo-Calc the compositions of the oxide layers were calculated as a function of oxygen partial pressure, and the oxidation mechanisms in different atmospheres were discussed. Regarding the oxidation of H11 steel, Pieraggi et al. [43] explained the oxidation behaviour of H11 steel in dry and wet air at 600 °C and 700 °C. It was found that the oxidation kinetics follows the parabolic law and is also quite sensitive to the presence of water vapour. Bruckel et al. [44] investigated the isothermal oxidation behaviour of an X38CrMoV5 hot-work tool steel at 600 °C and 700 °C. It was also found that the oxidation kinetics is very sensitive to the presence of water vapour. On the other hand, it was proved that the oxide layer is duplex consisting of an outer hematite sublayer and an inner (Fe, Cr)$_3$O$_4$ sublayer. It was mentioned that carbides could play an important role in the growth process of the oxide layer. These studies served as a first insight into what the formed oxide layers might consist of and how the oxidation kinetics might behave.

2. Materials and Methods

We investigated the H11 (EN X37CrMoV5-1) hot-work tool steel with the chemical composition given in Table 1, measured by wet chemical analysis and infrared absorption after combustion with ELTRA CS-800 (ELTRA GmbH, Haan, Germany).

Table 1. Chemical composition of investigated H11 hot-work tool steel given in weight percent.

Sample	C	Si	Mn	P	S	Cr	Ni	Mo	V	Fe
H11	0.36	0.97	0.54	0.015	0.002	5.05	0.09	1.22	0.38	Bal.

First the heat treatment of H11 hot-work tool steel was carried out. The austenitization temperature was 1020 °C and the soaking time was 30 min. Quenching was carried out in oil, followed by two-stage tempering, at 550 °C and 620 °C for 2 h, for reaching the hardness 42–44 HRC. The heat treatment was carried out in Bosio EUP-K 6/1200 (Bosio d.o.o., Celje, Slovenia) chamber furnace. Since air atmosphere was used, we had to mill off 2 mm of the steel surface, due to decarburization and oxidation during the heat treatment. Sample preparation using a CNC (Computer Numerical Control) Deckel Maho DMC 63V (Deckel Maho Pfronten GmbH, Pfronten, Germany) machine was next. three different types of samples were prepared, the first ones were cubes with dimensions of 10 mm × 10 mm × 10 mm (length × width × height), which were used during the high-temperature oxidation in the chamber furnace. These samples were specially prepared for further analysis by light optical microscope and SEM (Scanning Electron Microscope). The second set of samples were cuboids with dimensions 10 mm × 10 mm × 5 mm, which were also used during high-temperature oxidation in the chamber furnace. The dimensions are different because the height of the XRD (X-ray Diffraction) sample holder is limited. The last ones are cylinders with dimensions h = 4 mm and Φ = 4 mm, which were used in the high-temperature oxidation in NETZSCH STA (Simultaneous Thermal Analyzer) Jupiter 449C (NETZSCH Holding, Selb, Germany) instrument. All surfaces were polished with 1 μm polycrystalline diamond suspension, after polishing, the samples were ultrasonically cleaned with ethanol and dried on hot air, this procedure was also repeated before each oxidation test.

To get a first insight into the composition of the oxide layers, CALPHAD (CALculation of PHAse Diagrams) simulations were performed with the software Thermo-Calc (Thermo-Calc 2020a, Thermo-Calc Software AB, Stockholm, Sweden) using the thermodynamic database TCOX9 [45] (Metal Oxide Solutions Database). In the diagrams the amount of phase is shown on the Y-axis and the activity of O_2 is shown on X-axis. The composition of the oxide sublayers in the formed oxide layer in the studied temperature range (400–700 °C), was calculated. Thermo-Calc software and HSC Chemistry software (HSC Chemistry 9.0, Metso Outotec Finland Oy, Tampere, Finland) were also used to calculate the Gibbs free energy of the oxidation reactions.

High-temperature oxidation was studied at 400 °C, 500 °C, 600 °C, and 700 °C in air atmosphere for 100 h. The samples analysed by light optical microscope, SEM and XRD were oxidised in Bosio EUP-K 6/1200 (Bosio d.o.o., Celje, Slovenia) chamber furnace. The heating rate was 15 °C min^{-1} and the high-temperature oxidation was carried out in air atmosphere. To improve the oxidation conditions, additional air (79 vol% nitrogen, 21 vol% oxygen 0.9 vol% argon and 0.1 vol% of hydrocarbons and other inert gases) was injected into the furnace at a flow rate of 0.3 L min^{-1}. Cooling was performed in the chamber furnace at the cooling rate of 0.5 C min^{-1}, which was intentionally slow. The main reason was not to damage the oxide layer.

The samples used for the study of kinetics were oxidized in the NETZSCH STA Jupiter 449C (NETZSCH Holding, Selb, Germany) instrument. The heating and cooling rate were 10 °C min^{-1}, in addition air was injected into the furnace at a flow rate of 30 mL min^{-1} (during the heating/cooling and the isothermal section). Correction runs were also performed prior to sample runs with an empty Al_2O_3 pedestal on which the samples were located during oxidation. TGA (Thermogravimetric Analysis) was used to study the kinetics.

The samples for SEM analysis were ground and polished. An FEG-SEM (Field Emission Gun Scanning Electron Microscopy) ThermoFisher Scientific Quattro S (ThermoFisher Scientific, Waltham, MA, USA) was used to examine the composition and thickness of the oxide layers. The composition was investigated using EDS (Energy-Dispersive X-ray Spectroscopy) analysis (Ultim® Max, Oxford Instruments, Abingdon, UK). EBSD (Electron Backscatter Diffraction) analysis was also performed using ZEISS CrossBeam 550 (Carl Zeiss AG, Oberkochen, Germany) SEM with HikariSuper EBSD camera (EDAX, Mahwah NJ, USA). In addition, the samples were ground, polished, and etched with Vilella's reagent for optical microscopy. Microphot FXA, Nikon (Nikon, Minato City, Japan) with 3CCD video camera Hitachi HV-C20A (Hitachi, Ltd., Tokyo, Japan) was used for microstructure analysis. Vickers hardness was measured using Instron Tukon 2100B (Instron, Norwood, MA, USA). The hardness measurements were performed on the metallographically prepared samples (ground and polished), and we only measured the hardness of the steels. PANalytical XPert Pro PW3040/60 (Malvern Panalytical, Malvern, England, UK), X-ray diffractometer with Cu anode (λ = 0.15419 nm) using Bragg-Bretano geometry with a theta-theta goniometer with a radius of 240 mm was used to study the types of oxides in the oxide layers.

Since we were investigating samples in two different conditions soft annealed and hardened and tempered condition, we marked the first ones with H11 (soft annealed) and the second ones with H11-HT (hardened and tempered).

3. Results

3.1. CALPHAD

Figure 1 depicts the thermodynamically stable phases in oxide layer and the sublayers formed on the surface of H11 steel at 400 °C (Figure 1a) and 500 °C (Figure 1b). According to this, up to $a_O \approx 10^{-24}$ (Fe, Cr)$_2$O$_3$ is stable, which predominates (93 wt%), followed by (Fe, Cr, Mn, V)$_3$O$_4$ (3 wt%), Mo$_4$O$_{11}$ (1.9 wt%), (Cr, Fe, V, Mn)$_3$O$_4$ and (Fe, Mo)$_2$C$_3$ (both together 2.1 wt%). In the range $a_O \approx 10^{-24}$–10^{-34}, (Fe, Cr, Mn, V)$_3$O$_4$ is predominant (82.8 wt%), followed by (Cr, Fe, V, Mn)$_3$O$_4$ from $a_O \approx 10^{-26}$ onwards (Cr, Fe, Mn)$_3$O$_4$ (13.1 wt%) and (Fe, Mo)$_2$C$_3$. Ferrite also occurs, increasing with the decreasing a_O. At 500 °C there are no drastic changes, all thermodynamically stable phases are the same as at 400 °C, the only difference is the stability of (Fe, Cr)$_2$O$_3$ and (Fe, Cr, Mn, V)$_3$O$_4$, the former is stable from $a_O \approx 10^{-13}$–10^{-19} and the latter from $a_O \approx 10^{-19}$–10^{-29}.

On the other hand, the same changes take place at 600 °C (Figure 2a), (Fe, Cr)$_2$O$_3$ is stable from $a_O \approx 10^{-10}$–10^{-15} and (Fe, Cr, Mn, V)$_3$O$_4$ from $a_O \approx 10^{-15}$–10^{-25}. At both temperatures (500 °C and 600 °C), the calculations also showed that internal oxidation could occur in the steel directly under the oxide layer there could be stable (Fe, Cr, Mn, V)$_3$O$_4$ or (Cr, Fe, Mn, V)$_3$O$_4$. At $a_O < 10^{-29}$ internal oxidation could occur at 500 °C and $a_O < 10^{-25}$ at 600 °C. At 700 °C (Figure 2b), we can see that up to $a_O \approx 10^{-12}$ (Cr, V, Fe, Ti)$_2$O$_3$ is stable, which is predominant (91.7 wt%), followed by (Fe, Cr, Mo, Mn, V)$_3$O$_4$ (5.4 wt%), Mo$_4$O$_{11}$ (2.2 wt%) and (Fe, Cr, V) O (0.7 wt%). In the range $a_O \approx 10^{-12}$–10^{-22}, (Fe, Cr, Mo, Mn, V)$_3$O$_4$ is predominant (97.8 wt%), followed by ferrite, which occurs at $a_O \approx 10^{-16}$ and increases with decreasing a_O (up to a_O 10^{-12}–10^{-22}) and then (Fe, Cr, Mo, V, Mn)$_3$C, and Mo$_4$O$_{11}$ (the content of the latter two together depends on the ferrite content). It is also observed that wüstite appears in the oxide layer and reaches the highest contents at $a_O \approx 10^{-21}$–10^{-22}, up to 48 wt%. At a temperature of 700 °C, internal oxidation is also present, when a_O decreases below 10^{-22}, (Fe, Cr, Mo, Mn, V)$_3$O$_4$ appears just below the surface, and there is also more (Cr, V, Fe, Ti)$_2$O$_3$ present.

3.2. Microscopy

The microstructures after high-temperature oxidation at an individual temperature and the corresponding initial microstructure are shown in Figure 3. It can be seen that the initial microstructure in the soft annealed condition consists of a ferritic (highly decomposed martensite) matrix and spherical carbides. After oxidation at 400 °C, the microstructure consists of a ferrite matrix and spherical carbides. The carbides continue to grow and become even more spherical. After oxidation at 500 °C and 600 °C, the microstructure is very similar to the microstructure obtained after oxidation at 400 °C. There are no significant differences in the microstructure, it consists of a ferrite matrix and spherical carbides. This can also be confirmed by practically equal hardness (Table 2). The drop in hardness, compared to the hardness of the initial condition, is due to the growth of the spheroidized carbides, which become larger and more spherical. This reduces the effect of precipitation hardening. On the other hand, the microstructure of the sample oxidized at 700 °C consists of a ferrite matrix and spherical carbides. The microstructure is very similar to that of the initial sample. This is because this temperature is already in the range of spheroidization annealing.

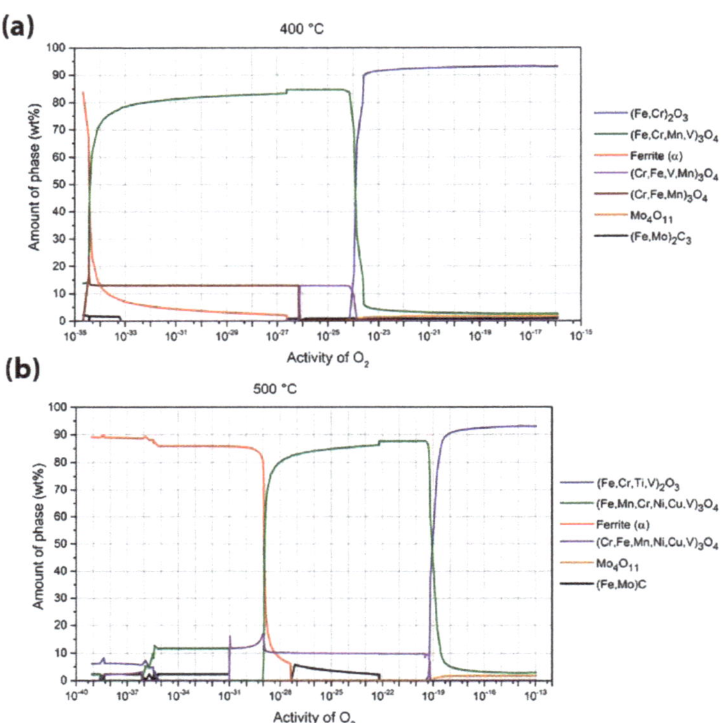

Figure 1. Amount of thermodynamically stable phases formed during high-temperature oxidation as a function of oxygen activity for H11 steel at (**a**) 400 °C and (**b**) 500 °C.

Figure 4 shows the microstructure after high-temperature oxidation at a single temperature and the corresponding initial microstructural condition. As can be seen from the microstructures, the initial microstructure is in a hardened and tempered state, which means that it consists of martensite and carbides. The microstructure after oxidation at 400 °C and 500 °C remains very similar to the initial microstructure. This indicates that the microstructure is stable in this temperature range. This is also confirmed by the hardness measurements (Table 2), as we see that they change very little, compared to the hardness of the initial state. The microstructure in both cases consists of martensite and carbides. Tempering is the reason that the microstructure is so stable, as the 2nd tempering was at the temperature of 620 °C. After oxidation at a temperature of 600 °C, additional softening occurs, which leads to a drop in hardness (Table 2). The microstructure continues to consist of martensite and carbides. In contrast, after oxidation at a temperature of 700 °C, the microstructure consists of a ferrite matrix and spherical carbides. This is because in this temperature range is already the range of spheroidization annealing.

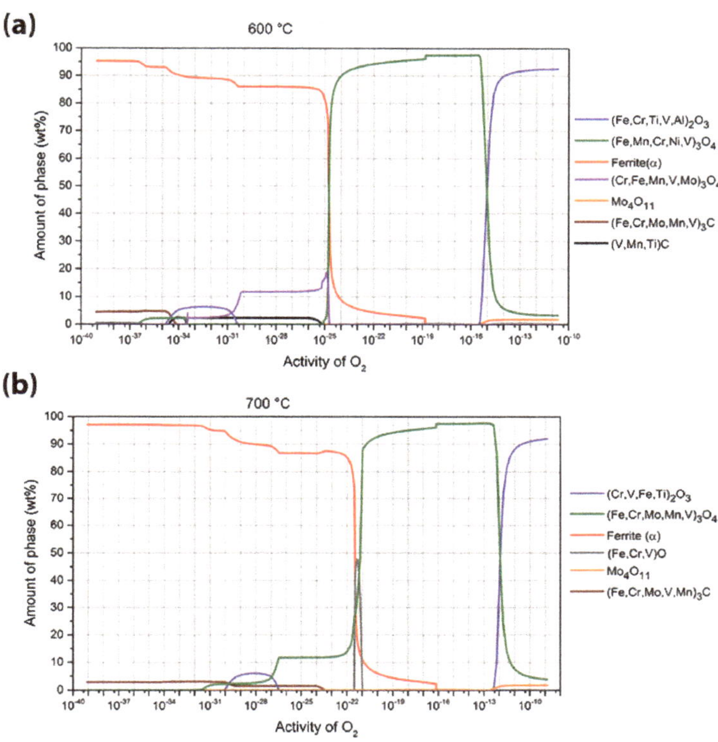

Figure 2. Amount of thermodynamically stable phases formed during high-temperature oxidation as a function of oxygen activity for H11 steel at (**a**) 600 °C and (**b**) 700 °C.

Figure 3. Etched microstructure of H11 hot-work tool steel in the soft annealed state in initial state and after oxidation at 400 °C, 500 °C, 600 °C and 700 °C for 100 h.

Table 2. Measured hardness of H11 hot-work tool steel in soft annealed (H11) and hardened and tempered (H11-HT) state before and after oxidation.

Sample	Before Oxidation	Hardness [HV 10]			
		After Oxidation			
		400 °C	500 °C	600 °C	700 °C
H11	180	166	163	164	164
H11-HT	478	460	466	294	243

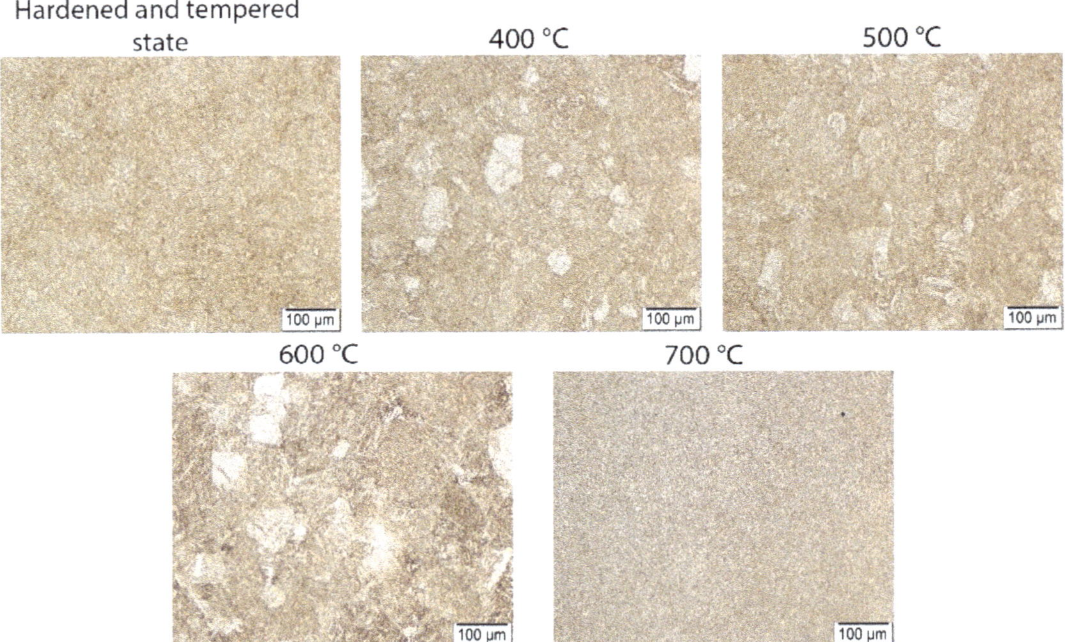

Figure 4. Etched microstructure of H11 hot-work tool steel in the hardened and tempered state in initial state and after oxidation at 400 °C, 500 °C, 600 °C and 700 °C for 100 h.

A thicker oxide layer is formed for the soft annealed samples, except at 600 °C, as evident from the figures showing the oxide layers (Figure 5) and the average of the measured thicknesses (obtained with SEM) of the oxide layers (Table 3). At 400 °C, the oxide layer was 3.9 µm thicker on the soft annealed sample and 1.8 µm thicker at 500 °C, compared to the thickness of the oxide layer on the hardened and tempered samples. On the other hand, the oxide layer formed on the hardened and tempered samples was 3.9 µm thicker at 600 °C. At 700 °C, the oxide layer on the soft annealed sample was thicker again by 22.8 µm.

Carbides are also seen in the microstructures of the soft annealed samples at 400 °C, 500 °C and 600 °C, which are not observed in the hardened and tempered samples because they are finer. The results show the presence of carbides in the inner oxide sublayer, while they are not detected in the outer oxide sublayer. In the soft annealed sample, the oxide layer is solid, and cracking is observed between the inner and middle oxide sublayers. The voids are clearly visible in the oxide layer of the soft annealed samples. They are also believed to form in the oxide layer of the hardened and tempered sample.

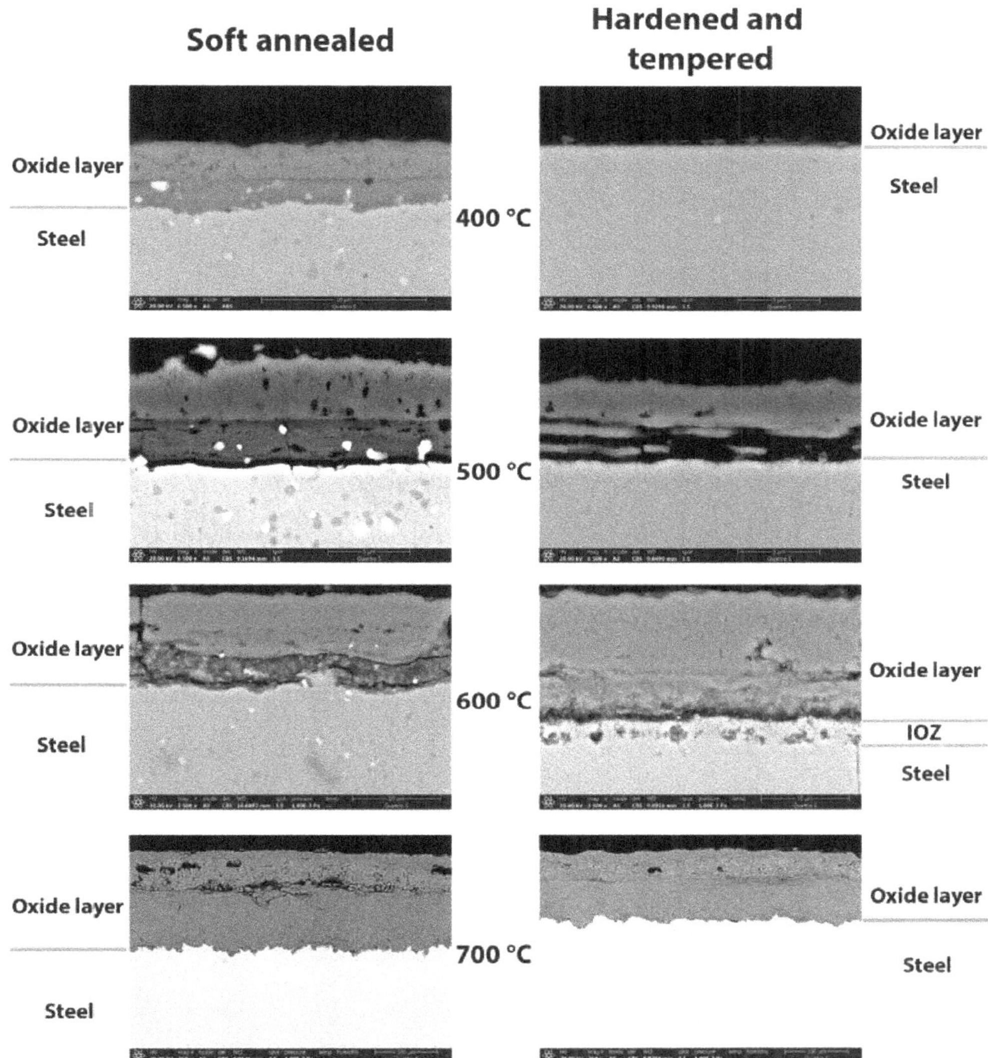

Figure 5. H11 hot-work tool steel in soft annealed and hardened and tempered state and oxide layer formed after high-temperature oxidation at 400 °C, 500 °C, 600 °C and 700 °C.

Table 3. Thickness of the oxide layer formed on H11 hot-work tool steel in the soft annealed (H11) and hardened and tempered (H11-HT) state after high-temperature oxidation at 400 °C, 500 °C, 600 °C and 700 °C.

Temperature [°C]	Thickness of Oxide Layer [μm]	
	H11	H11-HT
400	4.1	0.2
500	6.5	4.7
600	10.9	14.8
700	135.6	112.8

The results of EDS analysis of the oxide layer on the soft annealed sample oxidised at 400 °C are shown below (Figure 6a). Since the Fe:O ratio in hematite is 2:3 and the content of other alloying elements is very low compared to the inner oxide sublayer, it can be assumed that the outer oxide sublayer (Figure 6a)—Spectrum 13) consists of hematite. This is followed by two more oxide sublayers. Immediately behind the hematite is magnetite and some hematite (Figure 6a)—Spectrum 14), which form the middle oxide sublayer. Assuming that chromium and silicon are bound in the magnetite and considering the Fe:O ratio in the magnetite (3:4), it can be estimated that the inner oxide sublayer (Figure 6a)—Spectrum 15) is most likely spinel oxide ((Fe, Cr)$_3$O$_4$). In this case, the content of molybdenum, silicon, vanadium, and manganese is also increased, so that the mentioned alloying elements are bound in the spinel. The composition of the steel (Figure 6a)— Spectrum 16) was also analysed, and the matrix near the inner oxide sublayer already has a different chemical composition than the original one. The contents of some alloying elements are higher, these are mainly alloying elements that are also found in the inner oxide sublayer (chromium, molybdenum, and silicon).

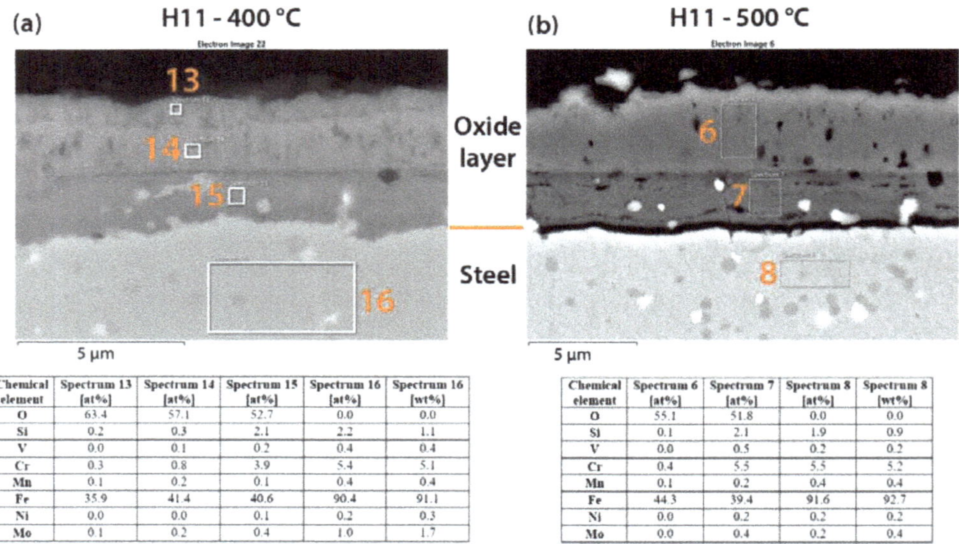

Chemical element	Spectrum 13 [at%]	Spectrum 14 [at%]	Spectrum 15 [at%]	Spectrum 16 [at%]	Spectrum 16 [wt%]
O	63.4	57.1	52.7	0.0	0.0
Si	0.2	0.3	2.1	2.2	1.1
V	0.0	0.1	0.2	0.4	0.4
Cr	0.3	0.8	3.9	5.4	5.1
Mn	0.1	0.2	0.1	0.4	0.4
Fe	35.9	41.4	40.6	90.4	91.1
Ni	0.0	0.0	0.1	0.2	0.3
Mo	0.1	0.2	0.4	1.0	1.7

Chemical element	Spectrum 6 [at%]	Spectrum 7 [at%]	Spectrum 8 [at%]	Spectrum 8 [wt%]
O	55.1	51.8	0.0	0.0
Si	0.1	2.1	1.9	0.9
V	0.0	0.5	0.2	0.2
Cr	0.4	5.5	5.5	5.2
Mn	0.1	0.2	0.4	0.4
Fe	44.3	39.4	91.6	92.7
Ni	0.0	0.2	0.2	0.2
Mo	0.0	0.4	0.2	0.4

Figure 6. The areas analysed with EDS are shown with the corresponding chemical composition in at% for oxide layers and wt% for steel. Results are shown for samples in the soft annealed state after high-temperature oxidation at (**a**) 400 °C and (**b**) 500 °C.

The results of EDS analysis of the oxide layer formed on the soft annealed sample oxidised at 500 °C are also shown (Figure 6b). Moreover, in this case, the outer oxide sublayer consists of hematite, which is a bright area on the surface of the oxide layer. The middle oxide sublayer consists of magnetite (Figure 6b)—Spectrum 6). Assuming that chromium and silicon are bound in magnetite and considering the Fe:O ratio in magnetite (3:4), it can be estimated that the inner oxide sublayer (Figure 6b)—Spectrum 7) is most likely spinel oxide ((Fe, Cr)$_3$O$_4$). In this case, the content of molybdenum, silicon, vanadium, and manganese is also increased, so the mentioned alloying elements are most likely bound in the spinel. The composition of the steel (Figure 6b)—Spectrum 8) was also analysed and it was found that the contents of some alloying elements are increased (chromium and nickel), while others are lower (silicon, vanadium, manganese, and molybdenum) compared to the original composition of the steel.

Carbides are observed in the soft annealed sample (Figure 7a), they are finer in the hardened and tempered sample (Figure 7b). The soft annealed sample contains mainly chromium and molybdenum rich carbides and some vanadium carbides. The results show that carbides are also present in the inner and middle sublayer, while they are not observed in the outer oxide sublayer. The concentration of iron is the lowest in the outer sublayer, higher in the inner sublayer and highest in the middle sublayer. On the other hand, the oxygen content is the highest in the outer oxide sublayer and then decreases towards the interior. It can also be seen that there is less chromium in the outer and middle oxide sublayers than in the inner one. Meanwhile, the concentration of molybdenum is greatly increased in the inner sublayer, while it is much lower in the middle and outer oxide sublayers. The nickel content is increased at the contact surface between the oxide layer and the steel. The concentration of silicon is increased in the inner oxide sublayer and then decreases towards the oxide surface.

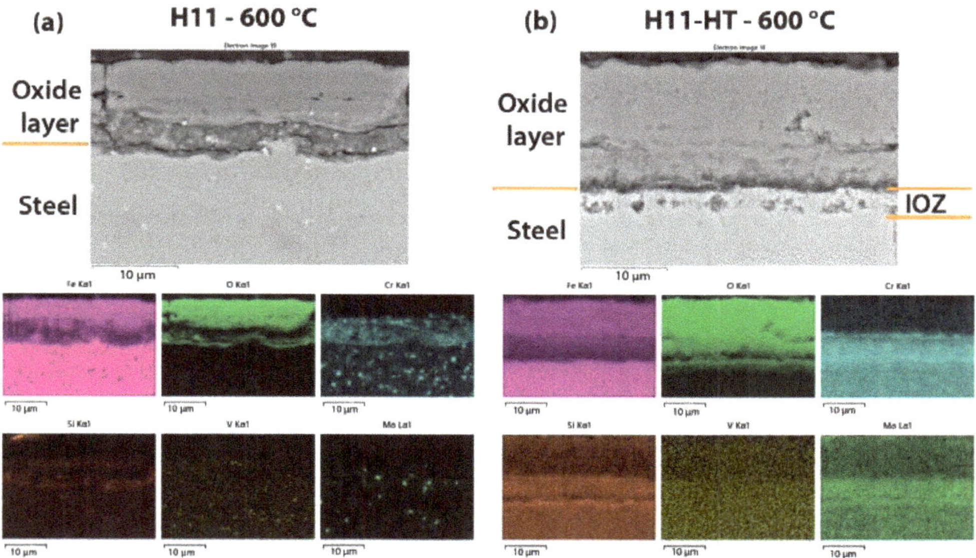

Figure 7. EDS element mapping images are shown for samples in the (a) soft annealed and (b) hardened and tempered state after high-temperature oxidation at 600 °C.

No major carbides are observed in the hardened and tempered sample (Figure 7b). Moreover, in this case the elements are very homogeneously distributed in the matrix. There are three oxide sublayers and internal oxidation of the steel is also observed i.e., an internal oxidation zone (Figure 7b—IOZ). IOZ occurs just in the case of hardened and tempered sample in the areas where the alloying elements concentrations are increased. The concentration of iron is lowest in the outer oxide sublayer, higher in the inner oxide sublayer and highest in the middle oxide sublayer. On the other hand, the oxygen content is the highest in the outer oxide sublayer and then decreases towards the interior. There are also some areas of increased oxygen content in the steel in the IOZ just below the oxide layer. Chromium concentration is lower in the outer and middle oxide sublayers, than in the inner one. In the case of molybdenum, the concentration is greatly increased in the inner oxide sublayer, compared to the middle and outer. On the other hand, the nickel content is increased at the contact surface between the oxide and the steel. While the silicon content is highest in the inner sublayer, it is slightly lower in the outer and lowest in the middle oxide sublayer.

The oxide layer formed at 700 °C was also investigated by EBSD analysis. First, the areas of the soft annealed sample examined by EBSD analysis are shown (Figure 8). The analysis was divided into four parts, namely: (1) Upper area of oxide layer—outer and part of middle oxide sublayer, (2) Middle area of oxide layer—inner oxide sublayer in contact area with middle oxide sublayer, (3) Middle area of oxide layer—middle area of inner oxide layer and (4) Steel/oxide layer—contact area between steel and inner oxide sublayer.

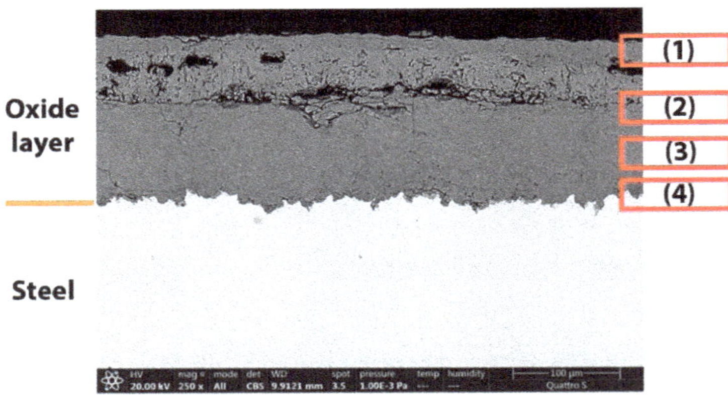

Figure 8. Areas of the oxide layer of the soft annealed sample after high-temperature oxidation at 700 °C analysed by EBSD analysis.

First, the results for the upper part of the oxide layer are shown (Figure 9). The outer oxide sublayer consists only of hematite. This is followed by a band of magnetite and hematite with low magnetite content, the middle oxide sublayer starts below the magnetite band. The magnetite content in the middle oxide sublayer increases towards the interior. Crystal orientations were coloured with IPF (Inverse Pole Figure) colour coding of orientation maps, IPF-Z standard colour triangle was used.

(1)

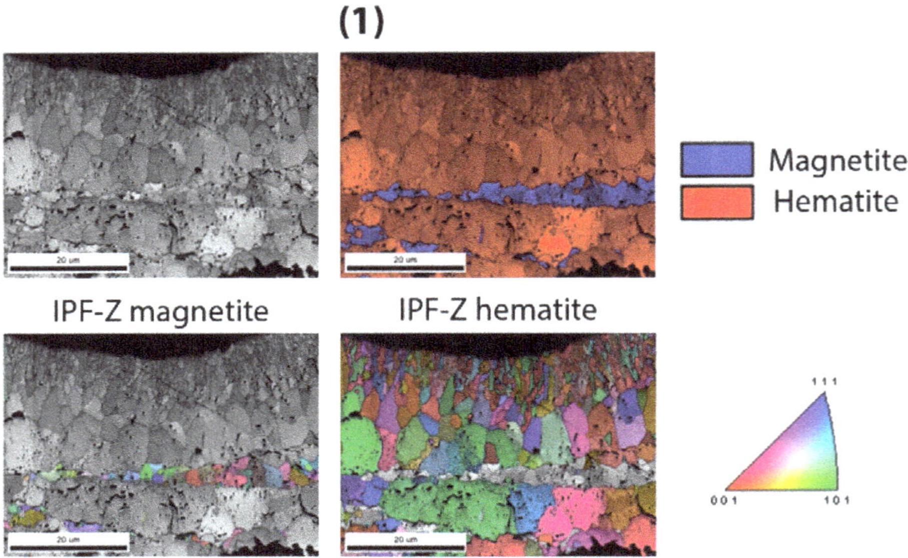

Figure 9. EBSD analysis of the oxide layer of the soft annealed sample after high-temperature oxidation at 700 °C (**1**) Upper part of the oxide layer—outer and part of the middle oxide sublayer.

The inner oxide sublayer (Figure 10) consists mainly of magnetite, together with small amounts of hematite. The composition of magnetite shows an increased concentration of chromium, molybdenum, and silicon. Therefore, along EBSD analysis an EDS analysis of the surface distribution of the elements was also performed. Chromium is distributed throughout the oxide layer (with visible increased concentrations at locations where previously carbides were present), and molybdenum is also concentrated where carbides were most likely previously present. This is already an indication that molybdenum has less influence on the oxidation kinetics than chromium. It can also be seen that a crack forms between the middle and inner oxide sublayers. It is also observed that the chromium content is significantly lower in the middle oxide sublayer. The results show that the inner oxide sublayer is mainly composed of magnetite and some hematite. This indicates that the inner oxide sublayer is very dense, and the magnetite grains are small, resulting in difficult diffusion of both oxygen and iron and other alloying elements.

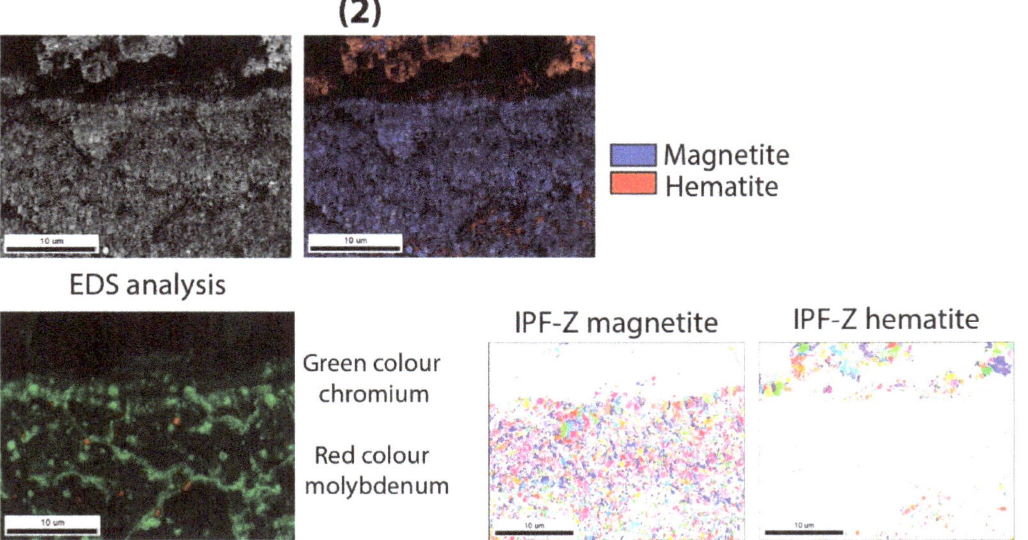

Figure 10. EBSD analysis of the oxide layer of the soft annealed sample after high-temperature oxidation at 700 °C (**2**) Middle oxide layer—inner oxide sublayer in contact area with middle oxide sublayer.

The analysis of middle area of inner oxide layer (Figure 11) reveals that some wüstite grains are also present, meaning that the inner oxide layer consists of magnetite, some hematite and very little wüstite. An EDS analysis was also performed, examining the concentrations of chromium and molybdenum in the inner oxide sublayer. Distribution is the same as at area (2) middle oxide layer.

The contact area between the oxide layer and the steel (4) was also investigated using EBSD analysis (Figure 12). First, a phase analysis is presented, where it can be observed that a mixed area of magnetite and hematite occurs at the contact area between the inner oxide sublayer and the steel. Carbides are also visible in this area. An analysis of the surface distribution of the elements was also carried out, it shows that the chromium content is increased, and the molybdenum is still present in the places where the carbides were before. This is the indication that molybdenum has less influence on the oxidation kinetics than chromium.

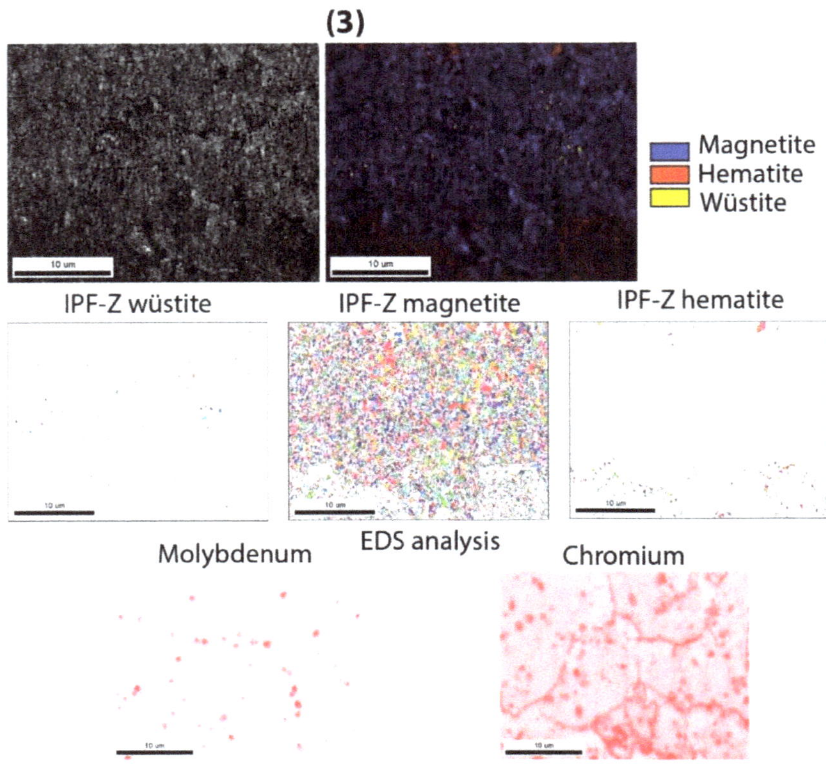

Figure 11. EBSD analysis of the oxide layer of the soft annealed sample after high-temperature oxidation at 700 °C (3) Middle oxide layer—middle area of inner oxide layer.

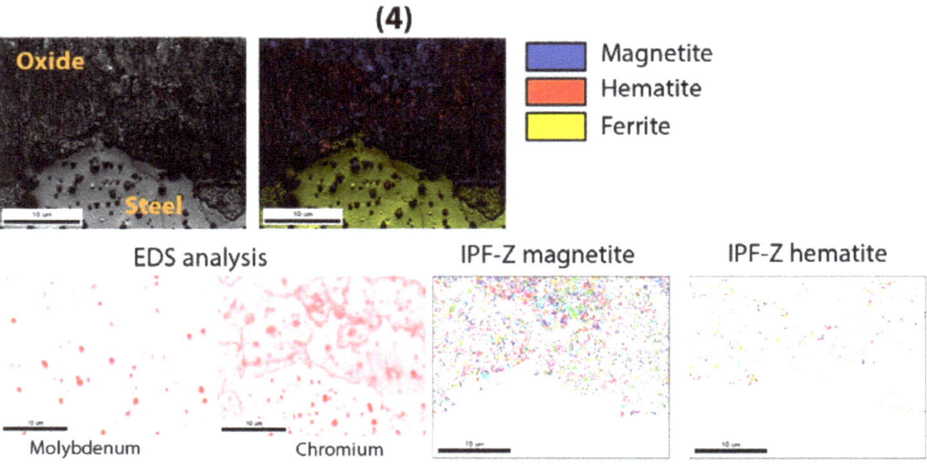

Figure 12. EBSD analysis of the oxide layer of the soft annealed sample after high-temperature oxidation at 700 °C (4) Steel/oxide layer—contact area between steel and inner oxide sublayer.

The areas of the hardened and tempered sample examined by EBSD analysis are shown (Figure 13). The analysis was divided into two parts, namely: (1) Upper part of oxide layer—outer and a part of middle oxide sublayer, (2) Middle oxide sublayer—inner oxide sublayer and contact area between steel and inner oxide sublayer.

Figure 13. Areas of the oxide layer of the tempered and hardened sample after high-temperature oxidation at 700 °C analysed by EBSD analysis.

First, the results for the upper part of the oxide layer are shown (Figure 14). The outer oxide sublayer consists only of hematite. This is followed by a middle oxide sublayer of magnetite and a small amount of hematite, which is stable around the voids and cracks. The cracks between the middle and inner oxide sublayer are also visible.

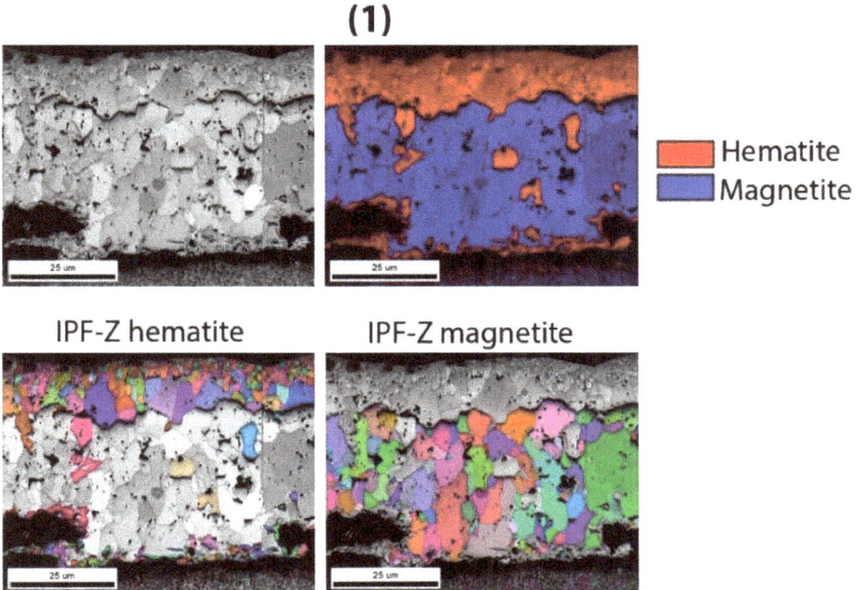

Figure 14. EBSD analysis of the oxide layer of the hardened and tempered sample after high-temperature oxidation at 700 °C (**1**) Upper part of oxide layer—outer and a part of middle oxide sublayer.

Meanwhile, the inner oxide sublayer (Figure 15) is mainly magnetite with low hematite content around voids and cracks. There is also a hematite band visible in the oxide layer/steel interface. In this case, no wüstite content was observed. The oxide grains in the inner oxide sublayer are much smaller compared to ones in the soft annealed sample (Figures 10–12).

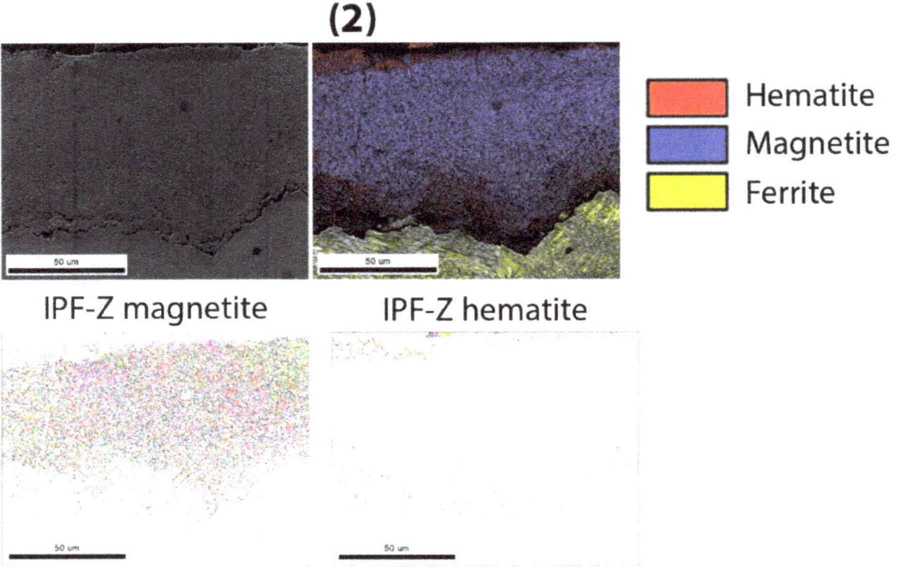

Figure 15. EBSD analysis of the oxide layer of the hardened and tempered sample after high-temperature oxidation at 700 °C (**2**) Middle oxide sublayer—inner oxide sublayer and contact area between steel and inner oxide sublayer.

3.3. Thermogravimetric Analysis

Thermogravimetric analysis was used to measure the change in mass of the samples during high-temperature oxidation. The high-temperature oxidation kinetics for the studied steel was described by mathematical functions. An iterative rectangular distance regression algorithm was used for the best fit of the actual curves with the mathematical functions. Equations describing the change in mass of the sample per unit area over time for each experimental temperature are also presented for each sample studied. Since the realistic conditions were simulated, air was blown into the furnace during the heating of the samples in the STA instrument, the curve in the heating segment differs from that under isothermal conditions. Therefore, each oxidation curve is described by two equations, the first giving the oxidation kinetics under heating and the second giving the oxidation kinetics under isothermal conditions. A two-phase and a single-phase exponential growth function (Equations (10) and (11)), a second-degree polynomial (Equation (12)—parabolic law) and a third-degree polynomial (Equation (13)—cubic law) were used to describe the curves.

$$W\left(\frac{\Delta m}{A}\right) = y = Ae^{\left(\frac{t}{B}\right)} + Ce^{\left(\frac{t}{D}\right)} + E \tag{10}$$

$$W\left(\frac{\Delta m}{A}\right) = y = Ae^{\left(\frac{t}{B}\right)} + E \tag{11}$$

$$W\left(\frac{\Delta m}{A}\right) = y = A + Bt + Ct^2 \tag{12}$$

$$W\left(\frac{\Delta m}{A}\right) = y = A + Bt + Ct^2 + Dt^3 \tag{13}$$

In all cases, t is the time in s, Δm is the mass change in mg, A is the specific surface area of the sample in cm^2, while the other coefficients depend on the temperature and chemical composition of the steel. In cases, where the kinetics can be described by a parabolic or cubic function, the rate constant (k_n) was determined. The rate constants were calculated for all the samples studied and for both parts of the curve, after equation (8) was derived using a simple linear regression. The basic equation for calculating the rate constant (k_n) is as follows:

$$\left(\frac{\Delta m}{A}\right)^n = k_n t \tag{14}$$

where k_n is the rate constant, where n can be 1, 2, or 3 in the case of a linear, parabolic, or cubic law, respectively. In the case of the exponential law, $1 < n < 3$. The subscript n was named according to the type of equation, so the exponential law has the subscript e, the parabolic p, and the cubic c. It should be added that the results of the rate constant of the first part of the curve are reliable. In the case of the second part, the results are less reliable up to 700 $°$ C, which can also be seen in the fluctuations of the curves.

Figure 16 shows TG results, the upper graph shows the curves of all the samples studied. It can be seen that the sample in the soft annealed state at 700 °C is different from the others, its final weight gain is 5.91 mg cm^{-2}. Since all the other samples did not gain more than 1.5 mg cm^{-2} in weight, the scale on the Y-axis was reduced to the 1.5 mg cm^{-2} in both graphs below (Figure 16b,c). The red rectangle marks the area enlarged in the left graph, where the fluctuations on the curves are visible, which are most pronounced on the curves at 400 °C and 500 °C. The mentioned effect decreases with increasing temperature of oxidation. The blue rectangle marks the area that is enlarged in the graph on the right, where the scale on the X-axis is reduced to 10 h. The reason for this is to show the difference between shape of the curves obtained from the heating section (1st part of the curve) and under isothermal conditions (2nd part of the curve). This will be explained in more detail in the discussion.

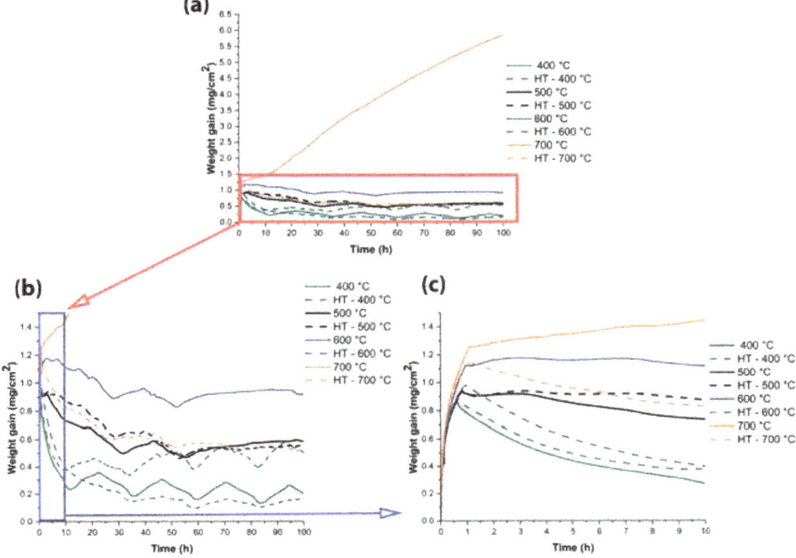

Figure 16. TG results for all the investigated samples of H11 hot-work tool steel (**a**) TG results, (**b**) enlarged Y-axis to the 1.5 mg cm^{-2} and (**c**) enlarged X-axis to the 10 h and Y-axis to the 1.5 mg cm^{-2}.

Table 4 shows the weight changes in % for the end of each part of the curve are presented. It is obvious that there is a weight loss at isothermal conditions (2nd part) for all the samples studied, except for those soft annealed at 700 °C. This is also evident from the oxidation curves in Figure 16.

Table 4. The weight changes in % for each part of the curve.

Temperature [°C]	Part of the Curve	TG [%]	
		H11	H11-HT
400	1.	0.168	0.175
	2.	0.038	0.095
500	1.	0.183	0.184
	2.	0.113	0.109
600	1.	0.218	0.191
	2.	0.178	0.034
700	1.	0.242	0.223
	2.	1.138	0.101

Each of the curves (Figure 16) was also described by the mathematical function that best fits the curves of the experimental data. The labels of part 1 (under heating conditions) and part 2 (under isothermal conditions) of the curve are used. It should be noted that oxidation itself under heating conditions affects the kinetics under isothermal conditions. First, the type of equation (Table 5) associated with each part of the curve is presented, and then the corresponding coefficients for each sample studied.

Table 5. The coefficients are given for each part of the oxidation curve the studied samples with the corresponding equation type.

Temperature [°C]	Sample	Part of Curve	Equation Type	Number of the Equation	Coefficients				
					A	B	C	D	E
400	H11	1.	exponential	10	−0.44	−192.54	−0.58	−1432.74	0.99
		2.	exponential	10	0.70	−9660.25	0.16	−230,346.79	0.17
	H11 HT	1.	exponential	10	−0.44	−200.67	−0.64	−1502.7	1.04
		2.	exponential	11	0.70	−6444.01	/	/	0.46
500	H11	1.	exponential	10	−0.48	−217.16	−0.64	−1948.81	1.09
		2.	cubic	13	0.94	$-5.73 \cdot 10^{-6}$	$2.41 \cdot 10^{-11}$	$-3.03 \cdot 10^{-17}$	/
	H11 HT	1.	exponential	10	−0.52	−220.61	−0.67	−2415.19	1.14
		2.	cubic	13	0.99	$-4.10 \cdot 10^{-6}$	$1.00 \cdot 10^{-11}$	$-5.22 \cdot 10^{-18}$	/
600	H11	1.	exponential	10	−0.48	−193.42	−0.94	−2639.44	1.38
		2.	cubic	13	1.25	$-4.61 \cdot 10^{-6}$	$1.92 \cdot 10^{-11}$	$-2.42 \cdot 10^{-17}$	/
	H11 HT	1.	exponential	10	−0.41	−242.47	−0.62	−2181.65	1.12
		2.	exponential	10	0.56	−11,080.79	0.49	−54,103.34	0.13
700	H11	1.	exponential	10	−0.42	−185.2	−0.81	−1759.68	1.27
		2.	parabolic	12	1.03	$1.62 \cdot 10^{-5}$	$-6.89 \cdot 10^{-12}$	/	/
	H11 HT	1.	exponential	10	−0.41	−186.54	−0.73	−1490.03	1.16
		2.	exponential	11	0.58	−56,349.86	/	/	0.54

Number of the equation: A two-phase exponential (10), a single-phase exponential (11), a second-degree polynomial (12—parabolic law) and a third-degree polynomial (13—cubic law).

The rate constants k (Table 6) are given for each sample studied and each part of the oxidation curve. In the case of the first part of the curve, the rate constants are accurate and there is not much variation between the results. It can also be determined, from the calculated constants, which sample had the fastest oxidation kinetics when heated, this was clearly H11 (soft annealed) at 700 °C.

However, for the second part they deviate greatly, which is logical as the curves also vary greatly and consequently the linear regression also deviates. However, comparing only the rate coefficients for the second part, a tendency can be seen, most of them are negative, which is the consequence of the mass loss during isothermal oxidation. The fastest oxidation kinetics was found for H11 at 700 °C, which was expected.

Table 6. The rate constants are given for each part of the oxidation curve for the studied samples with the corresponding equation type.

Temperature [°C]	Sample	Part of Curve	Equation Type	Rate Constants		
				k_e [mg^e cm^{-2e} s^{-1}]	k_p [mg^2 cm^{-4} s^{-1}]	k_c [mg^3 cm^{-6} s^{-1}]
400	H11	1.	exponential	$3.104 \cdot 10^{-4}$	/	
		2.	exponential	$-2.650 \cdot 10^{-7}$	/	
	H11 HT	1.	exponential	$3.475 \cdot 10^{-4}$	/	
		2.	exponential	$9.300 \cdot 10^{-8}$	/	
500	H11	1.	exponential	$2.998 \cdot 10^{-4}$	/	/
		2.	cubic	/	/	$-1.700 \cdot 10^{-7}$
	H11 HT	1.	exponential	$2.936 \cdot 10^{-4}$	/	/
		2.	cubic	/	/	$-1.57 \cdot 10^{-6}$
600	H11	1.	exponential	$4.089 \cdot 10^{-4}$	/	/
		2.	cubic	/	/	$-1.134 \cdot 10^{-7}$
	H11 HT	1.	exponential	$2.770 \cdot 10^{-4}$	/	/
		2.	exponential	$-4.983 \cdot 10^{-7}$	/	/
700	H11	1.	exponential	$4.983 \cdot 10^{-4}$	/	/
		2.	parabolic	/	$7.744 \cdot 10^{-5}$	/
	H11 HT	1.	exponential	$4.487 \cdot 10^{-4}$	/	/
		2.	exponential	$-1.526 \cdot 10^{-6}$	/	/

3.4. XRD

The results of the XRD analysis of the oxide layers at 400 °C, 500 °C, and 600 °C are shown in Figure 17. They indicate that the layers in both conditions consist of hematite, magnetite, and chromium spinel. Meanwhile, the oxide layer of the soft annealed sample at 700 °C consists of hematite, magnetite, wüstite and chromium spinel. On the other hand, the hardened and tempered sample does not contain wüstite.

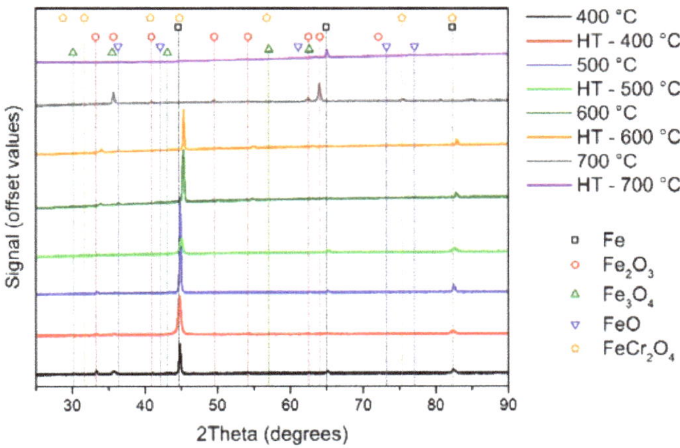

Figure 17. Results of XRD analysis of all samples examined.

4. Discussion

For discussion, we plotted a diagram of Gibbs free energies (Figure 18) for oxidation reactions of some carbides, alloying elements, and oxides, calculated using Thermo-Calc and HSC Chemistry software. All chemical reactions have been normalized to 1 mol O_2 for better comparability.

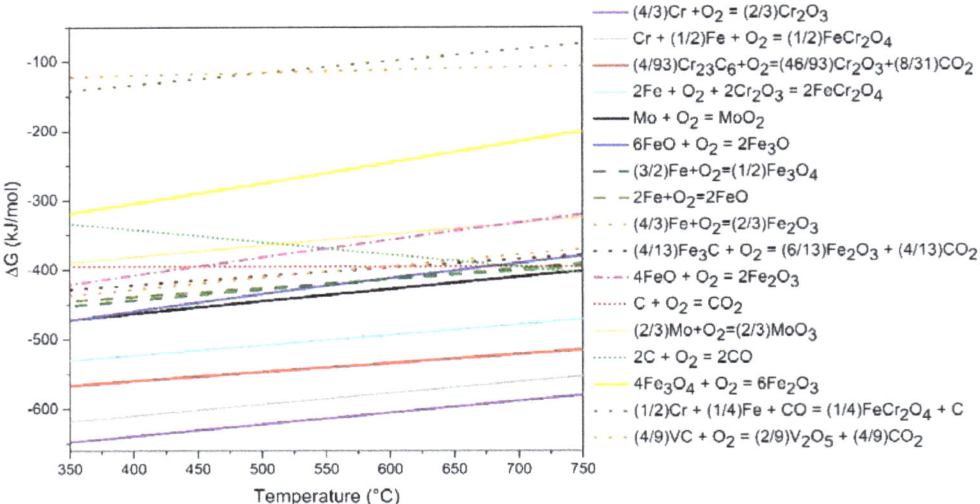

Figure 18. Calculated Gibbs free energies for oxidation reactions of selected carbides, alloying elements, and oxides.

The results of all analyses were used to explain the oxidation kinetics of H11 steel. The oxidation kinetics (Figure 16) is unstable below 700 °C, which can also be seen from the fluctuating oxidation curves. As evidenced by thermodynamic calculations and metallographic analysis, the microstructure of the steel in the soft annealed state consists of a ferrite matrix and spherical carbides (Cr, Mo, Fe)$_{23}$C$_6$ and VC, and some Ti(C,N). In the case of hardened and tempered state, the microstructure consists of a martensitic matrix and primary (VC and TiCN), and secondary (Cr, Mo, Fe)$_{23}$C$_6$) carbides. There are more of the latter, as they are precipitated during tempering.

The oxidation kinetics (Figure 16) during heating (Part 1) show in all cases the beginnings of the parabolic or cubic law (best described by the exponential growth function). Under isothermal conditions (part 2), the slope of the curves begins to decrease, except at 700 °C in the case of a soft annealed sample. As can be seen from the results presented, an oxide layer forms during heating and becomes thicker, as the oxidation temperature increases. The weight increases on average from 0.7 mg at 400 °C to 1.1 mg at 700 °C. The reason for this is that oxidation of the matrix and alloying elements and carbides occurs simultaneously during heating. It depends on the formed oxide layer, whether it protects the steel from further oxidation. In other words, whether a formed oxide layer is in thermodynamic equilibrium with the atmosphere under the given conditions or not.

Namely, in the transition from heating to the isothermal part of oxidation, it is crucial, whether the oxide layer is already in thermodynamic equilibrium with the atmosphere and protects the steel from further oxidation or not. The curves decrease at lower temperatures (400–600 °C) and increase at oxidation temperatures of 700 °C in the case of a soft annealed sample. Here it is essential, which has a greater influence on the oxidation rate or which process controls the oxidation kinetics. The oxidation of the iron matrix is on one side and the oxidation of carbides and other alloying elements in the matrix on the other side. In this case, mainly chromium, molybdenum, silicon, and vanadium. At lower temperatures (T < 700 °C), the oxidation of carbides and alloying elements dissolved in the matrix has been shown to have a greater effect on the oxidation kinetics than the oxidation of iron, even though the iron concentration in the matrix is highest. It has been shown that the oxide layer in this temperature range consists of three sublayers, namely the outer hematite, the middle magnetite, and the inner spinel. It is the latter that best protects the steel from further oxidation. Magnetite grows by the predominant transport of Fe^{2+} cations to wüstite (not yet thermodynamically stable) and Fe^{3+} to hematite. At the same time, grain boundaries in the oxide layer below 600 °C are very important for diffusion in both magnetite and hematite [12,46–48]. However, in this temperature range, diffusion is limited precisely because of the inner oxide sublayer, which consists of spinel oxide, so that the diffusion of iron and oxygen, and thus the growth of the magnetite oxide sublayer is also impeded. This is also contributed by the very fine crystalline grains of magnetite in the inner oxide sublayer (Figures 10–12), which further impede diffusion. As a result, oxygen is mainly used for the oxidation of carbides, which are most oxidized at the interface between the middle/inner oxide sublayers. From the calculated Gibbs free energies (Figure 18), chromium should oxidize first, then chromium spinels are formed, afterwards the oxidation of $M_{23}C_6$ carbide starts, followed by the oxidation of iron and finally carbon. The oxidation of the carbides themselves has been reported by several authors [49–53]. It was found that $M_{23}C_6$ carbides start to oxidize in the air atmosphere at elevated temperatures and thus can affect the oxidation kinetics, while MC carbides do not have a much effect on the oxidation kinetics. Recent research [49] has shown that $M_{23}C_6$ ($Cr_{23}C_6$) carbides in high-silicon ferritic/martensitic steel begin to oxidize under isothermal conditions in an atmosphere of air at 823 K. It was found that carbides and areas around (with increased alloying elements) can serve as potential sites for the formation of chromium oxides. In the case of $Cr_{23}C_6$, chromium begins to form chromium oxide (Cr_2O_3), which is a product of carbide oxidation. At the same time, the oxidation of the carbide itself also produces CO_2 [51], which is also a product of oxidation, the chemical reaction proceeds as follows:

$$\frac{4}{93}Cr_{23}C_6 + O_2 = \frac{46}{93}Cr_2O_3 + \frac{8}{31}CO_2 \tag{15}$$

The remaining molybdenum, silicon and iron also diffuse into the spinel oxide and partially into the magnetite. However, chromium, will not form a single oxide (Cr_2O_3), due to its low chromium content (5% by weight) [49,54,55]. As a result, it will react with iron and oxygen after a chemical reaction (16) to form a spinel-shaped oxide that forms the inner oxide sublayer.

$$2Fe + O_2 + 2Cr_2O_3 = 2FeCr_2O_4 \tag{16}$$

At the same time, MC (VC) carbides themselves do not have a significant effect on the oxidation kinetics because, as can be seen from the Gibbs free energy calculations (Figure 18), they oxidize last, slowly decompose into the inner oxide sublayer during oxidation, vanadium slowly diffuses into spinel oxides due to the decomposition of carbides and increases the density of the inner oxide sublayer. At the same time, it has been demonstrated [35,49] that silicon is also incorporated into the spinel oxide.

There are more causes of curve oscillations (Figure 16). As can be seen from the chemical reaction (15) and the oxidation reactions of other carbides (Figure 18), CO_2 is the product of carbide oxidation. This means that the local gas volume in the inner oxide sublayer will increase, especially at the interface with the middle sublayer, as carbide oxidation is most intense there. At the same time, the expansion coefficients (α) of the individual oxides formed must be considered. The CO_2 formed during the oxidation of carbides will leave the oxide layer, which is seen on the curve as a slope in the negative direction, as at the same time stresses are created in the oxide layer, causing cracking, and some oxide to fall off. It has been proven by other authors that if the resulting gas (CO or CO_2, as in our case) cannot escape through micro channels (micro pores and cracks), its pressure begins to increase, leading to the formation of new pores or even cracking or oxide layer fractures [21,56]. Mainly for the purpose of demonstrating that part of the oxide layer is crumbling, we also took an image of the morphology of the oxide layer (Figure 19—orange circles) with a SEM to explain the oscillations of the curves obtained with TG.

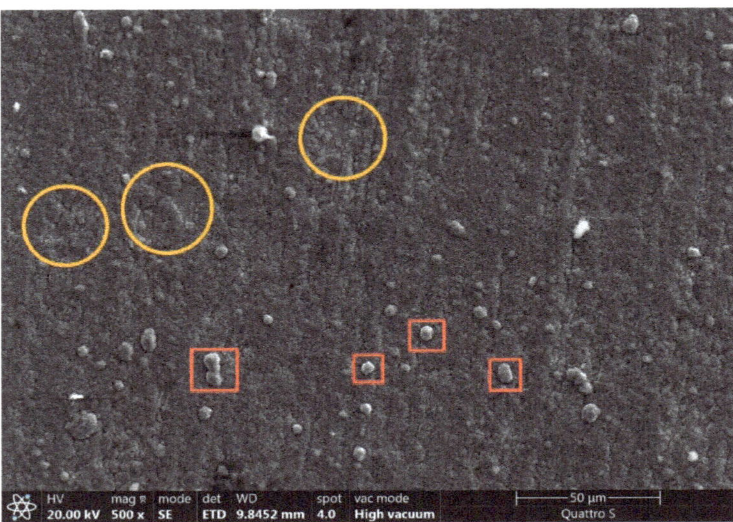

Figure 19. Oxide layer morphology for H11 steel after oxidation at 400 °C. The places where the oxide has fallen off (**orange circles**) and where the oxide has formed on the carbide (**red squares**) are marked.

At high-temperature oxidation at 700 °C, the aforementioned effect is still visible in a hardened and tempered sample, but in the soft annealed condition, it is only visible at the beginning of the oxidation, after which the oxidation rate depends mainly on oxygen and iron diffusion. This means that there is a faster diffusion of oxygen and iron in the soft annealed sample, and the resulting inner oxide sublayer and carbide oxidation no longer control the oxidation kinetics. In this case, chromium rich $M_{23}C_6$ carbides are no longer oxidized as intensively, but begin to decompose and thus the alloying elements begin to diffuse into the magnetite and form an inner oxide sublayer as is the case with other carbides. This can also be seen by the fact that both the inner and middle oxide sublayers are thicker than those that form in the temperature range between 400 °C and 600 °C. In addition, wüstite begins to form in the inner oxide sublayer, where iron diffusion is most rapid. In the hardened and tempered sample, the influence of carbide oxidation is still present, which is also evident from the curves. The curves at about 55 h oxidation slowly begin to flatten their shape (no oscillations are visible), indicating that the oxidation of the carbides no longer controls the oxidation kinetics.

In the soft annealed samples, the oxide grows throughout the crystal grain and on the carbides. In a hardened and tempered sample, the areas with increased contents of alloying elements and carbides oxidize first and then the matrix begins to oxidize.

Regarding the composition of the oxide layer, in all cases the oxide layer has the same composition and differs only in the thicknesses, which increase with temperature. Thus, the oxide layer consists of three sublayers, namely the outer hematite, the middle magnetite and the inner one of spinel oxide (Fe, Cr)$_3$O$_4$. At the oxidation temperature of 700 °C in the case of the soft annealed sample, the inner oxide sublayer consists of spinel and some wüstite. However, in the case of the hardened and tempered sample, no wüstite was detected. At the same time, a hematite band also appears on the contact surface between the steel and the inner oxide sublayer at the temperature of 700 °C. A similar composition of the oxide layer has also been reported by other researchers [40,57,58]. The thicknesses increase from 3.8 µm to 154.2 µm for the soft annealed sample and from 0.1 µm to 117.1 µm for the hardened and tempered sample (Table 3).

In the temperature range 400–600 °C, the oxide layer also consists of three sublayers (Figure 5). The inner oxide sublayer consists of spinel oxide ((Fe, Cr)$_3$O$_4$), which also contains increased concentrations of molybdenum, silicon, vanadium, and manganese. The middle oxide sublayer is also composed of magnetite, in which the concentrations of chromium, molybdenum, silicon, manganese and vanadium are much lower. The outer oxide sublayer, which is very thin, consists of hematite. This is also consistent with thermodynamic calculations, except in the case of molybdenum oxide, which was not formed. Thermodynamic calculations also indicated that internal oxidation should take place, which can be confirmed only in the case of the hardened and tempered sample, namely after oxidation at the temperature of 600 °C. Figure 20 shows the oxide layer of the H11 hot-work tool steel in soft annealed state, after oxidation at 500 °C. There are no differences in the composition of the oxide layer between soft annealed and hardened and tempered samples up to 700 °C. The only difference is the density of the inner oxide sublayer, which is denser in the hardened and tempered samples, as shown by the EBSD results.

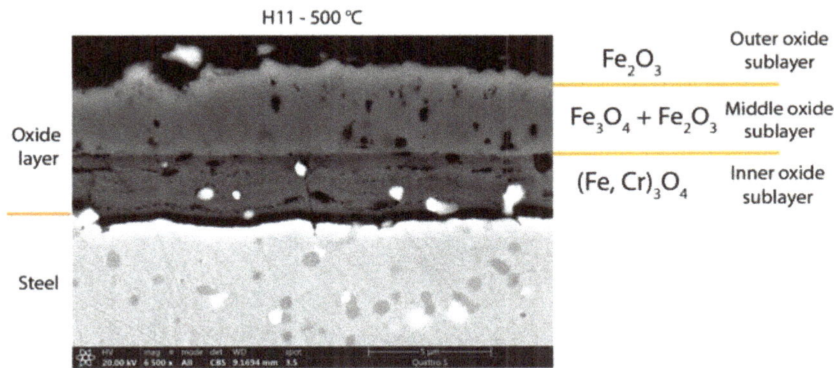

Figure 20. Oxide layer of H11 steel in the soft annealed state after oxidation at a temperature of 500 °C.

At 700 °C (Figure 21) in the case of the soft annealed state, the outer oxide sublayer also consists of hematite, the middle of magnetite and some hematite, and the inner of magnetite (Fe, Cr)$_3$O$_4$, hematite bands, and some wüstite. On the other hand, the oxide layer formed on the hardened and tempered sample after oxidation at a temperature of 700 °C was denser and did not contain any wüstite, as shown by the results of the EBSD analysis (Figure 15).

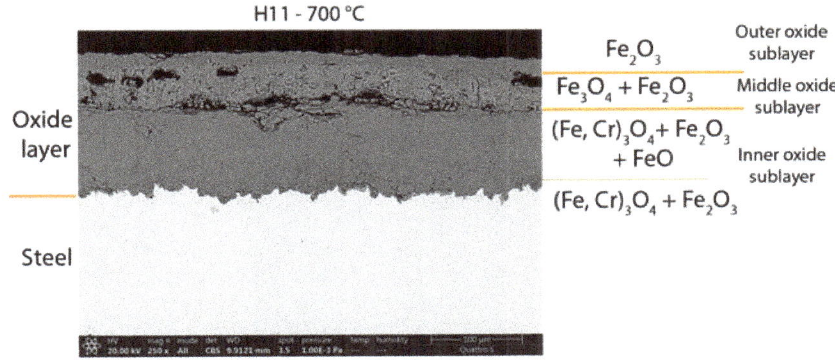

Figure 21. Oxide layer of H11 steel in the soft annealed state after oxidation at a temperature of 700 °C.

The morphology of the oxide layer formed in soft annealed (H11) and hardened and tempered (H11-HT) samples after oxidation at a temperature of 400 °C is shown in Figure 22. In the case of the soft annealed sample, the oxide starts to grow over the entire surface. On the other hand, in the case of the hardened and tempered sample, the areas with elevated concentrations of alloying elements and carbides begin to oxidize first.

H11   H11-HT

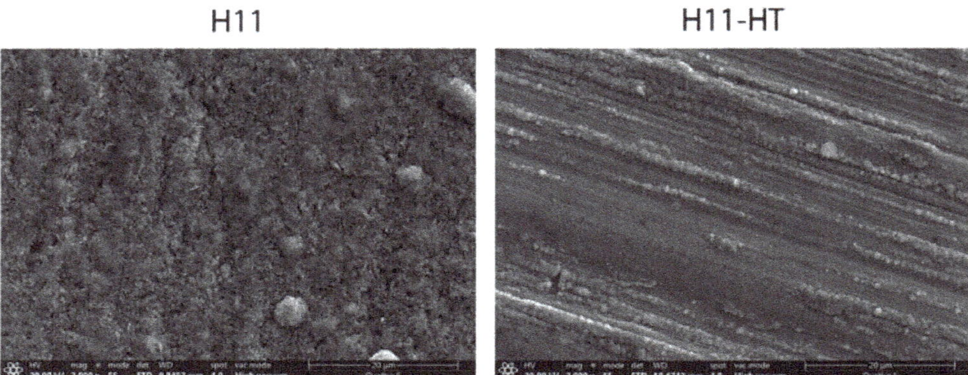

Figure 22. Morphology of the oxide layer of the soft annealed (**H11**) and hardened and tempered (**H11-HT**) sample after oxidation at a temperature of 400 °C.

5. Conclusions

Chemical composition affects high-temperature oxidation. It depends on the alloying elements, their content and affinity for oxygen, whether they form independent oxides or whether they are involved in oxidation processes at all. In the case of AISI H11 hot-work tool steel, silicon has the greatest affinity to oxygen, followed by chromium, molybdenum, carbon, and others. However, silicon did not have much influence on the high-temperature oxidation, since it was not present in the carbides (only in the matrix) and on the other hand soft annealed samples had faster oxidation kinetics. So, we assumed that chromium is the element that starts to oxidize first, since it was also present in the carbides.

Carbides and areas around (with increased concentration of alloying elements) can serve as potential sites for the formation of oxides. This is most evident at lower temperatures 400–500 °C, where oxidation of the $Cr_{23}C_6$ carbides controls the oxidation kinetics. Oxidation of the carbides causes stresses in the inner oxide sublayer, leading to pores and cracks, but can also lead to crumbling of the oxide layer, which is most evident in the oxidation kinetics at 400 °C. The higher the temperature, the smaller the effect, as diffusion is already so rapid that oxidation of the carbides no longer affects the oxidation kinetics of the steel, so that even $Cr_{23}C_6$ carbides ($\geq$700 °C) begin to slowly dissolve into the inner oxide sublayer at higher temperatures. Other carbides present, however, have no significant influence on oxidation.

The oxide layers formed during heating are in some cases sufficient protection to preserve the steel from further oxidation. The result is that the mass, which has increased during heating remains almost unchanged during further oxidation under isothermal conditions. This is observed at a temperature of 500 °C, and even more so at a temperature of 600 °C for both samples (soft annealed and hardened and tempered). Similarly, for the hardened and tempered sample at a temperature of 700 °C. In these cases, the change in mass is minimal, confirming the assumption that an oxide layer formed during heating provides sufficient protection to keep the steel from further oxidation.

Oxidation kinetics are also affected by the heat treatment of the steel. In general, the oxidation kinetics of the hardened and tempered samples are slower than the oxidation kinetics of the soft annealed samples. Small differences occur at a temperature of 400 °C and 500 °C. However, at 600 °C and 700 °C a difference is seen between soft annealed and hardened and tempered samples, with the oxidation kinetics of the latter being slower. This is largely influenced by the alloying elements dissolved in the matrix. More alloying elements are dissolved in the hardened and tempered samples and consequently the inner oxide sublayer formed is denser, preventing further diffusion of iron and oxygen.

Heat treatment affects the growth of the oxide layer and the size of the crystal grains in the oxide layer. As has been shown, the oxide layer, especially the inner sublayer, is denser in hardened and tempered samples. As a result, the growth of the oxide layer is slower because the diffusion of oxygen, iron and other alloying elements is impeded. Consequently, since the crystal grains in the matrix of the hardened and tempered samples are much smaller, the crystal grains in the oxide layer, especially in the inner sublayer, are also smaller than in the soft annealed samples.

Thermodynamic calculations give a first insight into the composition of the oxide layer. From the calculated thermodynamic results, one can conclude which oxides will form after oxidation at a given temperature. It has been shown that the calculations can be very accurate, provided we set all the input data correctly.

Author Contributions: Conceptualization and writing, T.B.; original draft preparation, T.B., J.B., M.V.; methodology and formal analysis, T.B., J.B., J.M., B.Š.B., M.V.; validation and formal analysis, T.B., J.B., B.Š.B., M.V.; supervision, J.M. and M.V. All authors have read and agreed to the published version of the manuscript.

Funding: Funding was provided by the Slovenian Research Agency ARRS program P2-0344 (B) and P2-0050 (C).

Data Availability Statement: Not applicable.

References

1. Mesquita, R.A. *Tool steels: Properties and performance*; CRC Press: Boca Raton, FL, USA, 2016.
2. Roberts, G.; Krauss, G.; Kennedy, R. *Tool Steels*, 5th ed.; ASM International: Materials park, OH, USA, 1998.
3. Zhou, Q.; Wu, X.; Shi, N.; Li, J.; Min, N. Microstructure evolution and kinetic analysis of DM hot-work die steels during tempering. *Mater. Sci. Eng. A* **2011**, *528*, 5696–5700. [CrossRef]
4. Medvedeva, A.; Bergström, J.; Gunnarsson, S.; Andersson, J. High-temperature properties and microstructural stability of hot-work tool steels. *Mater. Sci. Eng. A* **2009**, *523*, 39–46. [CrossRef]
5. Zhang, Z.; Delagnes, D.; Bernhart, G. Microstructure evolution of hot-work tool steels during tempering and definition of a kinetic law based on hardness measurements. *Mater. Sci. Eng. A* **2004**, *380*, 222–230. [CrossRef]
6. Mebarki, N.; Delagnes, D.; Lamesle, P.; Delmas, F.; Levaillant, C. Relationship between microstructure and mechanical properties of a 5% Cr tempered martensitic tool steel. *Mater. Sci. Eng. A* **2004**, *387–389*, 171–175. [CrossRef]
7. Jilg, A.; Seifert, T. Temperature dependent cyclic mechanical properties of a hot work steel after time and temperature dependent softening. *Mater. Sci. Eng. A* **2018**, *721*, 96–102. [CrossRef]
8. Caliskanoglu, D.; Siller, I.; Ebner, R.; Leitner, H.; Jeglitsch, F.; Waldhauser, W. Thermal Fatigue and Softening Behavior of Hot Work Tool Steels. *Proc. 6th Int. Tool. Conf.* **2002**, *2*, 707–719.
9. Markežič, R.; Mole, N.; Naglič, I.; Šturm, R. Time and temperature dependent softening of H11 hot-work tool steel and definition of an anisothermal tempering kinetic model. *Mater. Today Commun.* **2020**, *22*. [CrossRef]
10. Lai, G.Y. *High-Temperature Corrosion And Materials Applications*; ASM International: Materials park, OH, USA, 2007.
11. Revie, R.W. *Uhlig's corrosion handbook*; John Wiley & Sons, Inc.: Hoboken, NJ, USA, 2011.
12. Richardson, T.J.A.; Cottis, B.; Lindsay, R.; Lyon, S.; Scantlebury, D.; Stott, H.; Graham, M. *Shreir's Corrosion*; Elsevier Science: London, UK, 2009; Volume 1.
13. Cramer, S.D.; Covino Jr, B.S. *ASM Handbook Volume 13B: Corrosion: Materials*; ASM International: Materials Park, OH, USA, 2005; Volume 13.
14. Ellingham, H.J.T. Reducibility of oxides and sulphides in metallurgical processes. *J. Soc. Chem. Ind.* **1944**, *63*, 125–160. [CrossRef]
15. Popov, B.N. *Corrosion Engineering: Principles and Solved Problems*; Elsevier: London, UK, 2015.
16. Pedeferri, P. *Corrosion Science and Engineering*; Springer International Publishing: Cham, Switzerland, 2018.
17. Young, D.J. *High Temperature Oxidation and Corrosion of Metals*; Elsevier: London, UK, 2008.
18. Birks, N.; Meier, G.H.; Pettit, F.S. *Introduction to the High Temperature Oxidation of Metals*; Cambridge University Press: Cambridge, UK, 2006.
19. Hauffe, K. *Oxidation of metals*; Springer US: New York, NY, USA, 1965.

20. Davies, M.H.; Simnad, M.T.; Birchenall, C.E. On the Mechanism and Kinetics of the Scaling of Iron. *JOM* **1951**, *3*, 889–896. [CrossRef]

21. Chen, R.Y.; Yeun, W.Y.D. Review of the High-Temperature Oxidation of Iron and Carbon Steels in Air or Oxygen. *Oxid. Met.* **2003**, *59*, 433–468. [CrossRef]

22. Pilling, N.B.; Bedworth, R.E. Oxidation of metals at high temperatures. *J. Inst. Met.* **1923**, *29*, 529–539.

23. Ajersch, F. Scale formation in steel processing operation. In *Proceedings of the 34th Mechanical Working and Steel Processing Conference Proceedings*; KUHN, L.G., Ed.; Iron and Steel Society: Warrendale, PA, USA, 1993; pp. 419–437.

24. Caplan, D.; Cohen, M. Effect of cold work on the oxidation of iron from 400–650 °C. *Corros. Sci.* **1966**, *6*, 327–335. [CrossRef]

25. Caplan, D.; Sproule, G.I.; Hussey, R.J. Comparison of the kinetics ofhigh-temperature oxidation of Fe as influenced by metal purity and cold work. *Corros. Sci.* **1970**, *10*, 9–17. [CrossRef]

26. Caplan, D.; Cohen, M. Scaling of iron at 500 °C. *Corros. Sci.* **1963**, *3*, 139–143. [CrossRef]

27. Boggs, W.E.; Kachik, R.H. The Oxidation of Iron-Carbon Alloys at 500 °C. *J. Electrochem. Soc.* **1969**, *116*, 424–430. [CrossRef]

28. Abuluwefa, H.; Guthrie, R.I.L.; Ajersch, F. The effect of oxygen concentration on the oxidation of low-carbon steel in the temperature range 1000 to 1250 °C. *Oxid. Met.* **1996**, *46*, 423–440. [CrossRef]

29. Chen, R.Y.; Yuen, W.Y. Oxidation of low-carbon, low-silicon mild steel at 450–900 °C under conditions relevant to hot-strip processing. *Oxid. Met.* **2002**, *57*, 53–79. [CrossRef]

30. Caplan, D.; Sproule, G.I.; Hussey, R.J.; Graham, M.J. Oxidation of Fe-C alloys at 500 °C. *Oxid. Met.* **1978**, *12*, 67–82. [CrossRef]

31. Caplan, D.; Sproule, G.I.; Hussey, R.J.; Graham, M.J. Oxidation of Fe-C alloys at 700 °C. *Oxid. Met.* **1979**, *13*, 255–272. [CrossRef]

32. Malik, A.U.; Whittle, D.P. Oxidation of Fe-C alloys in the temperature range 600–850 °C. *Oxid. Met.* **1981**, *16*, 339–353. [CrossRef]

33. Cao, G.; Liu, X.; Sun, B.; Liu, Z. Morphology of oxide scale and oxidation kinetics of low carbon steel. *J. Iron Steel Res. Int.* **2014**, *21*, 335–341. [CrossRef]

34. Bak, S.-H.; Kim, M.-J.; Lee, J.-H.; Bong, S.-J.; Kim, S.-K.; Lee, D.-B. High-temperature oxidation kinetics and scales formed on Fe-2.3%Cr-1.6%W alloy. *J. Korean Ceram. Soc.* **2011**, *48*, 57–62. [CrossRef]

35. Hao, M.; Sun, B.; Wang, H. High-Temperature Oxidation Behavior of Fe–1Cr–0.2Si Steel. *Materials* **2020**, *13*, 509. [CrossRef]

36. Li, D.; Dai, Q.; Cheng, X.; Wang, R.; Huang, Y. High-Temperature Oxidation Resistance of Austenitic Stainless Steel Cr18Ni11Cu3Al3MnNb. *J. Iron Steel Res. Int.* **2012**, *19*, 74–78. [CrossRef]

37. Ghosh, S.; Kumar, M.K.; Kain, V. High temperature oxidation behavior of AISI 304L stainless steel - Effect of surface working operations. *Appl. Surf. Sci.* **2013**, *264*, 312–319. [CrossRef]

38. Kofstad, P. Oxidation of Metals: Determination of Activation Energies. *Nature* **1957**, *179*, 1362–1363. [CrossRef]

39. Browne, K.M.; Dryden, J.; Assefpour, M. Modelling Scaling and Descaling in Hot Strip Mills. In *Proceedings of the Recent Advances in Heat Transfer and Micro-Structure Modelling for Metal Processing*; Too, J.J.M., Guo, R.-M., Eds.; American Society of Mechanical Engineers: San Francisco, CA, USA, 1995; pp. 187–198.

40. Bidabadi, M.H.S.; Chandra-ambhorn, S.; Rehman, A.; Zheng, Y.; Zhang, C.; Chen, H.; Yang, Z.G. Carbon depositions within the oxide scale and its effect on the oxidation behavior of low alloy steel in low (0.1 MPa), sub-(5 MPa) and supercritical (10 MPa) CO_2 at 550 °C. *Corros. Sci.* **2020**, *177*, 108950. [CrossRef]

41. Zhang, X.; Jie, X.; Zhang, L.; Lui, S.; Zheng, Q. Improving the high-temperature oxidation resistance of H13 steel by laser cladding with a WC/Co-Cr alloy coating. *Anti-Corrosion Methods Mater.* **2016**, *63*, 171–176. [CrossRef]

42. Min, Y.; Wu, X.; Wang, K.; Li, L.; Xu, L. Prediction and analysis on oxidation of H13 hot work steel. *J. Iron Steel Res. Int.* **2006**, *13*, 44–49. [CrossRef]

43. Pieraggi, B.; Rolland, C.; Bruckel, P. Morphological characteristics of oxide scales grown on H11 steel oxidised in dry or wet air. *Mater. High Temp.* **2005**, *22*, 61–68. [CrossRef]

44. Bruckel, P.; Lamesle, P.; Lours, P.; Pieraggi, B. Isothermal oxidation behaviour of a hot-work tool steel. *Mater. Sci. Forum.* **2004**, *461*, 831–838. [CrossRef]

45. Thermo-Calc Software. *TCOX9 - TCS Metal Oxide Solutions Database*, 7. Database version 9; Thermo-Calc Software: Solna, Sweden, 1992.

46. Atkinson, A. Transport processes during the growth of oxide films at elevated temperature. *Rev. Mod. Phys.* **1985**, *57*, 437–470. [CrossRef]

47. Atkinson, A.; Taylor, R.I. Diffusion of 55Fe in Fe2O3 single crystals. *J. Phys. Chem. Solids* **1985**, *46*, 469–475. [CrossRef]

48. Channing, D.A.; Graham, M.J. A study of iron oxidation processes by Mössbauer spectroscopy. *Corros. Sci.* **1972**, *12*, 271–280. [CrossRef]

49. Ye, Z.; Wang, P.; Li, D.; Li, Y. M23C6 precipitates induced inhomogeneous distribution of silicon in the oxide formed on a high-silicon ferritic/martensitic steel. *Scr. Mater.* **2015**, *97*, 45–48. [CrossRef]

50. Gong, Y.; Young, D.J.; Kontis, P.; Chiu, Y.L.; Larsson, H.; Shin, A.; Pearson, J.M.; Moody, M.P.; Reed, R.C. On the breakaway oxidation of Fe9Cr1Mo steel in high pressure CO2. *Acta Mater.* **2017**, *130*, 361–374. [CrossRef]

51. Jung, K.H.; Kim, S.J. Role of M23C6 carbide on the corrosion characteristics of modified 9Cr-1Mo steel in N2-O2-CO2-SO2 atmosphere at 650 °C. *Appl. Surf. Sci.* **2019**, *483*, 417–424. [CrossRef]

52. Wu, S.; Fei, Y.; Guo, B.; Jing, L. Corrosion of Cr23C6 coated Q235 steel in wet atmospheres containing Na2SO4 at 750 °C. *Corros. Sci.* **2015**, *100*, 306–310. [CrossRef]

53. Li, Z.; Cao, G.; Lin, F.; Cui, C.; Wang, H.; Liu, Z. Phase transformation behavior of oxide scale on plain carbon steel containing 0.4 wt.% Cr during continuous cooling. *ISIJ Int.* **2018**, *58*, 2338–2347. [CrossRef]

54. Mrowec, S. On the mechanism of high temperature oxidation of metals and alloys. *Corros. Sci.* **1967**, *7*, 563–578. [CrossRef]
55. Wood, G.C. High-temperature oxidation of alloys. *Oxid. Met.* **1970**, *2*, 11–57. [CrossRef]
56. Sachs, K.; Brown, J.R. A theory of decarburization by scale. *J. Iron Steel Inst.* **1958**, *190*, 169–170.
57. Saunders, S.R.J.; Monteiro, M.; Rizzo, F. The oxidation behaviour of metals and alloys at high temperatures in atmospheres containing water vapour: A review. *Prog. Mater. Sci.* **2008**, *53*, 775–837. [CrossRef]
58. Zurek, J.; Wessel, E.; Niewolak, L.; Schmitz, F.; Kern, T.U.; Singheiser, L.; Quadakkers, W.J. Anomalous temperature dependence of oxidation kinetics during steam oxidation of ferritic steels in the temperature range 550–650 °C. *Corros. Sci.* **2004**, *46*, 2301–2317. [CrossRef]

Article

Microstructural Evolution of 9CrMoW Weld Metal in a Multiple-Pass Weld

Yu-Lun Chuang [1], **Chu-Chun Wang** [1], **Tai-Cheng Chen** [2], **Ren-Kae Shiue** [3] **and Leu-Wen Tsay** [1,4,*]

[1] Department of Optoelectronics and Materials Technology, National Taiwan Ocean University, Keelung 20224, Taiwan; aisaka8514@gmail.com (Y.-L.C.); welfare0717@laser-station.com (C.-C.W.)

[2] Nuclear Fuels and Materials Division, Institute of Nuclear Energy Research, Taoyuan 32546, Taiwan; tcchen@iner.gov.tw

[3] Department of Materials Science and Engineering, National Taiwan University, Taipei 10617, Taiwan; rkshiue@ntu.edu.tw

[4] Center of Excellence for Ocean Engineering, National Taiwan Ocean University, Keelung 20224, Taiwan

* Correspondence: b0186@mail.ntou.edu.tw; Tel.: +886-2-2462-2192 (ext. 6405)

Citation: Chuang, Y.-L.; Wang, C.-C.; Chen, T.-C.; Shiue, R.-K.; Tsay, L.-W. Microstructural Evolution of 9CrMoW Weld Metal in a Multiple-Pass Weld. *Metals* **2021**, *11*, 847. https://doi.org/10.3390/met11060847

Academic Editors: Stefano Spigarelli and Elisabetta Gariboldi

Received: 23 April 2021
Accepted: 16 May 2021
Published: 21 May 2021

Abstract: 9CrMoW steel tubes were welded in multiple passes by gas-tungsten arc welding. The reheated microstructures in the Gr. 92 weld metal (WM) of a multiple-pass weld were simulated with an infrared heating system. Simulated specimens after tempering at 760 °C/2 h were subjected to constant load creep tests either at 630 °C/120 MPa or 660 °C/80 MPa. The simulated specimens were designated as the over-tempered (OT, below A_{C1}, i.e., WT-820T) and partially transformed (PT, below A_{C3}, i.e., WT-890T) samples. The transmission electron microscope (TEM) micrographs demonstrated that the tempered WM (WT) displayed coarse martensite packets with carbides along the lath and grain boundaries. Cellular subgrains and coarse carbides were observed in the WT-820T sample. A degraded lath morphology and numerous carbides in various dimensions were found in the WT-890T sample. The grain boundary map showed that the WT-820T sample had the same coarse-grained structure as the WT sample, but the WT-890T sample consisted of refined grains. The WT-890T samples with a fine-grained structure were more prone to creep fracture than the WT and WT-820T samples were. Intergranular cracking was more likely to occur at the corners of the crept samples, which suffered from high strain and stress concentration. As compared to the Gr. 91 steel or Gr. 91 WM, the Gr. 92 WM was more stable in maintaining its original microstructures under the same creep condition. Undegraded microstructures of the Gr. 92 WM strained at elevated temperatures were responsible for its higher resistance to creep failure during the practical service.

Keywords: 9CrMoW; weld metal; reheated microstructure; creep resistance

1. Introduction

Increasing the steam pressure and operating temperature of a boiler in a coal-fired power plant can reduce the CO_2 emissions and save energy. Advanced 9–12 Cr ferritic steels are often applied to manufacture boilers in ultra-supercritical fossil-fired power stations [1]. The Gr. 92 steel is applied extensively to replace T/P91 steel in steam pipes, headers, reheaters and superheaters [1]. $M_{23}C_6$ carbides precipitate mainly at the prior austenite grain boundaries (PAGBs). The MX precipitates composed of V, Nb, C, and N are dispersed in tempered martensite laths of normalized and tempered T92 steel [2]. Moreover, after a long period of service, the microstructures of the Gr. 92 steel deteriorate, and the mechanical properties are degraded. After creep of Gr. 92 steel, the laths and $M_{23}C_6$ carbides increase in size and the dislocation density decreases near the crack tip, as compared with those of the virgin material [3]. Aging P92 weld metal (WM) at 625 °C/1000 h causes a decisive reduction in the impact energy, which is ascribed to the coarse solidified microstructure, Laves phase and $M_{23}C_6$ carbide [4].

Metals **2021**, *11*, 847. https://doi.org/10.3390/met11060847

Complex transformation proceeds in the heat-affected zone (HAZ) of an arc weld. Creep failure in the HAZ of advanced ferritic steel welds is identified as Type IV cracking [5–8]. The cracking of the Gr. 92 steel is attributed to the lack of precipitation-strengthening at PAGBs and block boundaries [9–11]. The coarsening of $M_{23}C_6$ carbides at the PAGBs can be inhibited by adding B at 130 ppm to 9CrWCo steel [12]. Moreover, increasing B and decreasing N contents can suppress grain refinement after welding through a larger pinning effect imposed by B-stabilized $M_{23}C_6$ carbides [13]. Among the different zones of the P92 weld, the fine-grained HAZ (FGHAZ), heated just above the temperature of A_{C3}, shows the highest degree of creep damage [14]. As reported in international journals, the microstructural evolution of an advanced ferritic steel weld generally focuses on the creep failure of the simulated HAZ, and less attention is paid to the WM. The creep growth rate of the P92 WM is faster than that of the base metal (BM) but slower that of the FGHAZ [15]. With the same tempering at 760 °C/2 h, re-austenization at 1050 °C can significantly increase the Charpy energy of the Gr. 92 WM [16]. Despite the finer structure of electron and laser beam welds, as compared to gas-tungsten arc welds, the variation in hardness in the Gr. 92 WM with the three welding processes is not significant [17].

In the open literature, great concern has been paid to the degradation of creep resistance in the FGHAZ of the 9–12 Cr steel welds. It is noticed that few studies have clarified the effects of the solidified microstructure and/or welding thermal cycle on the creep resistance of Gr. 92 WM. In a multiple-pass weld, the prior deposits will be reheated by the following passes, producing a reheated microstructure. The microstructures of the reheated zone in the WM will depend on the peak temperature of the imposed thermal cycles. In this study, the reheated microstructure in a Gr. 92 WM was simulated by an infrared heating system. The goal of this study was to inspect the simulated over-tempered and partially transformed microstructures in the WM of a multiple-pass weld. Furthermore, the impacts of welding thermal cycles on the microstructures of the Gr. 92 WM were related to creep rupture life.

2. Material and Experimental Procedures

Gr. 92 tubes with 50 mm in diameter and 9.6 mm in wall thickness were applied for the butt joint in this work. The chemical composition of the tube was 0.003 B, 0.12 C, 8.89 Cr, 0.43 Mn, 0.47 Mo, 0.038 N, 0.05 Nb, 0.24 Ni, 0.014 P, 0.002 S, 0.23 Si, 0.003 Ti, 1.76 W, 0.20 V and balanced Fe in wt.%. The wire filler composition was 0.002 B, 0.09 C, 9.66 Cr, 0.65 Mn, 0.44 Mo, 0.037 N, 0.12 Nb, 0.42 Ni, 0.013 P, 0.002 S, 0.29 Si, 0.0002 Ti, 0.19 V, 1.46 W and balanced Fe. A single V groove was machined for the gas-tungsten-arc welding of the T92 tube in multiple passes. Due to the size restriction of the T92 tube, a Gr. 91 steel plate of 12.5 mm cut into a double V groove was used as the substrate for the preparation of the creep sample of all WM. The schematic dimensions of the tested sample sliced from the butt-welded groove weld are shown in Figure 1. A dilatometer preset at distinct heating and cooling rates was examined to determine the A_{C1}, A_{C3}, M_s and M_f temperatures of the WM for comparison with the T92 substrate [18]. The tempered WM (WT) was heated using an infrared heater to generate the reheated microstructures present in a multiple-pass weld. Reheated microstructures were produced by heating the tempered WM to 820 °C (WT-820) or 890 °C (WT-890) for 60 s. These temperatures were slightly below either the A_{C1} or A_{C3} temperatures of the WM, respectively. The simulated specimens were tempered at 760 °C for 2 h and labeled the WT-820T or WT-890T, according to the prior heating temperature.

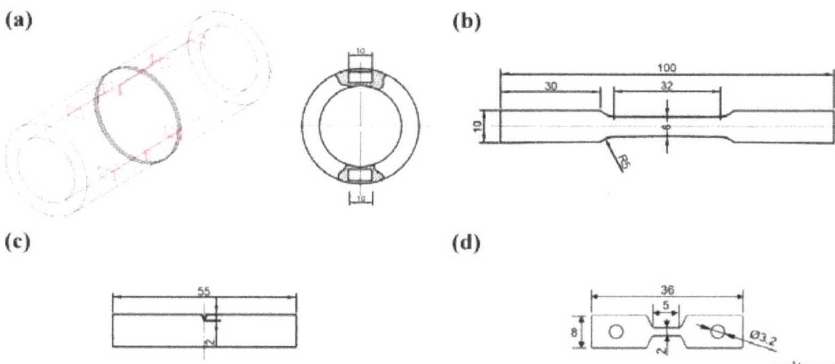

Figure 1. (**a**) The butt-welded groove weld; the dimensions of the (**b**) tensile, (**c**) Charpy V-notch and (**d**) creep samples cut from the finished weld.

The sample hardness was obtained with a MVK-G1500 micro-Vickers hardness tester (Mitutoyo, Kawasaki, Japan) under 300 gf loading for 15 s. To assess the influence of the reheated microstructures on the creep resistance, simulated specimens after tempering were subjected to the short-term creep tests with a constant load at either 630 °C/120 MPa or 660 °C/80 MPa. As shown in Figure 1b,c, the tensile and impact specimens of 5 mm thickness satisfied the ASTM E8 and E23 specifications, respectively. The tensile strain rate was kept at 1.6×10^{-4}/s at room temperature. The creep tests were terminated if no cracking occurred after straining for 4000 h. The results shown in this work are the average of three samples for each test. Their microstructures were explored with the aid of a JSM-7100F scanning electron microscope (SEM, JEOL, Tokyo, Japan). The minute microstructures in the specimens were investigated using a Tecnai G2 F30 transmission electron microscope (TEM, FEI, Hillsboro, OR, USA). A NordlysMax^2 electron backscatter diffraction (EBSD, Oxford Instruments, Abingdon, UK) detector was used to examine the grain size, high-angle and subgrain boundaries of the sample.

3. Results

3.1. Determination of Transformation Temperatures

Table 1 lists the measured A_{C1}, A_{C3}, M_s and M_f temperatures of the Gr. 92 WM at selected heating/cooling rates and compared them with those of the T92 substrate [18]. Transformation temperatures of A_{C1} and A_{C3} increased slightly with an increase in the heating rates. For the heating/cooling rates of 15 °C/s, the A_{C1} and A_{C3} temperatures of the Gr. 92 WM were 862 °C and 898 °C, relative to 886 °C and 934 °C for the T92 substrate. The A_{C1} and A_{C3} temperatures of the Gr. 92 WM were 20–30 °C lower than those of the T92 substrate. This implied that the phase transformation of Gr. 92 WM would occur at a lower temperature range than that of the T92 substrate if multiple welding passes were performed. In contrast, the M_s temperature was around 380 °C and the M_f temperature dropped to about 250 °C at the high cooling rate (45 °C/s). Thus, martensite would be formed completely after the solidified WM cooled to room temperature.

Table 1. Transformation temperatures of Gr. 92 WM determined by a dilatometer at specific heating/cooling rates.

Rates (°C/s)	Heating Rate (°C/s)				Cooling Rate (°C/s)			
Temperature (°C)	5	15	30	45	5	15	30	45
A_{C1}	851	862	869	894	-	-	-	-
A_{C3}	880	898	904	933				
M_s	-	-	-	-	388	385	384	377
M_f					285	284	280	249

3.2. Microhardness Distribution in the Cross Welds

The cross-sectional appearance of a T92 weld and microhardness curves in the as-welded (AW) and tempered conditions are shown in Figure 2. The WM microhardness fell to a range of HV 380 to 420 in the AW condition. The maximum microhardness above HV 450 appeared in the site near to the fusion boundary (FB). In the AW condition, the HAZ showed high hardening at a depth within 3 mm from the FB, and then a steep drop in microhardness further away from the FB. After tempering at 720 °C/2 h, the WM hardness decreased to below HV 350 and the HAZ microhardness dropped obviously, as compared with that of the AW condition. With an increase in tempering temperature to 760 °C, continuous decreases in WM and HAZ microhardness were observed, but the softening effect was weaker. The results reveal that the BM microhardness was less affected, regardless of the tempering conditions. It was noticed that a narrow soft zone developed ahead of the BM and it was more obvious in the weld tempered at 760 °C. Overall, after tempering at 760 °C, the WM microhardness was a little higher than those of the BM and HAZ.

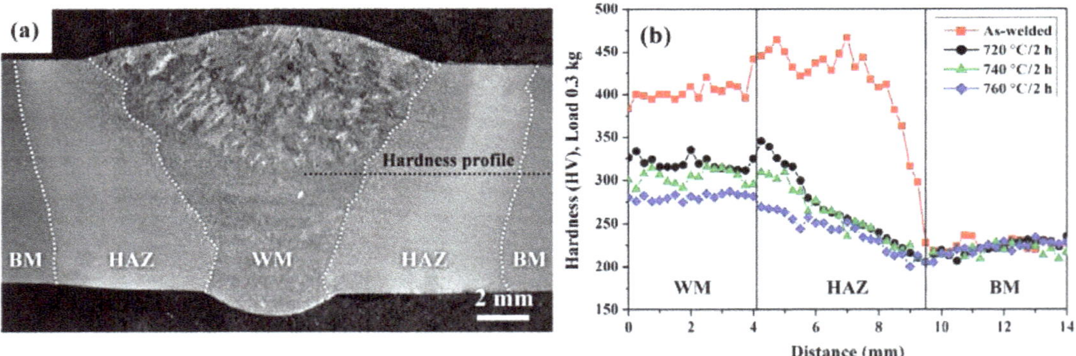

Figure 2. (**a**) Macroscopic appearance of a T92 weld in cross-sectional view cut from the butt-welded tube, (**b**) microhardness curves of the welds determined from the centerline of the WM across the heat-affected zone (HAZ) to the base metal (BM) in the as-welded and tempered conditions.

3.3. Mechanical Properties of a Cross Weld

Table 2 lists tensile properties of the Gr. 92 cross weld and the impact toughness of the WM in the AW or tempered conditions at 25 °C. For the T92 BM cut from the steel tube, the ultimate tensile strength (UTS) and yield strength (YS) were about 711 and 554 MPa, respectively. The elongation of the T92 BM was as high as 29%. In the AW condition, the UTS of the cross weld (707 MPa) was about the same as that of the BM, but the YS of the weld (526 MPa) was about 30 MPa lower than that of the BM. Moreover, the elongation of the cross weld (17%) in the AW condition was obviously lower than that of the BM, which could be attributed to the minimal deformation of the hardened HAZ during straining. Moreover, the fracture location of the AW condition was at the

BM. In the cross weld tempered at 740 °C, the UTS and YS of the weld were 667 and 500 MPa, respectively, together with 21% elongation. Increasing the tempering temperature to 760 °C, the UTS and YS of the weld decreased to 659 and 472 MPa, respectively, and the elongation improved to 24%. After tempering, the fracture locations of all the cross welds were at the HAZ near to the BM. The results indicate that increasing the tempering temperature would reduce the tensile strength but improve the ductility of the cross weld, as compared to the AW condition. A sub-size Charpy V-notch specimen of 5 mm thickness was used to determine the impact energy of the WM. The impact energy of T92 BM was about 80 J, which was obviously higher than that of the WM (12 J) in the AW condition. After tempering at 740 °C, the impact energy of the tempered WM was about 54 J, which was slightly lower than that of the BM. The impact energy of the WM raised to 80 J after tempering at 760 °C. The results indicate that the WM had very high brittleness in the AW condition, whereas the WM tempered at 760 °C exhibited the same resistance to impact fractures as the BM.

Table 2. Typical mechanical properties of the Gr. 92 cross welds in the as-welded and post-weld tempered conditions.

Specimen	Yield Strength (MPa)	Ultimate Tensile Strength (MPa)	Elongation (%)	CVN Impact Toughness (J)
AW [1]	506	707	17	12
PT 1 [2]	500	667	21	54
PT 2 [3]	471	659	24	80
BM [4]	554	711	29	80

[1] AW: as-welded. [2] PT 1: post-weld tempering at 740 °C/2 h. [3] PT 2: post-weld tempering at 760 °C/2 h. [4] BM: base metal.

3.4. Microhardness of the Simulated Sample

The micro-Vickers hardnesses of the WM specimens after different thermal treatments were determined. The peak WM microhardness in the AW condition (W sample) was about HV 450 and showed a certain degree of fluctuations in the microhardness of WM owing to the imposed thermal cycles of subsequent passes. After tempering at 760 °C/2 h, the microhardness of the WT sample was reduced to HV 266. The WT specimens reheated to 820 °C (WT-820) demonstrated little change in microhardness (HV 263) as compared with that of the WT sample. It was indicated that the short-term over-tempering at 820 °C induced less variation in the microstructures. Subjecting the WT-820 sample to tempering treatment, the specimen's microhardness decreased slightly to HV 254 (WT-820T sample). In the case of the WT sample reheated to 890 °C (WT-890 sample), the sample microhardness rose to HV 354, which was the result of incomplete hardening. Tempering the WT-890 sample was found to lower its microhardness to about HV 259, i.e., the WT-890T sample. The results reveal that the WT, WT-820T and WT-890T samples were all similar in microhardnesses but might have different microstructures.

3.5. Microstructural Observations and IPF Identifications

SEM micrographs and the inverse pole figure (IPF) maps, illustrating individual grain orientations of the WT, WT-820T and WT-890T samples in different colors, are displayed in Figure 3. Different colors in the IPF map represent the different orientations of the grains. The martensite packets of the WT sample were very coarse and mostly aligned in the same direction (Figure 3a,b). Coarser martensite packets separated by grain boundaries were seen in the WT-820T sample (Figure 3c,d). Those observations indicate that the short-term over-heating had little effect on the Gr. 92 WM microstructure. However, dislocation recovery, break-down of the lath structure, polygonization and carbide coarsening can occur during over-tempering of Gr. 92 steel [18,19]. Moreover, significant changes in granular morphology occurred in the WT-890T sample (Figure 3e,f). The SEM micrograph (Figure 3e) revealed that the WT-890T sample seemed to be composed of many fine grains.

The IPF map showed that the coarse martensite packets in the WT and WT-820T samples were replaced by short and fine ones in the WT-890T sample (Figure 3f). The fine-grained zone in the WM could be found after reheating to below the A_{C3} temperature in a multiple pass weld.

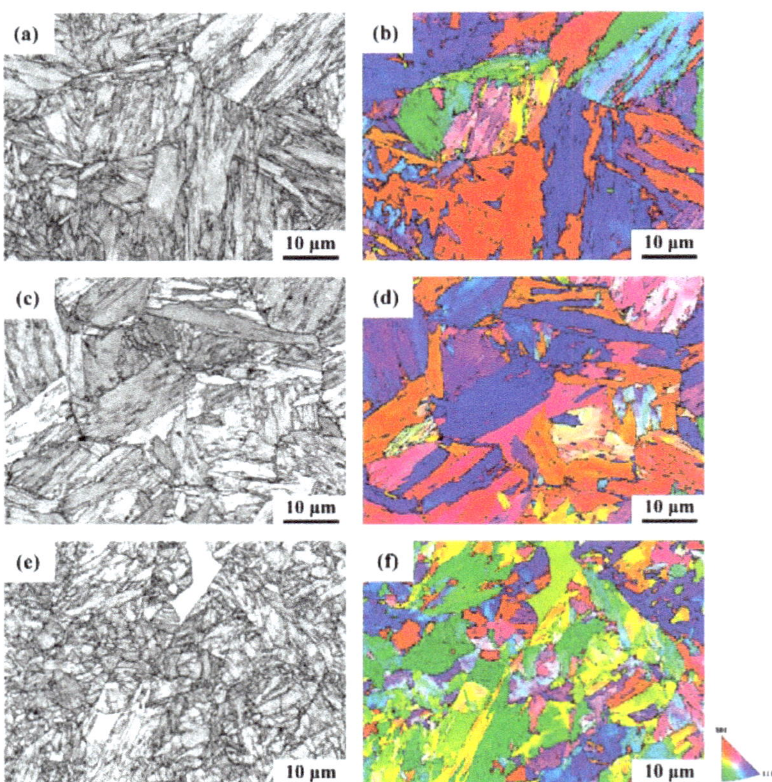

Figure 3. (**a**,**b**) WT, (**c**,**d**) WT-820T, (**e**,**f**) WT-890T. (**a**,**c**,**e**) Scanning electron micrographs; (**b**,**d**,**f**) the corresponding inverse pole figure (IPF) maps.

TEM micrographs of the Gr. 92 WM tempered under different conditions are shown in Figure 4. In the AW condition, the WM consisted of lath martensite with a high dislocation density and few precipitates in the sample (Figure 4a). After tempering at 760 °C, the dislocation density of the WT sample decreased, and the lath martensite boundaries and grain boundaries were decorated by precipitates (Figure 4b). The main precipitates present in the WT specimens were $M_{23}C_6$ carbides, as confirmed by the diffraction pattern. It has been reported that some fine precipitates, possibly carbides or carbonitrides (NbC, VC, and (NbV)(CN)), could be dispersed in the tempered martensite matrix [2]. For the WT-820T sample, the TEM micrograph showed a low dislocation density and fine cell subgrains instead of parallel lath martensite (Figure 4c), as compared to the WT sample. Furthermore, some carbides in the WT-820T sample aggregated to a very coarse size of about 400 nm and above. As shown in Figure 4d, the degraded lath morphology and the precipitation of numerous carbides of different sizes were seen in the WT-890T sample. The carbides in the WT-890T sample were greater in amount than those in the WT and WT-820T samples, but the exact reason for this is not known at this time.

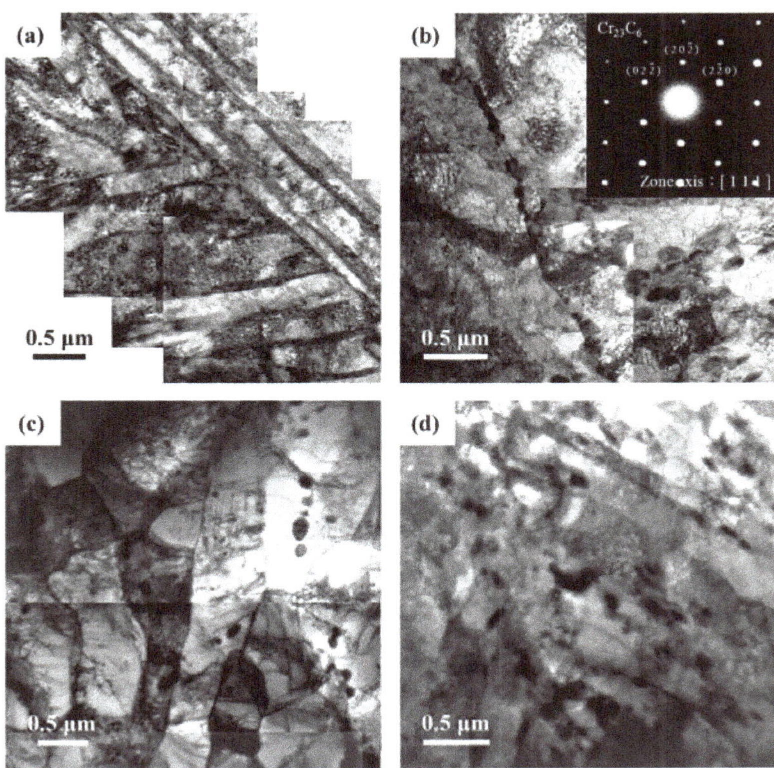

Figure 4. TEM micrographs of the (**a**) W, (**b**) WT, (**c**) WT-820T and (**d**) WT-890T samples.

3.6. Grain Boundary Characteristics

Figure 5 presents grain boundary maps showing the grain boundary characteristics of distinct samples. The PAGBs are high-angle grain boundaries (HAGBs), whereas the lath boundaries and block/packet boundaries are low-angle grain boundaries (LAGBs). As shown in Figure 5, the LAGBs are indicated by red (1°~5°) and green (5°~15°) lines, whereas the HAGBs (15°~62.5°) are indicated by black. The IPF map (Figure 3b) demonstrates that the WT sample consisted of coarse martensite packets. The grain boundary map of the WT sample shows that the coarse martensite packets were outlined by HAGBs (Figure 5a). Furthermore, the appearance of the HAGBs within a coarse grain (Figure 5a) was attributed to different orientations of the martensite packets, which was supported by the IPF map. The grain boundary feature (Figure 5b) of the WT-820T sample was similar to that of the WT one; both had coarse granular structures and coarse martensite packets. Moreover, few fine grains were present at the HAGBs of the WT-820T sample (Figure 5b). It was found that the HAGBs of the WT-890T sample occupied a large portion of the map (Figure 5c). The coarse-grained structure of the WT sample was replaced by numerous irregular fine grains in the WT-890T sample (Figure 5c). This revealed that the WM reheated to the temperature slightly below the A_{C3} temperature had a fine-grained structure. Those fine-grained structures were anticipated to deteriorate Gr. 92 WM creep resistance.

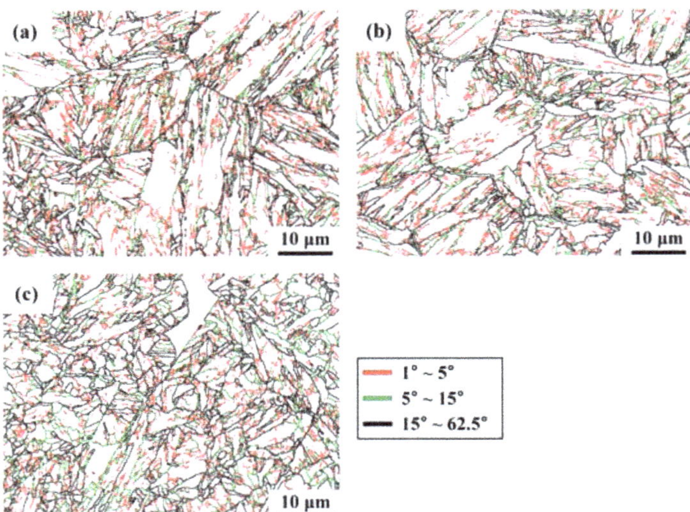

Figure 5. Grain boundary maps of the (**a**) WT, (**b**) WT-820T and (**c**) WT-890T samples.

3.7. Short-Term Creep Tests

Short-term creep tests were performed under the conditions either at 630 °C/120 MPa or 660 °C/80 MPa. Specimen elongation was also used to rank the creep resistance of the specimens. The WT samples did not fracture during the test period, despite the testing conditions. The WT-820T samples were resistant to rupture during the creep tests, but one of the samples fractured at 3472 h under the 630 °C/120 MPa conditions (Figure 6a). Moreover, the creep life of the WT-890T sample was much shorter than those of other samples under the 630 °C/120 MPa condition. As revealed in prior works [18–21], simulated thermal treatment will shorten the creep life of the sample, as compared to that of the original substrate, particularly in the case of the partially transformed samples. The results demonstrate that the reheated WM, which had been heated to the two-phase region, had the lowest creep resistance among the distinct zones of the WM. Under the 660 °C/80 MPa conditions, none of the samples fractured within the testing period, as shown in Figure 6b.

3.8. Fracture Features

The fracture features of the tensile and impact samples in the AW or tempered conditions were inspected, as shown in Figure 7. Before the tensile test, the sample was polished and slightly etched to ensure that the fracture location could be distinguished more accurately after tensile straining. The fracture of tensile specimens was located at the outer edge of the HAZ, regardless of the tempered conditions. In addition, the WM was resistant to deformation during straining (Figure 7). The slightly necked zone, symmetrically located with respect to the fracture site, was used to identify the weak zone in a cross weld. Tensile fracture of the weld in the AW condition was located at the BM (Figure 7a). The tempered weld was prone to rupture in the over-tempered zone just ahead of the BM (Figure 7c,e). The high thickness reduction in the tensile fractured zone was associated with extensive ductile dimple fracture (not shown here). The macro-fractured appearance of impact welds is shown in Figure 7b,d,f. The AW samples, as expected due to their low impact toughness, exhibited a flat fracture surface (Figure 7b). By contrast, the tempered welds showed a high distortion and change in sample profile (Figure 7d,f), which was related with the high resistance to impact rupture. The wide extent of the cleavage fracture was seen in the impact-fractured sample in the AW conditions, whereas ductile dimples were seen extensively in the tempered samples.

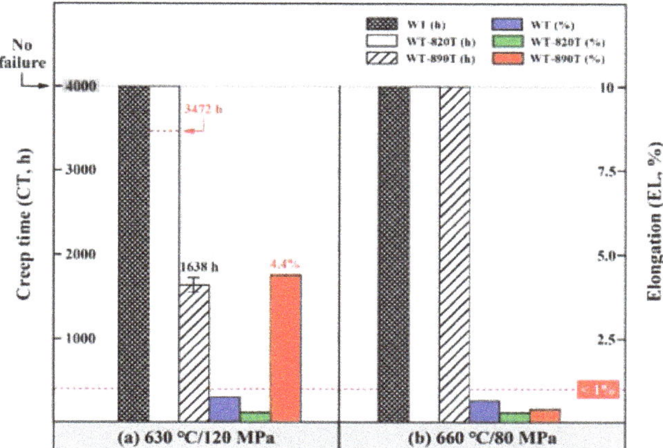

Figure 6. Short-term creep tests of the WT, WT-820T, WT-890T samples at (**a**) 630 °C/120 MPa, (**b**) 660 °C/80 MPa.

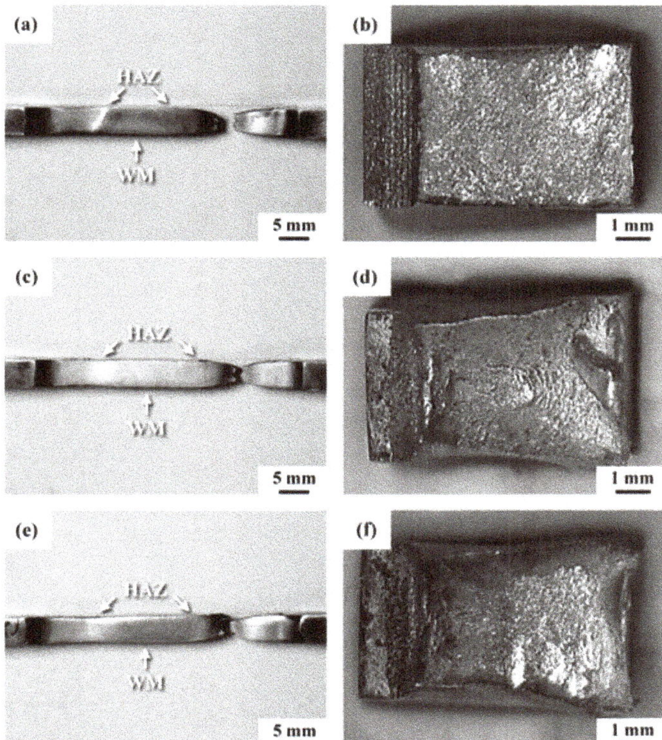

Figure 7. Appearance of the fracture zone in the welds in the (**a,b**) AW, (**c,d**) PT 1 and (**e,f**) PT 2 conditions. (**a,c,e**) the side view of the tensile fractured samples; (**b,d,f**) impact fractured samples.

The typical fracture features of crept samples were examined with an SEM, as shown in Figure 8. The crept fracture appearance of the WT-890T sample macroscopically showed obvious plastic deformation before rupture (Figure 8a) and microscopically mainly displayed fine dimples under the 630 °C/120 MPa conditions (Figure 8b). The occurrence of intergranular fracturing (as indicated by the arrow) at the specimen corner of the WT-890T sample (Figure 8b) was seen. By contrast, the fracture features of the WT-820T samples (creep life: 3472 h) showed mixed mode fracturing under the 630 °C/120 MPa conditions (Figure 8c). Intergranular fracturing (indicated by the arrows in Figure 8d) was observed predominantly at the corners of the WT-820T specimen after creep-straining. Moreover, the fast fracture (FF) zone displayed predominantly ductile dimple fracture in the tested sample. With increasing the creep life of the tested sample, intergranular fracturing was more likely to be observed.

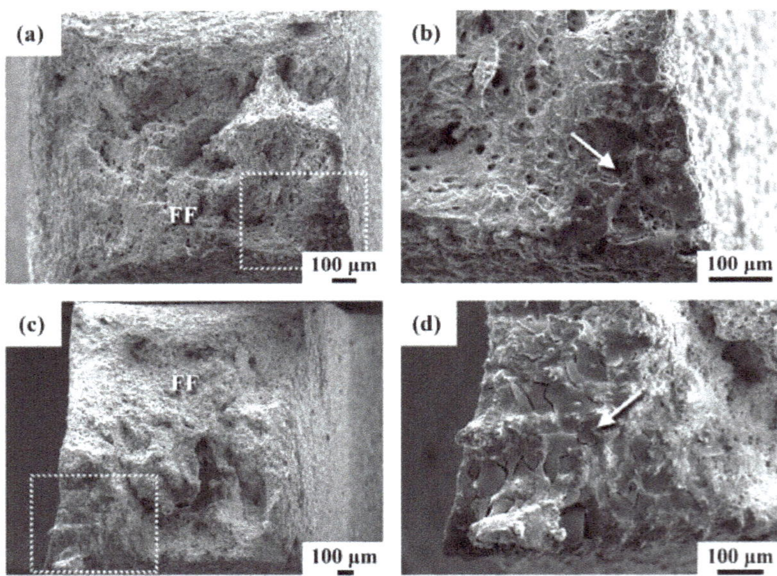

Figure 8. Fracture features of the crept samples: (**a**,**b**) WT-890T; (**c**,**d**) WT-820T at 630 °C/120 MPa.

Although not all of the tested samples fractured during the creep tests, creep defects could be also induced in the samples. The investigated specimen was cut from the gage section of the sample crept under the 660 °C/80 MPa conditions and subjected to metallographic preparation before inspection. The trace of carbides exhibited the profile of PAGBs. Figure 9 presents the microstructures of the samples before and after the creep loading. Before the creep-straining (Figure 9a,c,e), the precipitates in the original samples were very fine when inspected by SEM, especially in the WT sample. As compared with the initial state, a decrease in carbide density and an increase in carbide size occurred after creep-straining. Coarse precipitates in the WT-820T and WT-890T samples were observed after creep-loading, which implied the degraded creep strength (Figure 9d,f). The high density of carbides decorating the grain boundaries of the WT-890T sample revealed its fine-grained structure (Figure 9e,f). Under the creep-straining, microcracks were prone to be induced at the corners of the tested sample (Figure 9b,d,f), which propagated mainly along the grain boundaries. According to the carbide distribution, the coarse-grained structure was maintained in the WT and WT-820T samples (Figure 9b,d), but a fine-grained structure was found in the WT-890T sample (Figure 9f) after the creep tests.

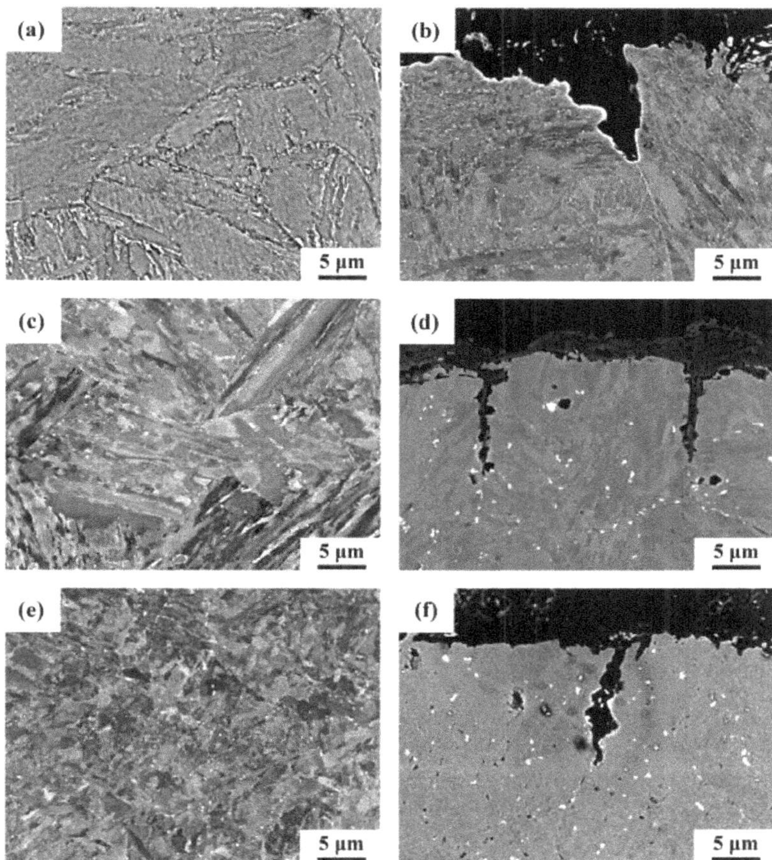

Figure 9. Changes in the microstructures of the samples before and after the creep-straining at 660 °C/80 MPa: (**a**,**b**) WT, (**c**,**d**) WT-820T, (**e**,**f**) WT-890T samples.

4. Discussion

As mentioned previously, the A_{C1} and A_{C3} temperatures of the Gr. 92 WM were about 20–30 °C lower than those of the Gr. 92 tube. At the heating/cooling rates of 15 °C/s, the A_{C1} and A_{C3} temperatures of the Gr. 92 WM were 862 °C and 898 °C, respectively. This indicated that over-tempering or fully austenitic transformation of the Gr. 92 WM would occur at lower temperatures, relative to the T92 substrate, if welding thermal cycles were applied. The microhardness distribution showed that the HAZ ahead of the BM had the lowest hardness among the distinct regions of a T92 cross weld, which could be attributed to the over-tempering induced by welding thermal cycles. The WM in the AW condition had very low impact toughness (12 J) relative to that of the steel substrate (80 J). After tempering at 740 °C, the impact energy of the tempered WM (54 J) was slightly lower than that of the BM. The impact energy of the WM could be raised to that of the substrate after post-weld tempering at 760 °C/2 h. With an increase in the post-weld tempering temperature, a gradual reduction in the tensile strength but an increase in ductility of the cross weld occurred. After the post-welding tempering, the tensile fracture of a cross weld occurred in the over-tempering zone, which was slightly ahead of the BM. The over-tempering zone in the T92 cross weld accounted for the inferior tensile properties of the weld, as compared to the T92 substrate.

The grain boundary and IPF maps of the WT and WT-820T samples revealed that coarse martensite packets were inter-dispersed in a coarse-grained structure and outlined by HAGBs. By contrast, the WT-890T sample comprised numerous irregular fine grains. The fine-grained structure of the WT-890T sample was proven to degrade its creep resistance. The creep life of the WT-890T sample was found to be much shorter than those of other samples under the 630 °C/120 MPa conditions. The relatively high elongation of the WT-890T sample also confirmed its inherent low creep strength at elevated temperature. The crept fracture appearance of the WT-890T sample revealed obvious plastic deformation macroscopically and mainly fine dimples microscopically under the 630 °C/120 MPa conditions. Moreover, intergranular cracks were found predominantly at the corners of the WT-820T sample after creep-straining for 3472 h.

Although none of the samples fractured within the testing period under the 660 °C/80 MPa conditions, creep defects still could be found in the tested samples. Creep voids were occasionally seen at the interior grain boundaries of the creep samples. The results indicate that the intergranular fracture was more likely to be seen in the crept samples with an increase in creep life (Figure 8d). It was deduced that the high surface strain combined with the high stress concentration at the specimen corners led to intergranular separation. It was deduced that the linking of surface creep voids into surface microcracks caused the creep fracture in the tested samples. Softening during the creep could result from two factors: the coalescence of subgrains and dislocation recovery, or coarsening of fine $M_{23}C_6$ carbides that hindered the subgrains and boundary motion. In prior work [19,21,22], the IPF maps and grain boundary maps around the fracture zones of creep-ruptured Gr. 91 and Gr. 92 samples exhibited a fine-grained structure and texture, which was related to dynamic recrystallization under straining at elevated temperature. As compared to the Gr. 91 steel or Gr. 91 WM, the Gr. 92 WM was more stable in maintaining the initial coarse microstructures under the same creep conditions, as shown in Figure 9. Undegraded microstructures of the Gr. 92 WM during straining at elevated temperature were responsible for its higher resistance to creep failure during the practical service.

5. Conclusions

An infrared heating system was used to simulate the reheated microstructures of Gr. 92 WM. The short-term creep life of the simulated samples was determined, and the results were related with the inherent microstructures of the samples.

(1)   Despite the tempering condition, the HAZ ahead of the BM had the lowest hardness among the distinct regions of a T92 cross weld. The AW WM was very brittle and showed low impact toughness relative to that of the substrate. After tempering at 740 °C, the impact energy of the tempered WM was slightly lower than that of the BM. The impact energy of the WM could be increased to that of the substrate after tempering at 760 °C/2 h. When increasing the post-weld tempering temperature, a gradual reduction in the tensile strength, but an increase in the ductility of the cross weld, occurred. The over-tempering zone in a T92 cross weld accounted for the inferior tensile properties of the weld, as compared to the T92 substrate.

(2)   The WT and WT-820T samples displayed coarse martensite packets inter-dispersed in a coarse-grained structure. By contrast, the WT-890T sample comprised numerous irregular fine grains. The creep life of the WT-890T sample was much shorter than those of the other samples under the 630 °C/120 MPa condition. Intergranular fracture was more likely to be found in the crept samples with increases in creep life. The high surface strain combined with the high stress concentration in the specimen corners assisted intergranular separation therein. By contrast, ductile dimple fracture was associated with the fast rupture zone in the sample. It was concluded that the fine-grained structure of the WT-890T sample played a crucial role in deteriorating its creep resistance. The formation of refined grains in the WM of the multiple-pass weld was not avoidable, but could be mitigated by lowering the heat input during welding.

(3)   As compared to the Gr. 91 steel substrate or Gr. 91 WM, the Gr. 92 WM was more stable in maintaining the original microstructures under the same creep conditions. Undegraded microstructures of the Gr. 92 WM during straining at elevated temperatures were responsible for its higher resistance to creep failure during the practical service.

Author Contributions: Experiment, Y.-L.C., C.-C.W. and T.-C.C.; formal analysis, R.-K.S.; writing—original draft preparation, L.-W.T. and R.-K.S.; writing—review and editing, L.-W.T.; funding acquisition, L.-W.T. All authors have read and agreed to the published version of the manuscript.

Funding: This research was funded by the Ministry of Science and Technology, R.O.C. (Contract No. MOST 106-2221-E-019 -060 -MY3). The authors also acknowledge MOST for research support of TEM examinations (Tecnai G2 F30).

Institutional Review Board Statement: Not applicable.

Informed Consent Statement: Not applicable.

Data Availability Statement: Not applicable.

Conflicts of Interest: The authors declare no conflict of interest.

Nomenclature

WM	weld metal
PT	partially transformed
OT	over-tempered
TEM	transmission electron microscope
WT	tempered weld metal
PAGBs	prior austenite grain boundaries
HAZ	heat-affected zone
FGHAZ	fine-grained heat-affected zone
BM	base metal
SEM	scanning electron microscope
EBSD	electron backscatter diffraction
AW	as-welded
FB	fusion boundary
PT	post-weld tempering
UTS	ultimate tensile strength
YS	yield strength
IPF	inverse pole figure
HAGBs	high-angle grain boundaries
LAGBs	low-angle grain boundaries
FF	fast fracture

References

1. Viswanathan, R.; Henry, J.F.; Tanzosh, J.; Stanko, G.; Shingledecker, J.; Vitalis, B.; Purgert, R.U.S. Program on Materials Technology for Ultra-Supercritical Coal Power Plants. *J. Mater. Eng.* **2005**, *14*, 281–292. [CrossRef]
2. Wang, S.S.; Peng, D.L.; Chang, L.; Hui, X.D. Enhanced mechanical properties induced by refined heat treatment for 9Cr–0.5Mo–1.8W martensitic heat resistant steel. *Mater. Des.* **2013**, *50*, 174–180. [CrossRef]
3. Sawada, K.; Hongo, H.; Watanabe, T.; Tabuchi, M. Analysis of the microstructure near the crack tip of ASME Gr.92 steel after creep crack growth. *Mater. Charact.* **2010**, *61*, 1097–1102. [CrossRef]
4. Výrostková, A.; Homolová, V.; Pecha, J.; Svoboda, M. Phase evolution in P92 and E911 weld metals during ageing. *Mater. Sci. Eng. A* **2008**, *480*, 289–298. [CrossRef]
5. Abson, D.J.; Rothwell, J.S. Review of type IV cracking of weldments in 9–12%Cr creep strength enhanced ferritic steels. *Int. Mater. Rev.* **2013**, *58*, 437–473. [CrossRef]
6. Liu, Y.; Tsukamoto, S.; Shirane, T.; Abe, F. Formation Mechanism of Type IV Failure in High Cr Ferritic Heat-Resistant Steel-Welded Joint. *Met. Mater. Trans. A Phys. Met. Mater. Sci.* **2013**, *44*, 4626–4633. [CrossRef]
7. Xu, X.; West, G.D.; Siefert, J.A.; Parker, J.D.; Thomson, R.C. The Influence of Thermal Cycles on the Microstructure of Grade 92 Steel. *Met. Mater. Trans. A Phys. Met. Mater. Sci.* **2017**, *48*, 5396–5414. [CrossRef]

8. Xu, X.; West, G.D.; Siefert, J.A.; Parker, J.D.; Thomson, R.C. Microstructural Characterization of the Heat-Affected Zones in Grade 92 Steel Welds: Double-Pass and Multipass Welds. *Met. Mater. Trans. A Phys. Met. Mater. Sci.* **2018**, *49*, 1211–1230. [CrossRef]
9. Abe, F.; Tabuchi, M.; Tsukamoto, S.; Shirane, T. Microstructure evolution in HAZ and suppression of Type IV fracture in advanced ferritic power plant steels. *Int. J. Press. Vessel. Pip.* **2010**, *87*, 598–604. [CrossRef]
10. Abe, F.; Tabuchi, M.; Tsukamoto, S. Metallurgy of Type IV fracture in advanced ferritic power plant steels. *Mater. High Temp.* **2011**, *28*, 85–94. [CrossRef]
11. Liu, Y.; Tsukamoto, S.; Sawada, K.; Abe, F. Role of Boundary Strengthening on Prevention of Type IV Failure in High Cr Ferritic Heat-Resistant Steels. *Met. Mater. Trans. A Phys. Met. Mater. Sci.* **2014**, *45*, 1306–1314. [CrossRef]
12. Yan, P.; Liu, Z.; Bao, H.; Weng, Y.; Liu, W. Effect of microstructural evolution on high-temperature strength of 9Cr–3W–3Co martensitic heat resistant steel under different aging conditions. *Mater. Sci. Eng. A* **2013**, *588*, 22–28. [CrossRef]
13. Matsunaga, T.; Hongo, H.; Tabuchi, M.; Sahara, R. Suppression of grain refinement in heat-affected zone of 9Cr–3W–3Co–VNb steels. *Mater. Sci. Eng. A* **2016**, *655*, 168–174. [CrossRef]
14. Xue, W.; Pan, Q.; Ren, Y.; Shang, W.; Zeng, H.; Liu, H. Microstructure and type IV cracking behavior of HAZ in P92 steel weldment. *Mater. Sci. Eng. A* **2012**, *552*, 493–501. [CrossRef]
15. Zhao, L.; Jing, H.; Xiu, J.; Han, Y.; Xu, L. Experimental investigation of specimen size effect on creep crack growth behavior in P92 steel welded joint. *Mater. Des.* **2014**, *57*, 736–743. [CrossRef]
16. Burgos, A.; Svoboda, H.; Zhang, Z.; Surian, E. Alternative PWHT to Improve High-Temperature Mechanical Properties of Advanced 9Cr Steel Welds. *J. Mater. Eng. Perform.* **2018**, *27*, 6328–6338. [CrossRef]
17. Paul, V.T.; Saroja, S.; Albert, S.K.; Jayakumar, T.; Kumar, E.R. Microstructural characterization of weld joints of 9Cr reduced activation ferritic martensitic steel fabricated by different joining methods. *Mater. Charact* **2014**, *96*, 213–224. [CrossRef]
18. Peng, Y.Q.; Chen, T.C.; Chung, T.J.; Jeng, S.L.; Huang, R.T.; Tsay, L.W. Creep Rupture of the Simulated HAZ of T92 Steel Compared to that of a T91 Steel. *Materials* **2017**, *10*, 139. [CrossRef]
19. Wu, T.J.; Liao, C.C.; Chen, T.C.; Shiue, R.K.; Tsay, L.W. Microstructural Evolution and Short-Term Creep Rupture of the Simulated HAZ in T92 Steel Normalized at Different Temperatures. *Metals* **2019**, *9*, 1310. [CrossRef]
20. Hsiao, D.H.; Chen, T.C.; Jeng, S.L.; Chung, T.J.; Tsay, L.W. Effects of simulated microstructure on the creep rupture of the modified 9Cr-1Mo steel. *J. Mats. Eng. Perf.* **2016**, *25*, 4317–4325. [CrossRef]
21. Wu, H.W.; Wu, T.J.; Shiue, R.K.; Tsay, L.W. The Effect of Normalizing Temperature on the Short-Term Creep Rupture of the Simulated HAZ in Gr. *91 Steel Welds. Metals* **2018**, *8*, 1072. [CrossRef]
22. Liao, C.C.; Wang, C.C.; Chen, T.C.; Shiue, R.K.; Tsay, L.W. Effects of Thermal Simulation on the Creep Fracture of the Mod. 91 Weld Metal. *Metals* **2020**, *10*, 1181. [CrossRef]

Article

Weldability and Damage Evaluations of Fresh-to-Aged Reformer Furnace Tubes

Chengming Fuyang [1,2], Yang Zhou [1,2], Bing Shao [1,2], Tianyu Zhang [1,2], Xiaofeng Guo [3], Jianming Gong [1,2,*] and Xiaowei Wang [1,2]

[1] School of Mechanical and Power Engineering, Nanjing Tech University, Nanjing 211816, China; fycm@njtech.edu.cn (C.F.); zhouyang0903@foxmail.com (Y.Z.); 18852990717@163.com (B.S.); zhangtianyu@njtech.edu.cn (T.Z.); xwwang@njtech.edu.cn (X.W.)

[2] Key Laboratory of Design and Manufacture of Extreme Pressure Equipment, Nanjing 211816, China

[3] School of Mechanical Engineering, Inner Mongolia University of Science & Technology, Baotou 014010, China; guoxiaofeng@njtech.edu.cn

[*] Correspondence: gongjm@njtech.edu.cn; Tel.: +86-025-5813-9361

Abstract: The microstructures and tensile properties of fresh and aged reformer furnace tubes and a fresh-to-aged welded joint were investigated to assess the weldability of fresh-to-aged reformer furnace tubes. Damage evaluation of the fresh-to-aged welded joint was also carried out using the modified Kachanov–Rabotnov model. The experimental results showed that M_7C_3 carbide transforms into $M_{23}C_6$ carbide and secondary carbides precipitate in the matrix after aging treatment. With continuous exposure, the interdendritic precipitates coalesced and coarsened and the number of secondary carbides reduced gradually. Microdefects were absent in the fresh-to-aged welded joint, and the tensile properties of the welded joint were close to the as-cast alloy, which confirms the weldability of fresh-to-aged furnace tubes. According to the results of the simulation, stress redistribution occurred during the creep process and the peak damage of the welded joint was located in the aged tube. The maximum damage of the fresh-to-aged welded joint reached 34.01% at 1.5×10^5 h.

Keywords: reformer furnace tube; weldability; damage; fresh-to-aged welded joint

Citation: Fuyang, C.; Zhou, Y.; Shao, B.; Zhang, T.; Guo, X.; Gong, J.; Wang, X. Weldability and Damage Evaluations of Fresh-to-Aged Reformer Furnace Tubes. *Metals* **2021**, *11*, 900. https://doi.org/10.3390/met11060900

Academic Editors: Elisabetta Gariboldi and Stefano Spigarelli

Received: 6 May 2021
Accepted: 22 May 2021
Published: 31 May 2021

Publisher's Note: MDPI stays neutral with regard to jurisdictional claims in published maps and institutional affiliations.

1. Introduction

A steam reformer furnace tube is the key component for producing hydrogen and carbon monoxide in a steam reformer furnace [1–3]. The designed lifetime of reformer furnace tubes is 10^5 h according to API 530 [4]. Due to the elevated temperature and inner pressure, the tube mainly suffers from creep damage under service conditions, which contributes greatly to the lifetime of the furnace tube [2,5–7]. Furthermore, the outer surface of the tube is also subjected to oxidation damage [8,9]. Therefore, it is necessary to develop a heat resistance alloy to withstand severe service conditions so as to extend the lifetime of the furnace tube. In the past decades, heat-resistant HP40Nb (25Cr35Ni1Nb) alloy has been developed by adding a small amount of niobium to HP40 alloy [10–13]. The alloy is being widely used for steam reformer furnace tubes due to superior oxidation and creep resistance as well as mechanical properties at elevated temperatures, including strength and toughness. Actually, the wall temperature in the whole furnace tube is heterogeneous, which leads to inhomogeneous damage distribution [7]. The maximum temperature in the lower segment can reach as high as 950 °C, and the microstructures are seriously deteriorated after long-term thermal exposure. The tubes should be replaced for the safety of steam reformers. Nevertheless, the damage level in the upper segment is comparatively lower and the upper tubes can continue to service. Additionally, the reformer tubes comprise a significant fraction of the petrochemical reforming plants cost, due to the high alloy content [7,14]. Therefore, there is a cost-driven motivation of local repairment in contrast to replacing the entire set of tubes, where the lower portion with the deteriorated microstructures is replaced with an as-cast tube [7,15]. Unfortunately, welding as-received tubes to aged ones may bring about weldability issues due to the diversities in microstructures and mechanical properties.

As is known to all, the microstructures of as-cast HP40Nb alloy are composed of a supersaturated austenite matrix and interdendritic primary carbides (M_7C_3, $M_{23}C_6$, and NbC). The primary NbC and M_7C_3 carbides transform into the G phase [3,9–11,16] and $M_{23}C_6$ carbide after thermal exposure, respectively [5,11,16,17]. The secondary $M_{23}C_6$ carbide precipitates in the matrix. Meanwhile, the inter- and intradendritic phases are coarsen and the morphology of precipitates changes gradually with continuous exposure. The phase transformation and precipitate coarsening both result in a reduction in mechanical properties of the alloy, leading to low weldability. Studies suggest that the G phase can decrease the creep strength [11], while the weldability is not affected and is correlated to the coarsening precipitates [14]. In terms of the weldability of the component, there are several assessment methods. Weldability can be evaluated by some standard weldability tests and the corresponding test specimen, including a Tekken test specimen and a CTS specimen [18,19]. After welding, non-destructive tests are conducted, including visual, penetrant, radiographic, and ultrasonic tests. If welding cracks and other defects are absent, further destructive tests are performed, such as tensile, bending, impact, and metallographic tests. As mentioned above, the microstructures of the reformer tube change with continuous service exposure and the degradation of the microstructure leads to a reduction in mechanical properties, especially tensile properties. Therefore, weldability can be primarily judged by the absence of post-welding macro-cracks for an HP40Nb reformer tube [20,21]. In addition, the elongation of aged tubes at room temperature equal to or greater than 4% are considered to be weldable according to technical guidelines [14,22]. Therefore, the weldability of the reformer tube can be assessed by non-destructive tests followed by destructive tests, including microstructural characterization and tensile tests. Although investigations on the weldability of a serviced reformer tube are sufficient, studies on the weldability of fresh-to-aged reformer tubes in view of microstructures and mechanical properties are rare. Additionally, the damage evaluation of the fresh-to-aged welded joint should also be implemented to predict the lifetime of reformer tubes after local repairment. For the life prediction of components controlled by creep, it is necessary to analyze their creep behavior and evaluate the damage evolution. Continuum damage mechanics is usually used to reflect creep damage and to describe creep deformation behavior, which can be categorized into empirical-based models [23–27] and physical-based models [28–31]. Empirical-based damage models define a single damage variable to quantify the average loss of strength, while physical-based models usually couple different damage variables into a sinh function to study the influence of different damage mechanisms on creep deformation. The empirical-based Kachanov–Robatnov (KR) model [23,24] has been widely used owing to its accurate prediction and the available parameters. In our previous investigation [15], the damage evolution of the service furnace tube was deduced based on the modified KR model [26]. However, the issues in damage assessment of fresh-to-old reformer tubes have not been investigated.

In this paper, accelerated aging treatments were carried out for as-cast HP40Nb alloy to simulate the service microstructure. The weldability evaluation of fresh-to-aged reformer tubes was conducted based on their microstructures and tensile properties. Furthermore, the damage evolution of fresh-to-aged welded joints was also predicted using the modified Kachanov–Robatnov damage model based on the creep data of HP40Nb alloy.

2. Materials and Experimental Section

The material investigated in this study is centrifugally cast HP40Nb alloy. The outer diameter and wall thickness are 126 and 13 mm, respectively. The chemical composition of the as-cast HP40Nb alloy was measured using an optical emission spectrometer (OES) and is shown in Table 1. Long-term thermal exposure experiments on the as-cast alloy samples were conducted at 990 °C with different aging periods (500, 2000, and 4000 h) to model the microstructural evolution in long-term service exposure. The relationship between aging time and operating time according to the Larson Miller equation is illustrated in Table 2 [32]. To study the weldability and damage evolution of the upper reformer tube at 900 °C after local repairment, an aged tube (990 °C, 2000 h) corresponding to the service tube for 1.2×10^5 h at 900 °C was selected to be welded to the original tube by Gas Tungsten Arc Welding (GTAW) using a TIG 35CW filler metal. The lengths of the reformer tubes used for welding were both 150 mm. The chemical composition of the filler metal is also present in Table 1. A V-joint groove with a 30° half-angle and a root opening of 2 mm were used. The welding current and voltage were 130 A and 10–18 V, respectively. The welding speed was about 9 cm/min. Therefore, the maximum heat input was calculated as 1.56 kJ/mm. Argon with 99.9% purity was used as a shielding and backing gas with a 14 L/min flow. The preheating temperature and the maximum interpass temperature were both 150 °C. Six filling passes were applied to fill the seam. Dye penetration inspection and radiographic testing were carried out in sequence after covering welding. The results suggested that macro-cracks and other welding defects were absent in the fresh-to-aged welded joint. Therefore, further microstructural characterization and mechanical tests can be implemented to assess the weldability of fresh-to-aged reformer tubes.

Table 1. Chemical composition of base metal and filler metal (wt %).

Material	C	Ni	Cr	Nb	Si	Mn	P	S	Fe
HP40Nb	0.45	34.13	25.47	0.70	1.35	1.03	0.029	0.008	Bal.
TIG 35CW	0.41	34.95	25.12	1.28	1.49	1.30	0.025	0.006	Bal.

Table 2. Aging time versus operating time of HP40Nb alloy.

Temperature/°C	900	910	920	930	940	950	990
	3×10^4	1.5×10^4	1×10^4	6000	4000	3000	500
Time/h	1.2×10^5	8×10^4	5×10^4	3×10^4	2×10^4	1.2×10^4	2000
	2.5×10^5	1.5×10^5	1×10^5	6×10^4	4×10^4	2.4×10^4	4000

To identify the precipitates in the as-cast and aged alloys, the precipitates were extracted from the matrix using Berzelius solution [33]. The extracted residues were identified using a Rigaku Smartlab diffractometer at 40 kV and 30 mA from 30° to 60° with a step size of 0.02°. The microstructural morphology of samples with various aging times was observed by Zeiss AXIO Imager A1m optical microscopy (OM) (Oberkochen, Germany) and Zeiss Ultra 55 field emission–scanning electron microscopy (FE-SEM) (Oberkochen, Germany). Element chemical analysis was conducted using an energy-dispersive spectrometer (EDS). The microstructures and microdefects of the fresh-to-aged welded joint were also detected using OM. The specimens for OM observation and FE-SEM examination were prepared by standard metallographic methods, such as cutting, grinding, and polishing, and then the samples were tested after electrolytic etching at 0.5 A for 4 s using saturated oxalic acid solution.

The tensile properties of the as-cast alloy, aged alloy (990 °C, 2000 h), and fresh-to-aged welded joint were determined by tensile tests at room and elevated (950 °C) temperatures. The tensile tests at room and elevated temperatures were conducted according to the Chinese standards GB/T 228.1-2010 and GB/T 228.2-2015, respectively, performed on a Hydraulic Servo 4830 instrument (Kyoto, Japan) equipped with a 100 KN load cell and a multifunctional Hydraulic Servo MTS Landmark 370.10 machine (Minneapolis, MI, USA), respectively. Monotonic creep tests of the as-cast alloy, aged alloy, and fresh-to-aged welded joint were conducted on a mechanical CTM304 creep-testing machine (Shenzhen, China) in a wide range of stresses at 950 °C, and creep tests were conducted according to the standard GB/T 2039-2012. The specimens for tensile and creep tests were taken from the middle of the furnace tube. The configuration and dimensions of the specimen for creep tests are shown in Figure 1.

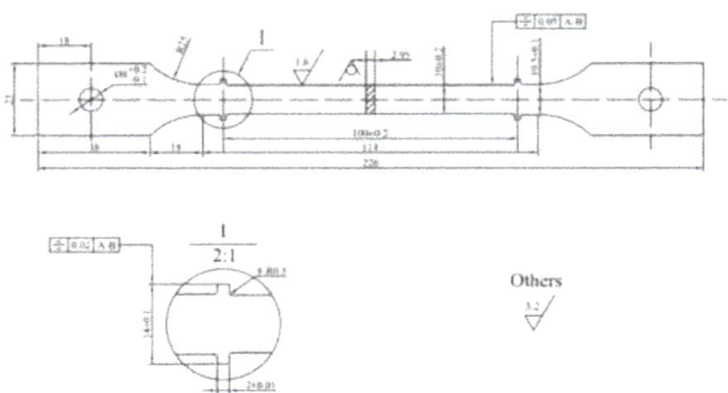

Figure 1. Configuration and dimensions of the specimen for constant-load creep tests (Units: mm).

3. Experimental Results and Discussion

3.1. Precipitate Identification

XRD patterns of the extracted powder of the as-cast and aged HP40Nb alloy samples are shown in Figure 2. In the as-cast condition, the main precipitates of the alloy were niobium carbide (NbC) and chromium-rich carbides (M_7C_3 and $M_{23}C_6$). Niobium first combines with carbon, forming NbC at higher temperature during solidification [11]. The atomic ratio of Nb to C acts as the driving force for the formation of NbC. A higher value of the atomic ratio leads to the greater driving force for forming NbC. The value (0.20) of the ratio lower than 0.5 in HP40Nb alloy is insufficient for Nb to combine with all the free carbon to form NbC [34]. As a result, chromium attracts the free carbon left in the matrix to form chromium carbides at lower temperature during solidification. Due to the high cooling rate, M_7C_3 carbide cannot transform into $M_{23}C_6$ carbide completely. Hence, M_7C_3 and $M_{23}C_6$ carbides coexist in the as-cast alloy. Compared with fresh material, the diffraction intensity of the M_7C_3 peaks was not seen on the X-ray spectrum in the aging state, which indicates that M_7C_3 carbide transforms to $M_{23}C_6$ carbide after long-term exposure. Moreover, the NbC peaks that still existed on the X-ray spectrum and in other phases, such as the G phase, were not detected in the aged alloy.

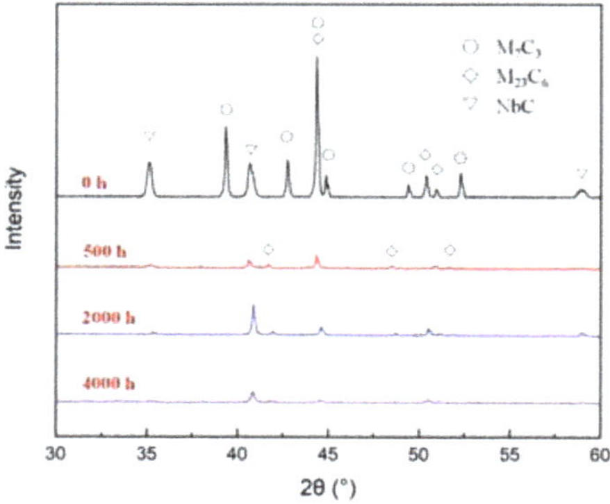

Figure 2. Representative XRD patterns of extracted residues in as-cast HP40Nb alloy and the specimens after aging at 990 °C.

3.2. Microstructural Characterization

Figure 3 presents the OM images of as-cast HP40Nb alloy and the specimens with various aging times at 990 °C. The original microstructures are composed of an austenite matrix with skeletal-shaped primary eutectic carbides, as can be seen in Figure 3a. During the aging process, although the austenite matrix remains, microstructural evolution occurs concerning the type, morphology, and distribution of the precipitates. Compared with the microstructure in the initial state, secondary carbides nucleate within the austenite matrix after aging for 500 h, as depicted in Figure 3b. The interdendritic phases were dominated by a skeletal-shaped structure, while worm-like and blocky structures were also observed. With an increase in aging time, the number of secondary carbides within the austenitic matrix declined gradually, which suggests that the secondary carbides dissolve into the matrix during the long-term aging process. The interdendritic precipitates coarsen continuously and change into the blocky structure. When the aging time increased to 4000 h, the skeletal-shaped precipitates were almost replaced by blocky and spherical structures.

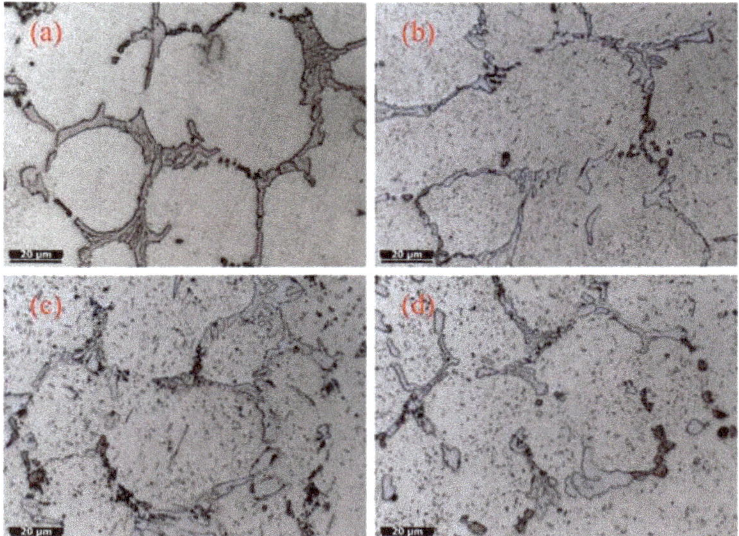

Figure 3. Typical optical microscopy images of (**a**) as-cast HP40Nb alloy and the specimens after aging at 990 °C for (**b**) 500, (**c**) 2000, and (**d**) 4000 h.

To investigate the types, morphology, distribution, and size of the intra- and interdendritic precipitates in HP40Nb alloy before and after thermal exposure, SEM observations were conducted in back-scattered electron (BSE) mode. Figure 4 exhibits the SEM micrographs of as-cast HP40Nb alloy and the samples after aging treatment at 990 °C. As illustrated in Figure 4a, two types of carbides were detected in the interdendritic region for the fresh material, which exhibit dark and bright contrast. Elemental chemical analysis using SEM elemental mapping technique indicated that the dark precipitate has a high chromium content, while the bright phase is highly rich in niobium, as seen in Figure 5. As analyzed above in Figure 2, the Cr-rich phases are regarded as $M_{23}C_6$ and M_7C_3 carbides, while the Nb-rich precipitate is deemed as NbC.

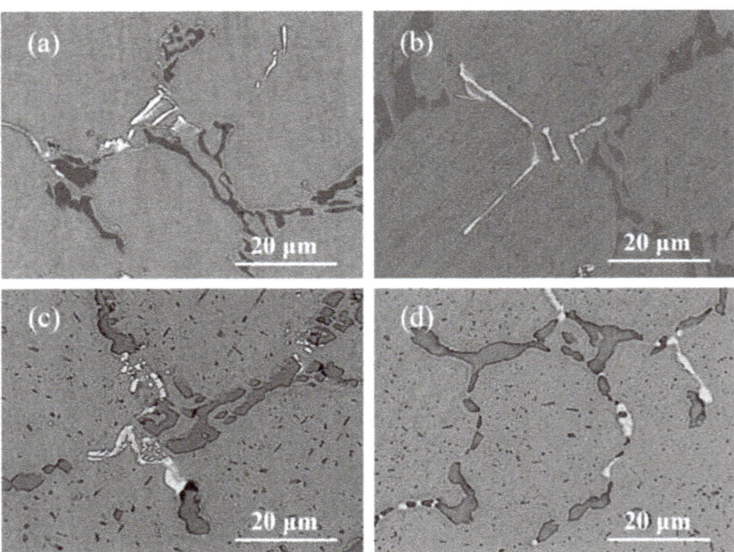

Figure 4. Typical back-scattered electron (BSE)–SEM images of (**a**) as-cast HP40Nb alloy and the samples after aging at 990 °C for (**b**) 500, (**c**) 2000, and (**d**) 4000 h.

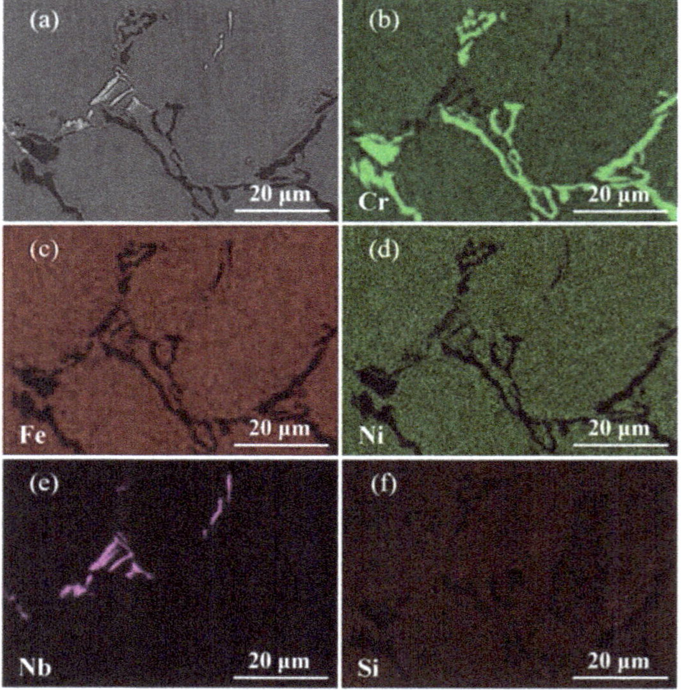

Figure 5. Precipitations of as-cast HP40Nb alloy detected using SEM elemental mapping: (**a**) SEM image, (**b**) Cr, (**c**) Fe, (**d**) Ni, (**e**) Nb and (**f**) Si elements.

For the aging samples, the interdendritic phases coarsened with continuous aging treatments, as can be seen in Figure 4b–d. The dark Cr-rich phase and bright Nb-rich carbide still existed in the interdendritic region, taking the aging specimen for 2000 h as an instance exhibited in Figure 6. According to the XRD patterns, the Cr-rich and Nb-rich phases are regarded as $M_{23}C_6$ carbide and NbC, respectively. It is noteworthy that M_7C_3 carbide is unstable and is prone to transform into $M_{23}C_6$ carbide at elevated temperature. Therefore, the Cr-rich phase is known as $M_{23}C_6$ carbide for the aging specimen. In various papers describing heat-resistant Fe-Cr-Ni alloys, the transformation from NbC into niobium nickel silicide ($Ni_{16}Nb_6Si_7$), identified as the G phase [35,36], occurs in the temperature range from 700 to 1000 °C [16,37]. However, the G phase was absent in the post-aged alloy at 990 °C in our study, as illustrated in Figures 2 and 6. The explanation for this discrepancy is the lower silicon content of HP40Nb alloy in this research. A higher silicon concentration contributes to the formation of the G phase. In addition, it can be observed from the SEM image that large amounts of secondary carbides precipitate within the austenite matrix after aging for 500 h. With continuous exposure, a gradual reduction in the number and density of secondary carbides occurs within the austenite matrix. SEM elemental mapping analysis showed that the precipitates are highly rich in chromium, as depicted in Figure 6c. This result indicates that the precipitated particles within the austenite matrix are $M_{23}C_6$ carbides. Due to the high cooling rate, carbon is supersaturated in the matrix for fresh alloy. When exposed to the aging condition, the supersaturated carbon tends to combine with chromium to precipitate dispersed $M_{23}C_6$ carbides in the matrix [34,38]. Nevertheless, the fine secondary $M_{23}C_6$ carbides are likely to dissolve and coalesce into larger carbides in order to reduce the surface energy [34], which leads to the reduction in secondary carbides with further thermal exposure. Meanwhile, the carbon in the interdendritic areas is inclined to combine with chromium to form $M_{23}C_6$ carbide. The formation of the carbide reduces the concentration of carbon in the interdendritic regions and creates a concentration gradient between the dendritic interface and the austenite matrix, which drives carbon to diffuse from the matrix into the interdendritic areas at elevated temperature. Owing to the carbon diffusion, the interdendritic carbides coalesce and coarsen with further aging treatment.

The optical microscopy images of the fresh-to-aged welded joint are illustrated in Figure 7, in which the welded joint is composed of a weld metal (WM) and a base metal (BM) as well as a fusion zone. However, the heat-affected zone (HAZ) was not apparent in our study. The fine grains in the weld metal grow perpendicular to the fusion zone. This is attributed to the direction with the fastest heat dissipation. Due to the different temperature gradients and cooling rates at various welding layers, the structures of the weld metal have a diverse morphology, including columnar and cellular structures. Moreover, continuous network carbides are located in the interdendritic region, and no secondary carbides are dispersed within the austenite matrix for the weld metal. As analyzed above, the morphology of interdendritic carbides of the fresh alloy (BM) is seen as a skeletal-shaped structure, while blocky interdendritic carbides and dispersed secondary carbides were observed in the aged alloy (BM) on account of the microstructural degradation after aging treatment for 2000 h. The carbides are partially and fully melted in the fusion zone, and the content of carbides in the fusion zone is greater compared to the base metal. The carbides of the fusion zone on the fresh side are mainly seen as spherical particles, while the ones on the aged side are seen as a skeletal-shaped and worm-like structure, as can be seen in Figure 8. The distinct morphology of carbides in the fusion zone is also attributed to the diversities in the heating temperature and cooling rate in the welding process. The carbides reacted with the surrounding austenite matrix in the heating cycle and melted at temperatures lower than the melting points of the matrix and the carbide. Afterward, these phases resolidified to the eutectic carbides in the cooling cycle. This phenomenon can be called constitutional liquation [20]. Furthermore, microcracks and other microdefects were absent in the fresh-to-aged welded joint in our investigation, which suggests the superior welding quality of fresh-to-aged reformer tubes.

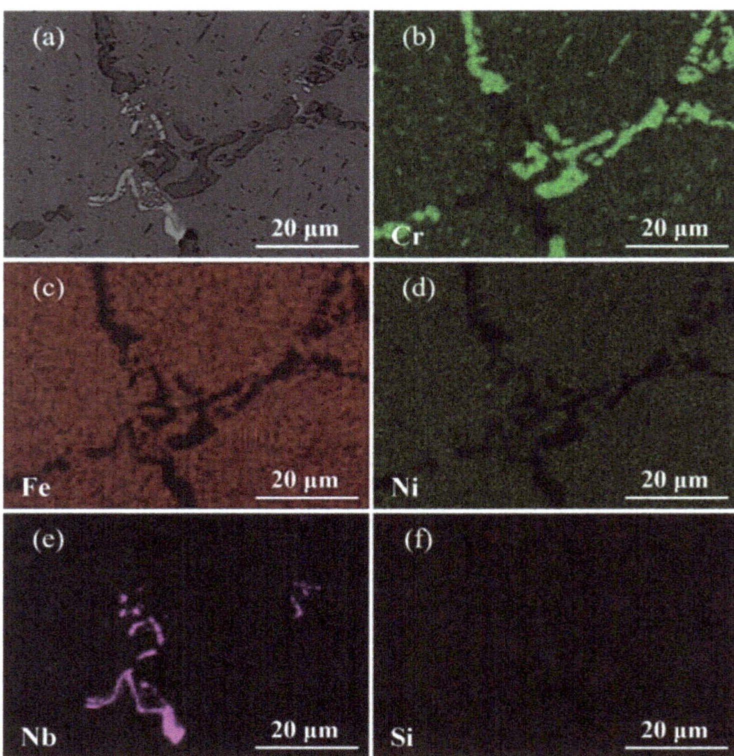

Figure 6. Precipitations in the sample after aging at 990 °C for 2000 h detected using SEM elemental mapping: (**a**) SEM image, (**b**) Cr, (**c**) Fe, (**d**) Ni, (**e**) Nb and (**f**) Si elements.

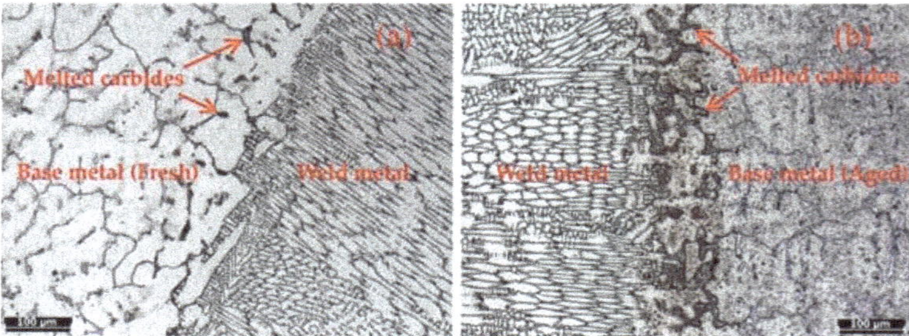

Figure 7. Optical microscopy images of the fresh-to-aged welded joint: (**a**) fresh alloy/weld metal and (**b**) aged alloy/weld metal.

3.3. Tensile Properties

Figure 9 presents the representative tensile stress–strain plots of the as-cast alloy, aged alloy (990 °C, 2000 h), and fresh-to-aged welded joint at room and elevated temperatures. The tensile properties of these samples are also listed in Table 3. The yield strengths of the fresh-to-aged welded joint at room and elevated temperatures were greater than those of fresh and aged alloys. The ultimate tensile strength of the welded joint was also the highest among these specimens at elevated temperature, while the ultimate tensile strength of the welded joint was almost the same compared with as-cast and aged alloys at room temperature. Overall, the tensile strengths of the fresh-to-aged welded joint were close to the as-cast alloy. Meanwhile, the ultimate tensile strength of the aged alloy was slightly lower than that of the as-cast alloy. The results suggest that microstructural deterioration reduces the tensile strengths limitedly and that the continuous network carbides of the weld metal may contribute to the enhancement of tensile strengths. The elongation of the fresh alloy at room temperature was apparently higher compared with the other two specimens, indicating that coarsening carbides contribute to the reduction in plasticity. It is noteworthy that the elongation of the aged alloy was slightly greater than 4%. In general, an elongation higher than the minimum value of 4% at room temperature is considered by industrial guidelines to avoid welding cold-cracking [14,22]. Therefore, the heating and cooling rates are supposed to be slowed down during the welding process to prevent cold-cracking. Furthermore, the elongations of the specimens at elevated temperature were much larger than those at room temperature, which shows the higher ductility at higher temperature for HP40Nb alloy.

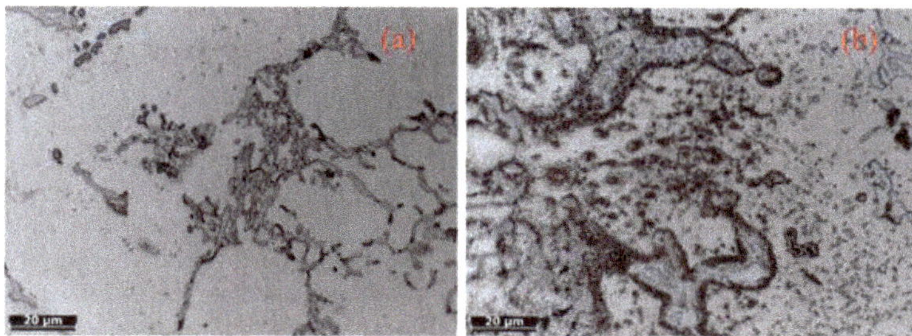

Figure 8. Optical microscopy images of melted carbides in the fusion zone for (**a**) as-cast alloy and (**b**) aged alloy.

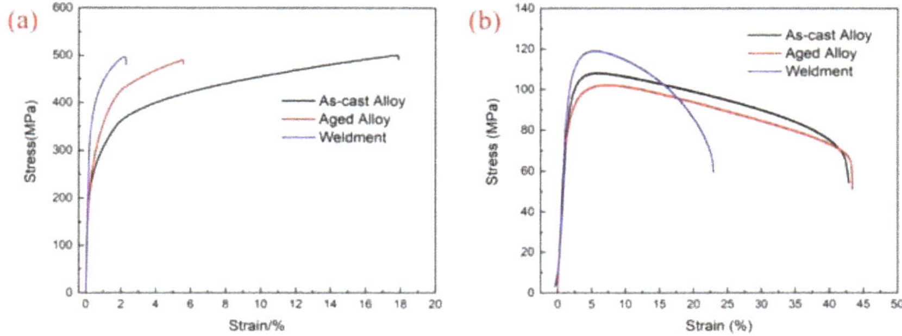

Figure 9. Typical stress–strain plots of the as-cast alloy, aged alloy, and fresh-to-aged welded joint at (**a**) room temperature and (**b**) elevated temperature (950 °C).

Table 3. Tensile properties of the as-cast HP40Nb alloy, aged alloy, and fresh-to-aged welded joint.

Temperature (°C)	Condition	Yield Strength (MPa)	Ultimate Tensile Strength (MPa)	Elongation (%)
20	As-cast alloy	248.35	500.46	26.67
	Aged alloy	258.39	490.69	4.17
	Welded joint	369.32	496.97	2.50
950	As-cast alloy	87.12	108.34	36.55
	Aged alloy	68.44	102.42	43.37
	Welded joint	92.17	119.29	21.05

In summary, the feasibility of welding an as-cast reformer tube to an aged tube for 2000 h can be verified in view of the absence of microdefects in the fresh-to-aged welded joint and the high tensile strengths of the welded joint at room and elevated temperatures. However, the heating and cooling rates in the welding process are supposed to be slowed down to avoid cold-cracking.

3.4. Creep Properties

Figure 10 exhibits the creep curves at various stress levels at 950 °C for the as-cast and aged alloys, along with the fresh-to-aged welded joint. The stress dependence of rupture time and the minimum creep strain rate is indicated in Figure 11. In general, similar creep deformation behaviors were obtained in these samples. They all exhibited well-defined primary, secondary, and tertiary creep stages under various creep conditions, while the rapid accumulation of strain in the tertiary creep stage prior to fracture was much larger than that in primary and secondary stages. In addition, the rupture time declined and the fracture strain increased with the growth of creep stress. Owing to the microstructural degradation after long-term thermal exposure, the rupture time of the aged alloy and the fresh-to-aged welded joint was obviously lower compared to the as-cast alloy at the same stress level. Nevertheless, the fracture strain of the fresh material was the lowest among these specimens, which may result from the softening of the aged microstructure during creep exposure. Furthermore, the fracture surfaces of the welded joint were located in the aged alloy, which originated from the coarsening carbides in the intra- and interdendritic regions after aging treatment. Stress dependence of rupture time and the minimum creep strain rate was linear in double-logarithmic plots, and the minimum creep strain rate values increased with the growth of stress levels. The minimum creep strain rates of the aged alloy and the fresh-to-aged welded joint were larger compared with the fresh alloy, which is attributed to the microstructural deterioration after long-term thermal exposure, as seen in Figure 11b.

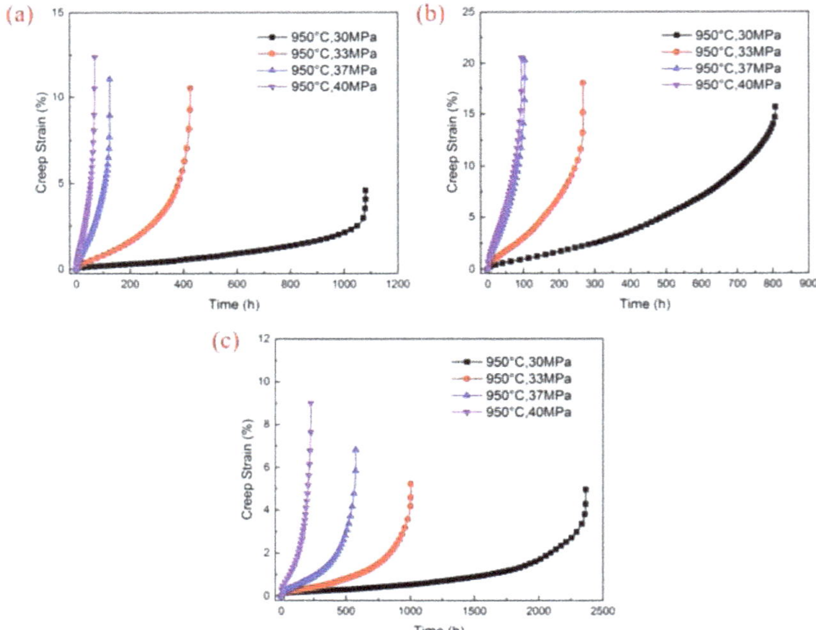

Figure 10. Standard creep time–strain curves of (**a**) as-cast alloy, (**b**) aged alloy, and (**c**) fresh-to-aged welded joint at 950 °C.

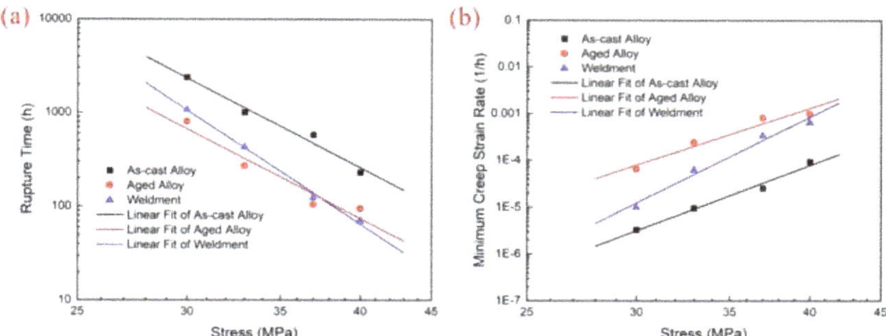

Figure 11. Stress dependence of (**a**) rupture time and (**b**) minimum creep strain rate at 950 °C for the as-cast alloy, aged alloy, and fresh-to-aged welded joint.

4. Numerical Simulation

In our investigation, the empirically based, modified Kachanov–Rabotnov model considering the heterogeneity of creep damage was used to analyze the creep damage evolution of the fresh-to-aged welded joint and can be expressed in multi-axial form as follows [26]:

$$\frac{d\varepsilon_{ij}^{c}}{dt} = \frac{3}{2}B\sigma_{e}^{n-1}S_{ij}\left[(1-\rho) + \frac{\rho}{(1-D)^{n}}\right],\tag{1}$$

$$\frac{dD}{dt} = gA\frac{[\alpha\sigma_{1} + (1-\alpha)\sigma_{e}]^{v}}{(1+\varphi)(1-D)^{\varphi}},\tag{2}$$

$$D_{cr} = 1 - (1-g)^{\frac{1}{\varphi+1}},\tag{3}$$

where ε_{ij}, S_{ij}, σ_{e}, and σ_{1} are multi-axial strain components, deviatoric stress components, the von Mises equivalent stress, and the maximum principal stress, respectively. α represents the effect of the multi-axial stress state behavior of the material. D and D_{cr} are the damage variable and the critical damage, respectively. B, n, A, and v are material constants, which can be determined according to the stress dependence of the minimum creep strain rate and rupture time, as shown in Figure 11. In addition, g and ρ are constants accounting for the inhomogeneity of the material damage and φ is the damage parameter, all of which can be obtained by the optimization algorithm based on the Newton iteration. The material constants in the creep constitutive equations for the as-cast and aged alloys are summarized in Table 4. The parameters of the weld metal are considered to be consistent with those of the as-cast alloy for simplicity in view of the similar chemical compositions. According to Equation (3), the value of critical damage for HP40Nb alloy can be calculated and is shown in Table 4. The critical damage of the fresh-to-aged welded joint corresponds to that of the aged alloy on account of the higher damage level in the aged alloy.

Table 4. Material constants in the damage constitutive equations for HP40Nb alloy at 950 °C.

Material	B	n	A	v	g	ρ	φ	α	D_{cr}
As-cast alloy	8.18×10^{-20}	9.22	2.14×10^{-15}	7.65	0.92	0.18	2.9	0.25	0.477
Aged alloy	1.75×10^{-22}	11.91	8.24×10^{-15}	7.62	0.88	0.06	3.8	0.25	0.357

Considering the dimensional feature and loading profile of the fresh-to-aged welded joint, a 2D axisymmetric model composed of a weld metal and a base metal (as-cast tube and aged tube) was used, as shown in Figure 12. The length and thickness of the fresh-to-aged welded joint were 50 and 13 mm, respectively. The as-cast tube and the aged tube were located in the upper and lower parts, respectively. The element type was CAX4. The number of nodes and elements was 4132 and 4011, respectively. The boundary conditions of the top and bottom sides were both $Y = 0$. The applied internal pressure was 3.5 MPa. User subroutines programmed with FORTRAN for the finite element ABAQUS code were used to investigate the stress distribution and damage evolution.

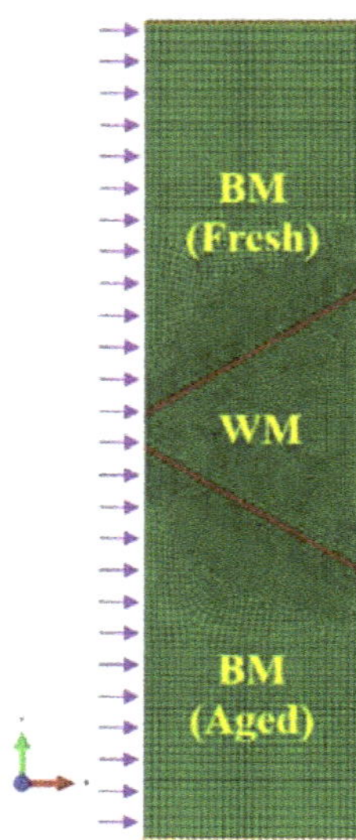

Figure 12. FE model and element division of the welded joint.

Figure 13 shows the Mises stress distribution of the fresh-to-aged welded joint during creep exposure. The peak Mises stress was situated on the inner surface of the welded joint, and the stress descended toward the outer surface before creep. Stress redistribution occurred during the creep process. The value of the peak Mises stress reduced to 14.98 MPa, and the position of the peak Mises stress shifted to the interface of the weld metal and the aged tube when the creep time increased to 1×10^4 h. In addition, the Mises stress in the aged tube was lower compared to that in the as-cast tube. With further exposure, the peak Mises stress increased to 15.27 MPa and then remained constant. The Mises stress evolution of the fresh-to-aged welded joint turned insignificant after creep for 1×10^5 h. The contour of the damage distribution for the fresh-to-aged welded joint is also exhibited in Figure 14. The damage value rose continuously with service exposure, and the location of peak damage changed gradually. The maximum damage was located on the inner surface of the aged tube at 1×10^4 h, with the value of 1.86%. With further exposure, the position of the peak damage shifted to the region adjacent to the outer wall at 5×10^4 h and the damage value increased to 9.179%. The peak damage reached 34.01% in the outer wall of the aged tube when the creep time increased to 1.5×10^5 h, while the highest damage was about 20.82% in the weld metal and the as-cast tube. In other words, the peak damage lies in the aged reformer tube during the creep process, which suggests that the microstructural degradation results in more serious damage evolution in the aged tube. However, the welding residual stress and the complex microstructures of the fresh-to-aged welded joint were not considered in numerical simulation. Meanwhile, void nucleation and growth occur in sequence for the service reformer tube, while the voids were absent in the aged alloy. Actually, the chemical composition of the filler metal differs from that of the fresh alloy. As mentioned above, the HAZ is not significant and the boundary between base metal and weld metal is mainly composed of a fusion zone. The parameters of the filler metal and the fusion zone were both unavailable. For simplicity, the material parameters of the filler metal corresponded to the as-cast alloy and the parameters in the fusion zone were ignored in this study. Therefore, the prediction of damage evolution for the fresh-to-aged welded joint is comparatively conservative in the finite simulation, and the stress and damage distribution are not continuous at the conjunction of the weld metal and the aged tube (BM).

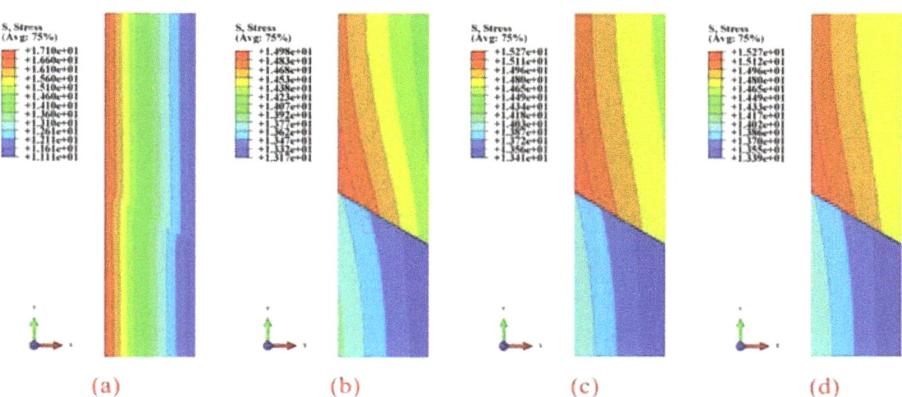

Figure 13. Mises stress distribution of the fresh-to-aged welded joint at (**a**) 0, (**b**) 1×10^4, (**c**) 1×10^5, and (**d**) 1.5×10^5 h.

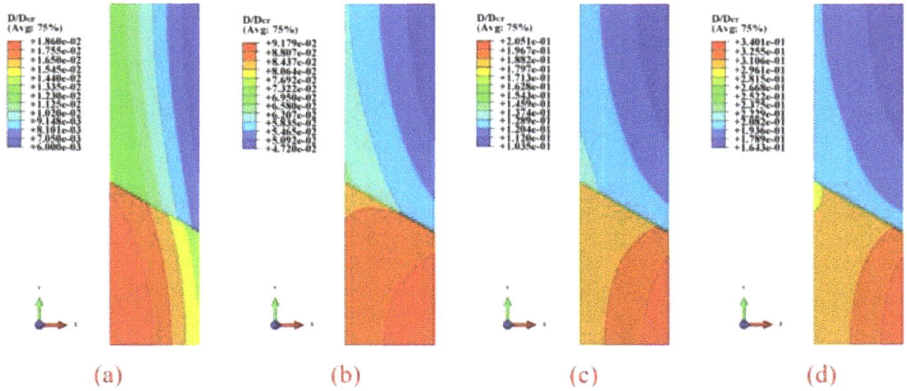

Figure 14. Damage distribution of the fresh-to-aged welded joint at (**a**) 1×10^4, (**b**) 5×10^4, (**c**) 1×10^5, and (**d**) 1.5×10^5 h.

5. Conclusions

The weldability of fresh-to-aged reformer furnace tubes was investigated based on microstructural characterization and tensile tests. Damage evaluations of a fresh-to-aged welded joint were carried out using the modified Kachanov–Rabotnov model. The main conclusions are as follows:

(1) The interdendritic M_7C_3 carbide transformed into $M_{23}C_6$ carbide after aging treatment at 990 °C, and interdendritic $M_{23}C_6$ carbide and NbC coarsened gradually with continuous exposure. Moreover, the number of precipitated $M_{23}C_6$ secondary carbides reduced by degrees.

(2) Micro-defects were absent in the fresh-to-aged welded joint, and the tensile strength of the welded joint was greater compared with the as-cast and aged alloys, which demonstrates the weldability of fresh-to-aged reformer tubes. Nevertheless, the heating and cooling rates in the welding process are supposed to be slowed down to inhibit cold-cracking.

(3) The creep rupture time of the as-cast alloy was greater compared with the aged alloy and the fresh-to-aged welded joint, while the fracture strain and the minimum creep strain rate were the lowest among the creep specimens.

(4) The stress and damage distributions in the fresh-to-aged welded joint changed continuously during the creep process. The highest Mises stress shifted to the interface of the weld metal and the aged tube with further exposure, while the predicted peak damage of the welded joint reached 34.01% in the outer wall of the aged tube at 1.5×10^5 h.

Author Contributions: Software, C.F. and T.Z.; formal analysis, C.F. and Y.Z.; investigation, C.F. and B.S.; resources, B.S.; data curation, C.F. and Y.Z.; writing—original draft preparation, C.F.; writing—review and editing, C.F., X.G. and J.G.; supervision, J.G. and X.W.; funding acquisition, X.W. All authors discussed the results and contributed greatly to the final manuscript. All authors have read and agreed to the published version of the manuscript.

Funding: The authors gratefully acknowledge the support provided by the National Key Research & Development (R&D) Program of China (no. 2018YFC0808800) and the National Natural Science Foundation of China (nos. 51805274 and 51905261).

Institutional Review Board Statement: Not applicable.

Informed Consent Statement: Not applicable.

Data Availability Statement: Not applicable.

Conflicts of Interest: The authors declare no conflict of interest.

References

1. Bonaccorsi, L.; Guglielmino, E.; Pino, R.; Servetto, C.; Sili, A. Damage analysis in Fe-Cr-Ni centrifugally cast alloy tubes for reforming furnaces. *Eng. Fail. Anal.* **2014**, *36*, 65–74. [CrossRef]
2. Bahrami, A.; Taheri, P. Creep failure of reformer tubes in a petrochemical plant. *Metals* **2019**, *9*, 1026. [CrossRef]
3. Abbasi, M.; Park, I.; Ro, Y.; Ji, Y.; Ayer, R.; Shim, J.-H. G-phase formation in twenty-years aged heat-resistant cast austenitic steel reformer tube. *Mater. Charact.* **2019**, *148*, 297–306. [CrossRef]
4. API. *Calculation of Heater Tube Thickness in Petroleum Refineries: API Recommended Practice 530*, 3rd ed.; API: Washington, DC, USA, 1988.
5. Alvino, A.; Lega, D.; Giacobbe, F.; Mazzocchi, V.; Rinaldi, A. Damage characterization in two reformer heater tubes after nearly 10 years of service at different operative and maintenance conditions. *Eng. Fail. Anal.* **2010**, *17*, 1526–1541. [CrossRef]
6. Haidemenopoulos, G.N.; Polychronopoulou, K.; Zervaki, A.D.; Kamoutsi, H.; Alkhoori, S.I.; Jaffar, S.; Cho, P.; Mavros, H. Aging Phenomena during In-service creep exposure of heat-resistant steels. *Metals* **2019**, *9*, 800. [CrossRef]
7. Gong, J.M.; Tu, S.T.; Yoon, K.B. Damage assessment and maintenance strategy of hydrogen reformer furnace tubes. *Eng. Fail. Anal.* **1999**, *6*, 143–153. [CrossRef]
8. Kondrat'ev, S.Y.; Anastasiadi, G.P.; Ptashnik, A.V.; Petrov, S.N. Kinetics of the high-temperature oxidation of heat-resistant statically and centrifugally cast HP40NbTi alloys. *Oxid. Met.* **2019**, *91*, 33–53. [CrossRef]
9. Silveiraa, R.; Arenas, M.P.; Pacheco, C.J.; Rocha, A.C.; Eckstein, C.B.; Bruno, A.C.; Pereira, G.R.; Almeida, L.H. Characterization of the oxide scale formed on external surface of HP reformer tubes. *J. Mater. Res. Technol.* **2018**, *7*, 578–583. [CrossRef]
10. Tancret, F.; Laigo, J.; Christien, F.; Gall, R.L.; Furtado, J. Phase transformations in Fe-Ni-Cr heat-resistant alloys for reformer tube applications. *Mater. Sci. Technol.* **2019**, *34*, 1333–1343. [CrossRef]
11. Almeida, L.H.D.; Ribeiro, A.F.; May, I.L. Microstructural characterization of modified 25Cr-35Ni centrifugally cast steel furnace tubes. *Mater. Charact.* **2003**, *49*, 219–229. [CrossRef]
12. Andrade, A.; Bolfarini, C.; Ferreira, L.; Vilar, A.; Filho, C.S.; Bonazzi, L. Influence of niobium addition on the high temperature mechanical properties of a centrifugally cast HP alloy. *Mater. Sci. Eng. A* **2015**, *628*, 176–180. [CrossRef]
13. Attarian, M.; Taheri, A.K.; Jalilvand, S.; Habibi, A. Microstructural and failure analysis of welded primary reformer furnace tube made of HP-Nb micro alloyed heat resistant steel. *Eng. Fail. Anal.* **2016**, *68*, 32–51. [CrossRef]
14. Abbasi, M.; Park, I.; Ro, Y.; Nam, J.; Ji, Y.; Kim, J.; Ayer, R. Microstructural evaluation of welded fresh-to-aged reformer tubes used in hydrogen production plants. *Eng. Fail. Anal.* **2018**, *92*, 368–377. [CrossRef]
15. Fuyang, C.-M.; Zhu, R.; Zhang, P.-P.; Guo, X.-F.; Li, H.; Geng, L.-Y.; Gong, J.-M. Feasibility assessment of local repairment for reformer furnace tubes in service exposure. *Int. J. Pres. Ves. Pip.* **2020**, *179*, 104032. [CrossRef]
16. Fuyang, C.-M.; Chen, J.-Y.; Shao, B.; Zhou, Y.; Gong, J.-M.; Guo, X.-F.; Jiang, Y. Effect of microstructural evolution in thermal exposure on mechanical properties of HP40Nb alloy. *Int. J. Pres. Ves. Pip.* **2021**, *191*, 104391. [CrossRef]
17. Mostafaei, M.; Shamanian, M.; Purmohamad, H.; Amini, M.; Saatchi, A. Microstructural degradation of two cast heat resistant reformer tubes after long term service exposure. *Eng. Fail. Anal.* **2011**, *18*, 164–171. [CrossRef]

18. Tomków, J.; Janeczek, A. Underwater in situ local heat treatment by additional stitches for improving the weldability of steel. *Appl. Sci.* **2020**, *10*, 1823. [CrossRef]
19. Tomków, J.; Janeczek, A.; Rogalski, G.; Wolski, A. Underwater local cavity welding of S460N steel. *Materials* **2020**, *13*, 5535. [CrossRef] [PubMed]
20. Mostafaei, M.; Shamanian, M.; Purmohamad, H.; Amini, M. Increasing weldability of service-aged reformer tubes by partial solution annealing. *J. Mater. Eng. Perform.* **2016**, *25*, 1291–1303. [CrossRef]
21. Bhaumik, S.; Rangaraju, R.; Parameswara, M.; Bhaskaran, T.; Venkataswamy, M.; Raghuram, A.; Krishnan, R. Failure of reformer tube of an ammonia plant. *Eng. Fail. Anal.* **2002**, *9*, 553–561. [CrossRef]
22. Branza, T.; Deschaux-Beaume, F.; Sierra, G.; Lours, P. Study and prevention of cracking during weld-repair of heat-resistant cast steels. *J. Mater. Process. Technol.* **2009**, *209*, 536–547. [CrossRef]
23. Kachanov, L.M. Rupture time under creep condition. *Izvestia Akademi Nauk SSSR Otd. Tekhn Nauk.* **1958**, *8*, 26–31.
24. Rabotnov, Y.N. *Creep Problems in Structural Members*; Elsevier (North Holland Publishing, Co.): Amsterdam, The Netherlands, 1969.
25. Hayhurst, D. Creep rupture under multi-axial states of stress. *J. Mech. Phys. Solids.* **1972**, *20*, 381–382. [CrossRef]
26. Liu, Y. *A Localized Creep Damage Theory and Its Application to Creep Crack Growth*; Southwestern Jiaotong University: Chengdu, China, 1990. (In Chinese)
27. Liu, Y.; Murakami, S. Damage localization of conventional creep damage models and proposition of a new model for creep damage analysis. *JSME Int. J.* **1998**, *41*, 57–65. [CrossRef]
28. Othman, A.M.; Hayhurst, D.R.; Dyson, B.F. Skeletal point stresses in circumferentially notched tension bars undergoing tertiary creep modeled with physically based constitutive equations. *Proc. R. Soc. Lond. A* **1993**, *441*, 343–358.
29. Kowalewski, Z.L.; Hayhurst, D.R.; Dyson, B.F. Mechanisms-based creep constitutive equations for an aluminium alloy. *J. Strain. Anal. Eng. Des.* **1994**, *29*, 309–316. [CrossRef]
30. Perrin, I.J.; Hayhurst, D.R. Creep constitutive equations for a 0.5Cr-0.5Mo-0.25V ferritic steel in the temperature range 600 °C to 675 °C. *J. Strain Anal. Eng Des.* **1996**, *31*, 299–314. [CrossRef]
31. Dyson, B. Use of CDM in materials modeling and component creep life prediction. *J. Pres. Ves. Technol. ASME.* **2000**, *122*, 281–296. [CrossRef]
32. Larson, F.R.; Miller, J. A time-temperature relationship for rupture and creep stress. *Trans. ASME* **1952**, *74*, 765–771.
33. Burke, K.E. Chemical extraction of refractory inclusions from iron- and nickel-base alloys. *Metallography* **1975**, *8*, 473–488. [CrossRef]
34. Shi, S.; Lippold, J.C. Microstructure evolution during service exposure of two cast, heat-resisting stainless steels—HP-Nb modified and 20-32Nb. *Mater. Charact.* **2008**, *59*, 1029–1040. [CrossRef]
35. Dessolier, T.; McAuliffe, T.; Hamer, W.J.; Hermse, C.G.M.; Britton, T.B. Effect of high temperature service on the complex through-wall microstructure of centrifugally cast HP40 reformer tube. *Mater. Charact.* **2021**, 111070. [CrossRef]
36. Chen, Q.Z.; Thomas, C.W.; Knowles, D.M. Characterisation of 20Cr32Ni1Nb alloys in as-cast and ex-service conditions by SEM, TEM and EDX. *Mater. Sci. Eng. A* **2004**, *374*, 398–408. [CrossRef]
37. Soares, G.D.D.A.; Almeida, L.H.D.; Silveira, T.L.D.; May, I.L. Niobium additions in HP heat-resistant cast stainless steels. *Mater. Charact.* **1992**, *29*, 387–396. [CrossRef]
38. Guo, X.; Jia, X.; Gong, J.; Geng, L.; Tang, J.; Jiang, Y.; Ni, Y.; Yang, X. Effect of long-term aging on microstructural stabilization and mechanical properties of 20Cr32Ni1Nb steel. *Mater. Sci. Eng. A* **2017**, *690*, 62–70. [CrossRef]

Article

High Temperature Behavior of Al-7Si-0.4Mg Alloy with Er and Zr Additions

Elisabetta Gariboldi *, Chiara Confalonieri and Marco Colombo

Department of Mechanical Engineering, Politecnico di Milano, Via La Masa 1, 20156 Milano, Italy; chiara.confalonieri@polimi.it (C.C.); marco1.colombo@polimi.it (M.C.)
* Correspondence: elisabetta.gariboldi@polimi.it

Abstract: In recent years, many efforts have been devoted to the development of innovative Al-based casting alloys with improved high temperature strength. Research is often oriented to the investigation of the effects of minor element additions to widely diffused casting alloys. The present study focuses on Al-7Si-0.4Mg (A356) alloy with small additions of Er and Zr. Following previous scientific works on the optimization of heat treatment and on tensile strength, creep tests were carried out at 300 °C under applied stress of 30 MPa, a reference condition for creep characterization of innovative high-temperature Al alloys. The alloys containing both Er and Zr displayed a lower minimum creep strain rate and a longer time to rupture. Fractographic and microstructural analyses on crept and aged specimens were performed to understand the role played by eutectic silicon, by the coarse intermetallics and by α-Al matrix ductility. The creep behavior in tension of the three alloys has been discussed by comparing them to tension and compression creep curves available in the literature for Al-7Si-0.4Mg improved by minor elemental additions.

Keywords: Al-7Si-0.4Mg (A356 alloy); erbium; zirconium; creep; microstructural stability

check for updates

Citation: Gariboldi, E.; Confalonieri, C.; Colombo, M. High Temperature Behavior of Al-7Si-0.4Mg Alloy with Er and Zr Additions. *Metals* **2021**, *11*, 879. https://doi.org/10.3390/met11060879

Academic Editor: Andrey Belyakov

Received: 30 April 2021
Accepted: 25 May 2021
Published: 28 May 2021

Publisher's Note: MDPI stays neutral with regard to jurisdictional claims in published maps and institutional affiliations.

1. Introduction

A356 (Al-7Si-0.4Mg) is a casting alloy widely applied in automotive and aeronautical fields, as referred by Mishra et al. [1] and Jeong et al. [2], among others. The casting alloy is age-hardenable, and, thus, its mechanical properties depend on the combination of several microstructural features, from macroscopic and microscopic ones (including dendrite arm spacing, grain size, coarse equilibrium or non-equilibrium intermetallic phases, porosities and other possible discontinuities related to casting processes) down to the presence of nanometric precipitates with noticeable strengthening effect [3,4]. The alloy is classically solution heat treated (SHT) at about 540 °C [4] with multiple effects of homogenizing the alloy compositions, reducing the amount of coarse embrittling phases, bringing precipitates in solid solution and concurrently modifying silicon morphology. An artificial aging at temperatures 155–200 °C follows [3], during which the β-Mg2Si precipitation sequence evolves similarly to what observed in wrought Al-Mg-Si alloys, where also the Si or U2-Mg-Al-Si phase can be observed substituting β″ in alloys with high Si content [5,6].

The high temperature behavior of the Al-Si-Mg age-hardenable casting alloys is strictly connected to overaging processes, related to the instability and fast growth of strengthening particles of the abovementioned precipitation sequence. Several minor elemental additions have been proposed during the years to improve mechanical properties at high temperature and/or mechanical stability of the A356 alloy [2,7–13]. The adopted criteria were analogous to those used for other casting Al alloys, but also for wrought ones. As an example, the addition of small amount of copper, besides a minor refining effect, favors the concurrence of β and of the Q precipitation sequences, this latter characterized by slower coarsening kinetics [2,7]. Nevertheless, the requests of high-temperature stability cannot be met adding this element only. During the years, modified A356 alloys with minor additions of Mo [8],

V [9], Ho [10], Zr [11], Sc [12], Er [13], Ce [14] and Nb [15] have been proposed with improved aging resistance and high temperature strength. In many cases, the criteria proposed by Knipling et al. [16] were applied, exploiting the formation of relatively high volumes of secondary trialuminides phases (Al$_3$M), specifically the ones characterized by the most favorable L12 structure such as those where M corresponds to Sc or to the rare earths Er, Yb and Tm [7]. Additionally, Al$_3$M with less favorable D022 and D023 structures (such as for Ti, V, Zr, Mo, Hf and Ta) can be considered since their low diffusivity leads to slow coarsening rates (such as for Mo, Mn and Cr [17–19]). A different and favorable situation can in any case be observed for Zr: its additions, for example can lead to the precipitation of L12 metastable intragranular particles [16] (in addition to the D023 coarse intragranular intermetallics). During high temperature exposure, the fine L12 particles display a slow coarsening, due to low Zr diffusivity, until they eventually transform into the stable phase.

In wrought alloys [20–25], the combination of Zr with Sc or its partial replacement by Er have proved to be beneficial for the high-temperature strength and microstructural stability of Al alloys, since the segregation of Zr at the particle/matrix interface reduces the particle coarsening rate until the stable LI2 structure is kept [20,22]. The role played by minor additions of silicon to enhance particle nucleation has also been observed for these alloys [26].

The combined addition of Sc, Er and Zr was also proposed in literature to improve the structural properties of Al-Si casting alloys [13,26–28], typically heat treated in a conventional way (SHT + aging). Complementary effects were reported for these minor additions: Zr acts as grain refiner and inoculant [11,29,30] while Er, together with Sc and other rare-earth elements, have modifying effects on eutectic Si, similarly to the widely diffused but more volatile Sr [31–33].

The high temperature behavior of Al heat-treatable casting alloys has been generally evaluated by means of tensile tests at high temperature [17–19,26] and/or by means of hardness evolutions at different aging times. Additionally, creep tests, where the material deforms under constant stress at constant temperature, and thus in a progressively overaging material, have been considered in some cases [34–36]. Even if a proper creep characterization should require tests at different temperatures and applied stresses, the potentialities offered by minor alloying additions on the high temperature behavior of Al-Si alloys has been typically evaluated at 300 °C in order to show the suitability of innovative Al-Si-Mg-based alloys with improved high temperature strength for automotive, and more in general for industrial components for which the service temperature is continuously or cyclically close to it [34,37–40]. Further, specific stress levels have been considered from several authors [41–43]. Due to the significant role played by microscopic and submicroscopic features, a meaningful comparison with literature data is not easy and, thus, the effect of compositional or microstructural changes is typically evaluated by the same authors within the same or related research works [36,41,44].

The authors of the present paper have previously investigated the beneficial effect of Er addition to A356 casting alloy on its microstructure and identified 0.3 wt% Er as the content offering the best compromise between tensile strength and ductility at room and high temperature [45]. Similarly, an additional 0.5 wt% Zr led to a further improvement of tensile strength [30]. The present article focuses on the tensile creep behavior at 300 °C of the two abovementioned optimized alloys, comparing them to those of the reference A356 alloy.

2. Materials and Methods

A standard Al-7Si-0.4Mg (A356) alloy, whose chemical analysis is given in Table 1, was used as a reference alloy for the present research study. Different amounts of Al-15 mass%Er and Al-10 mass%Zr master alloys were added to the reference alloy, to obtain nominally A356 + 0.3Er and A356 + 0.3%Er + 0.5%Zr (% in mass). The alloys will be hereafter referred as A356, E3 and EZ35, respectively. The master and reference alloys were melted under Ar atmosphere in an induction furnace at 800 °C, where they were kept and stirred for 30 min. The alloys were poured in a permanent steel mold, preheated to 200 °C. Cylinders with 40 mm diameter and 120 mm in height were obtained. The chemical compositions of the three alloys measured by glow discharge optical emission spectroscopy (GDOES) are shown in Table 1. They are close to those of ingots with the same nominal composition that were used for heat treatments optimization and tensile characterization [30,45,46].

Table 1. Chemical composition (mass%) of the studied alloys.

	Si	Mg	Fe	Ti	Er	Zr	Al
A356	7.02	0.41	0.07	0.14		-	bal
E3	7.40	0.34	0.10	0.12	0.29	-	bal
EZ35	6.66	0.37	0.11	0.12	0.31	0.49	bal

Bars of about $12 \times 12 \times 100$ mm^3 size were sampled from the mid-low part of each cylinder. Each one was solution heat treated (SHT) at 540 °C for 5 h, quenched in water at room temperature and immediately aged at 200 °C in order to reach the peak-aged (PA) condition. Ageing time was about 2 h for the EZ35 alloy and 3 h for the others [30,45]. Cylindrical specimens for creep tests were machined from these bars. The specimens were characterized by threaded ends and collars for extensometer attachment at the ends of the cylindrical gauge length. This latter was 30 mm long, with a diameter of 6 mm. The specimen geometry was compatible with ISO standards for both high-temperature tension tests [47] and creep tests [48].

Creep test in tension were performed at 300 °C on samples in the PA condition. Tests were carried out in air, in a lever-arm machine with lever ratio 25:1 (AMSLER, Sciaffusa, Switzerland). The tests have been carried out at initial stress (σ_0) of 30 MPa. The temperature and stress conditions were selected in order to compare results with those available for other improved Al-Si based casting alloys proposed in the literature [37,40]. During each creep tests, the specimen elongation was monitored with a diametral telescopic extensometer, to which a couple of LVDT (Solartron Metrology, Leicester, UK) were attached in order to supply the average displacement of the specimen gauge length and, from it, the material strain within the specimen gauge length. From each creep test where final fracture was reached, time to rupture, elongation at rupture and reduction of area at rupture were measured and minimum strain rate ($d\varepsilon/dt$) was calculated from each creep curve (engineering strain (ε) vs. time (t)). For further analysis and comparisons with literature data, the true strain ($e = \ln(1 + \varepsilon)$), true stress ($\sigma_t = \sigma_0 \times (1 + \varepsilon)$) and true strain rates ($de/dt = (d\varepsilon/dt)/(1 + \varepsilon)$) were calculated under the assumption of constant volume and uniform strain along the gauge length.

In order to correlate the fracture mode to microstructural features of the alloys, SEM (FEG-SEM, Zeiss Sigma 500, Jena, Germany) fractographic analyses were carried out on the fracture surface, on longitudinal sections of crept specimens. In the latter case, SEM observations were combined to optical ones (Nikon Eclipse LVision NL, Tokyo, Japan) for better identification of Si particles. In both cases, the specimen surface was prepared by conventional metallographic polishing.

Lastly, a discoidal sample 2 mm thick and of about 12 mm diameter was cut from the ingot for each alloy. These samples were SHT at 540 °C and then immediately aged at 300 °C for 100 h, corresponding to a time shorter than that of the onset of creep damage in the creep test of shorter duration. SEM observations and intermetallic compositional EDS analyses were performed on them. Quantitative analysis of the area fraction of coarse intermetallic compounds was conducted with ImageJ software in the Fiji distribution [49].

3. Results

3.1. Alloy Microstructure

The microstructural features and aging behavior of the three alloys in the T6 condition will be here summarized anticipating both creep test and fractographic analyses results. The authors of the present paper have presented and discussed them in detail elsewhere [46].

The casting alloys are characterized by the dendritic grain size of 1462 ± 350 μm for A356 alloy, 1157 ± 186 μm for E3 alloy and 383 ± 98 μm for the Zr-containing alloy. Additionally, the secondary dendrite arm spacing decreased from 31.5 μm for A356 to 24.1 μm and 18.2 for E3 and EZ35, respectively. In PA conditions, the α-Al phase at the dendritic core is flanked by Si particles and by coarse phase containing Er and Zr in the modified alloys and also combinations of Fe, Cu and Mg and minor alloying additions in the reference alloy. In as-cast conditions, the eutectic Si length was longer and with higher aspect ratio for A356 (about 27 μm and 10.8, respectively), but shows a rapid decrease of both morphology parameters within the first dozen of minutes at 540 °C, reaching respectively 9.2 μm and 4.2 after 5 h at 540 °C. On the other hand, E3 and EZ35 form coral-like eutectic Si structures, resulting in shorter and lower aspect ratio Si particles in metallographic sections (of the order of 4 μm and 1.65, respectively) [46]. The 5 h SHT only slightly increased the average eutectic Si particle length, while decreasing the aspect ratio to about 1.35 for both alloys. Thus, the morphology of Si in these alloys was more related to the as-cast microstructure than to further exposure at high temperature. Further, the area fractions of coarse intermetallic phases for the A356, E3 and EZ35 alloys, which were 0.5, 0.9 and 1.3 vol% in as-cast alloys, after 5 h SHT reduced to 0.3, 0.6 and 1.1 vol% [46]. The Mg-containing π-Al$_8$Mg$_3$FeSi$_6$ Chinese script and Mg$_2$Si phases show the most significant changes during the 5 h SHT. Contrarily to the other alloys, part of the coarse intermetallics in the EZ35 alloy are located in intradendritic regions and those in interdendritic regions are of complex shape and composition.

Alpha phase strengthening in A356, E3 and EZ35 alloys, due to the presence of nanometric precipitates of Al$_3$Er, Al$_3$Zr and Al$_3$(Zr,Er) is discussed in [50]. After solution heat treatment the peak hardness obtained by overaging at 200 °C reached about 97, 109 and 133 HV for A356, E3 and EZ35 alloys, respectively, in all cases with the maximum standard deviation close to 4 HV. Close hardness values are expected for the creep specimens, tested after performing the same solution treatment and aging.

3.2. Creep Tests at 300 °C

Creep tests at 300 °C, carried out at the same stress level of 30 MPa, led to the creep curves (strain vs. time plot) shown in Figure 1a. While the alloy with only Er addition displayed slightly lower times to rupture with respect to the reference A356 alloy, the one containing both Er and Zr clearly shows a longer test duration. The alternative arrangement of creep test results in terms of minimum strain rate plotted vs. time in Figure 1b better highlights differences in the initial transient stage, in the value of the minimum creep strain rate and in its increase. During the initial loading stage and during the primary stage transient, the alloys with greater addition of Er and Zr display less overall strain and lower strain rates and lower minimum creep rates. The minimum strain rate was slightly higher for the reference alloy rather than for the Er-modified alloy (0.0142 and 0.0129%/h, respectively). On the other hand, EZ35 alloy was clearly the one displaying the higher creep resistance under the testing conditions, since its minimum creep rate decreased to 0.0071%/h, that is about one half of those of the A356 and E3 alloy.

Figure 1b further shows that the time and strain range characterized by the minimum creep strain rate was limited. Rather, after reaching the minimum value, the strain rate gradually increased for the three alloys, up to a time closer to the final fracture. Despite its relatively good creep resistance in terms of the minimum creep strain rate, the E3 alloy displays a relatively rapid increase of strain rate in Figure 1b compared to the reference alloy.

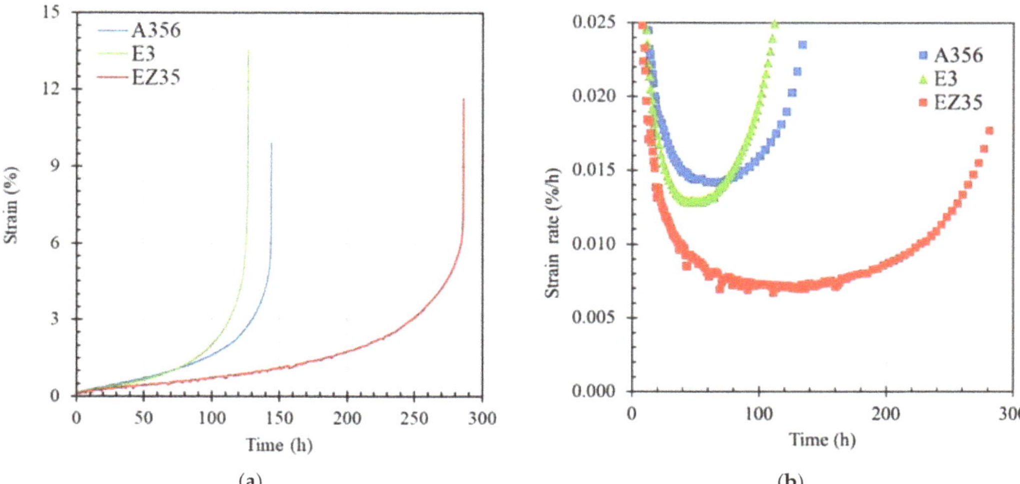

Figure 1. Results of creep tests carried out on the reference and modified alloys at 300 °C, 30 MPa. (**a**) Strain vs. time curves and (**b**) creep strain rate vs. time.

Both ductility indexes show the same trend for the alloys. As matter of facts, the strain at fracture was 9.93%, 13.54% and 11.68% for the A356, E3 and EZ35 alloy, while the reductions of area were 2.68%, 22.97% and 8.72% for the alloys in the same order.

3.3. Fractographic Analyses

Representative results from fractographic analyses of crept specimens are given in Figures 2–4. Figure 2a shows a low-magnification image of the fracture surface of A356 specimen. The big reliefs and depression on the specimen surface, also visible on the contour of the fracture surface, were, at least partly, associated to decohesion along dendrite grains, which were relatively large in this alloy. This agreed with the features observed on the external surface of the A356 crept sample (Figure 3a) at a few millimeters distance from the final fracture.

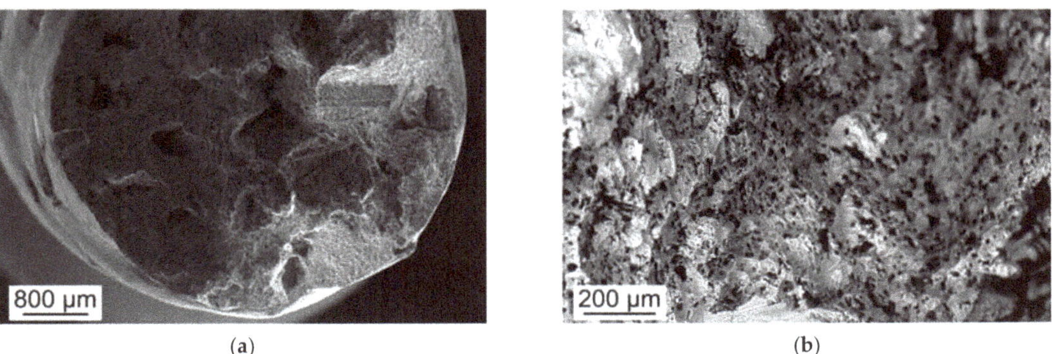

Figure 2. Fracture surface crept specimen in alloy A356 at low (**a**) and high (**b**) magnification, this latter taken with an electron backscattered probe to reveal the presence of phases with high Z-contrast.

At a higher magnification scale (Figures 2b and 4), the fracture surfaces of the three alloys were characterized by the presence of dimples, as a sign of the ductile behavior of the Al matrix, overaged after prolonged maintenance at 300 °C. In the A356 alloy dimples mainly included Si particles (Figure 2b). In the modified alloys E3 and EZ35, dimples also nucleated at Er- and/or Zr-rich coarse particles. These features can be clearly observed by comparing Figure 4a,b, taken in the same region of specimen E3 with secondary electrons and backscattered electrons (BSE), respectively. Due to their Z-contrast, Er containing particles are the brightest in BSE fractographs and micrographs.

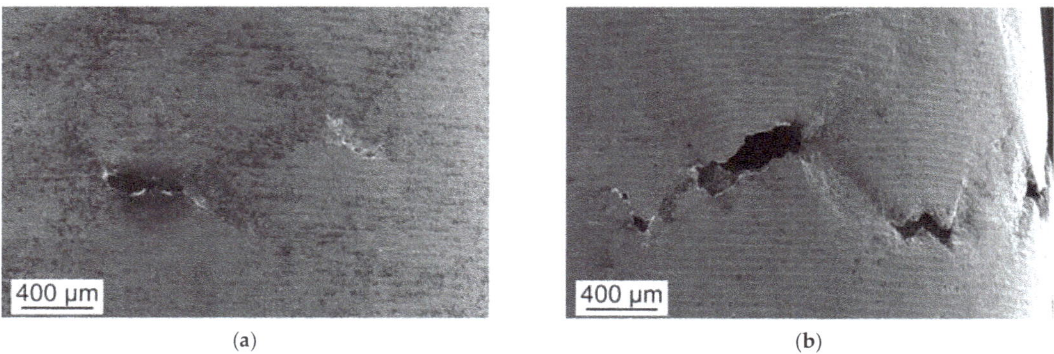

(a) (b)

Figure 3. Lateral surface of crept specimen in alloy A356 (**a**) and EZ35 (**b**).

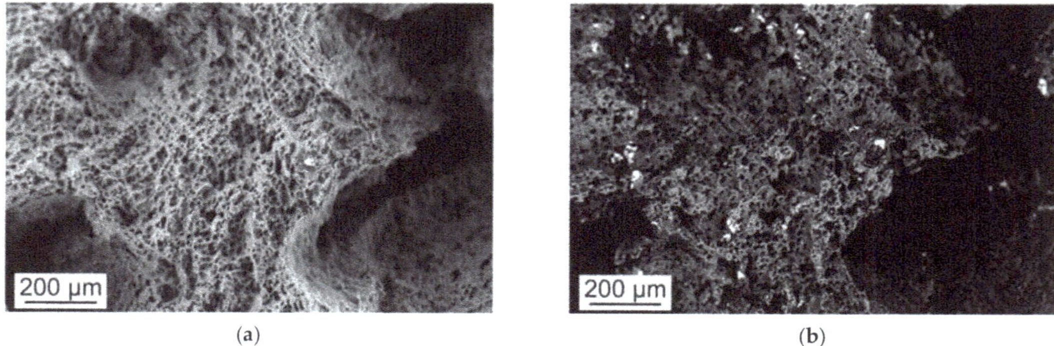

(a) (b)

Figure 4. Higher magnification images of fracture surface of specimen E3 taken by secondary electron probe (**a**) or backscattered electron probe (**b**), which clearly displays Er-containing phases, with higher Z-contrast.

Representative micrographs of longitudinal sections of the above specimens at the optical microscope are shown in Figures 5 and 6. Specifically, Figure 5a, taken on alloy A356, confirms that the fracture path generally follows the coarse dendritic grains of the alloy, where the eutectic Si and α-Al phase are present. The higher magnification image further shows that the creep fracture surface also follows interdendritic regions, which are underlined by the presence of the grey eutectic Si phase. In this alloy, eutectic Si particles have high aspect ratio (i.e., maximum to minimum length of the particle). In correspondence to these Si particles, decohesions at the particle/matrix interface rather than their fractures are observed.

The fracture path described above for the reference A356 alloy was essentially the same observed on longitudinal sections of Er- and Zr-containing alloys, with the difference that in them higher local strains at the final fracture were suggested by secondary cracks that follow the same interdendritic path. In Figure 6, refers to alloy E3, the high strain at fracture can be noticed in the low-magnification micrograph. In the same crept specimen, some discontinuities between matrix and Er-containing coarse interdendritic particles were observed. The SEM-BSE micrograph in Figure 7 of the longitudinal section of E3 sample further shows these as bright particles, with a simple, more or less elongated shape, which tended to orient along the vertical, loading and strain direction (vertical in Figure 7). In the modified alloys E3 and EZ35, the fracture path can also include some of the intergranular porosities related to the casting process.

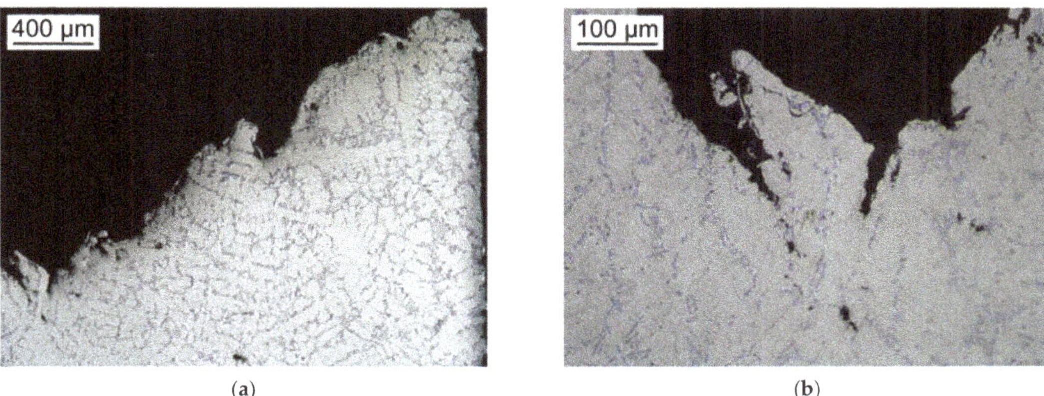

(a)   (b)

Figure 5. Optical micrographs of longitudinal sections of crept specimens A356 alloy at low (**a**) and high (**b**) magnification.

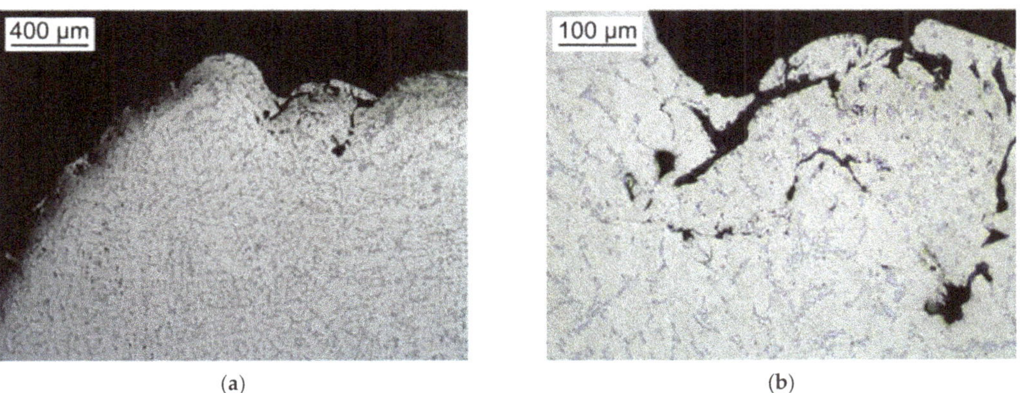

(a)   (b)

Figure 6. Metallographic features on the longitudinal section of the crept sample E3 in optical micrograph at low- (**a**) and high-magnification (**b**) clearly displaying the eutectic regions, the material deformation in vertical direction and the presence of discontinuities.

Even if macroscopic strains were noticed for the three alloys only close to the fracture surface, the first signs of microstructural creep damage were observed a few millimeters far from the final surface. This clearly appears on the outer surface of the crept specimens, where the fracture mechanism leading to the formation of intergranular and interdendritic fracture can be clearly observed from the presence of microcracks. Figure 3a,b depicts the ones for the A356 and EZ35 alloy, respectively. While the change of crack orientations corresponding to an interface between two dendritic grains can be observed on both alloys, grain boundary sliding is suggested by the change of levels of the horizontal turning lines only on the surface of EZ35, characterized by the finest dendritic grains (Figure 3b). Due to the small sliding extension (some tenth of μm on grains displaying it compared to the material grain size), the contribution of grain boundary sliding to the overall creep elongation is limited also for this alloy.

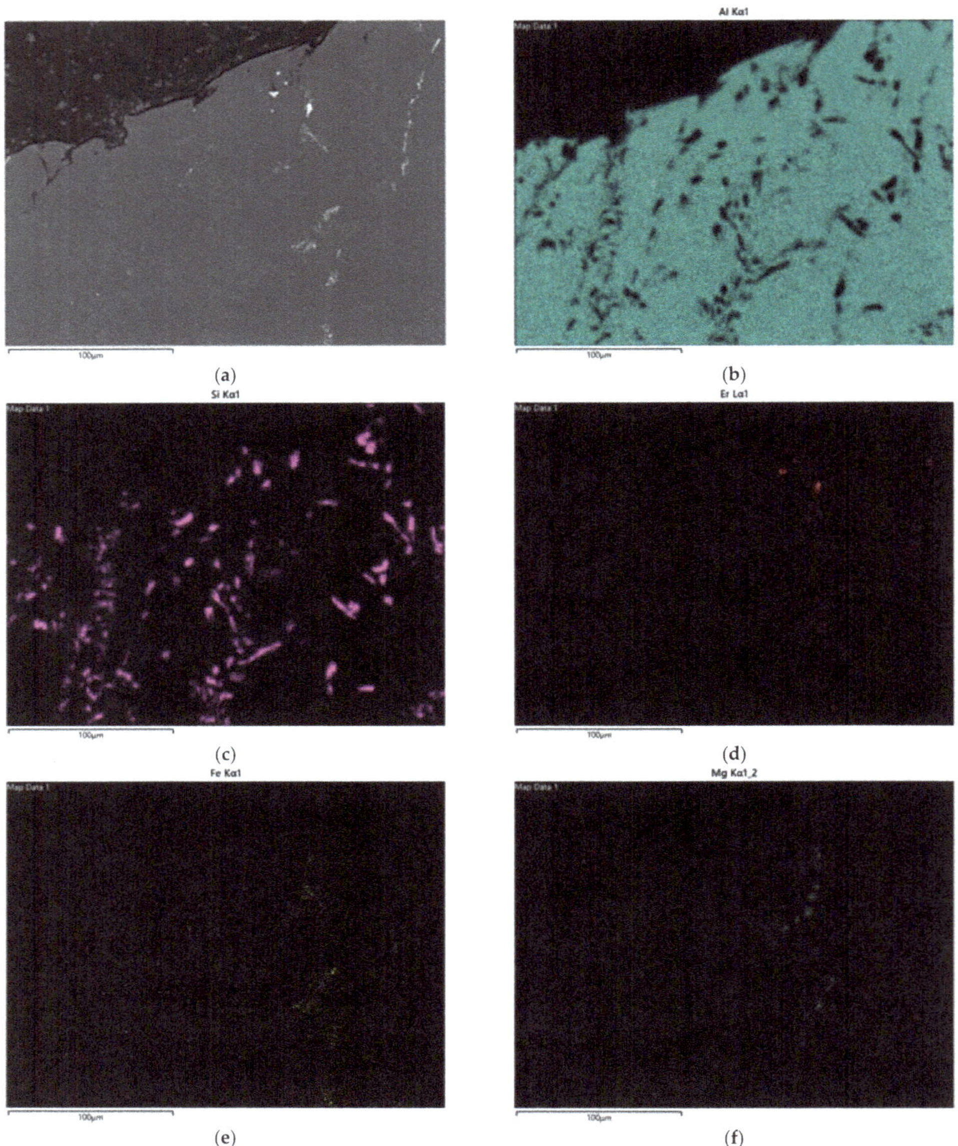

Figure 7. SEM BSE micrograph (**a**) and corresponding EDS map for the alloying elements (Al (**b**), Si (**c**), Er (**d**), Fe (**e**) and Mg (**f**)) in the longitudinal section of E3 alloy close to its final fracture.

Figure 7 shows an EDS map of element distributions taken close to the fracture surface of the E3 specimen. The combined analysis of Al and Si once again shows that the preferred fracture path was localized along the interdendritic eutectic regions, where Si was clearly well globularized. Er-containing particles can also be identified as the bright particles in the BSE micrograph. The corresponding maps show that Er can be found in elongated interdendritic particles (reasonably combined with Al). Er is also partly combined with Mg or Fe in particles whose morphology suggests the progressive dissolution of the coarse particles with complex composition and complex morphology.

Lastly, at higher magnification some extremely fine, Er-containing particles appear at the surface of eutectic silicon, at least partly related to the presence of Er in eutectic silicon in the as cast structure [50].

3.4. Metallographic Analyses of Specimens Aged at 300 °C

The analysis of the microstructure of samples aged for 100 h at 300 °C was mainly aimed at investigating the differences between the alloys in terms of coarse intermetallic structure. Thus, BSE images of the samples were taken and the amount of coarse intermetallics, appearing as bright particles, was measured in terms of area fraction on images (each one with an area equal to 0.196 mm^2). The micrographs E3 and EZ35 alloy are shown in Figure 8. The threshold of the counting method was set so that the counting of eutectic particles, slightly visible in micrographs, was avoided. The area fraction of coarse intermetallic phases in A356, E3 and EZ35 alloys resulted as 0.16%, 0.88% and 3.02%, respectively. In A356 alloys the minor amount of intermetallics consists in Fe-containing particles, while in E3 alloys the intermetallics are of a different type, characterized by the combination of Er with Fe or Mg. These particles were generally located in interdendritic regions (Figure 8a) and corresponded to those visible on crept specimen and shown in Figure 7. The Er-containing intermetallics are observed also in the sample of the EZ35 alloy. Clearly, in this latter alloy also Zr-containing particles were identified. Most of them were elongated Zr-containing particles clearly visible in Figure 8b.

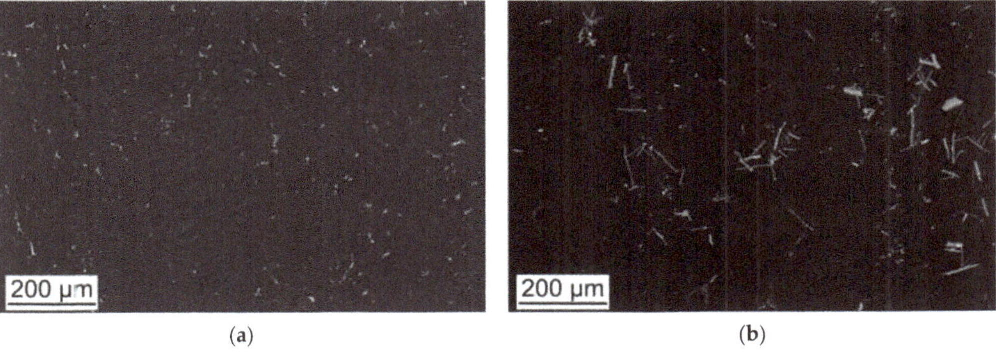

(a) (b)

Figure 8. SEM BSE micrograph of E3 (**a**) and EZ35 (**b**) alloys after SHT for 5 h at 540 °C and aging for 100 h at 300 °C.

Figure 9 shows a representative element mapping for the EZ35 alloy after 100 h at 300 °C. Similar to what observed on the crept sample in the E3 alloy shown in Figure 7, Er was distributed within different groups of particles. A first group of compact and bright particles (corresponding to Spectrum 12 in element maps and to the chemical composition in Table 2) includes Er, Al and possibly some Si (a eutectic silicon particle was very close to the analyzed one). The other Er-containing sets of particles were characterized by the concurrent presence of Fe or Mg, as clearly shown by elemental maps in Figure 9. The morphology of some of these Fe- or Mg-containing particles was similar to that observed in the E3 alloy in the metallographic sample (Figure 8b) and in the gauge length of the crept specimen (Figures 7 and 9) and suggests that they could be the residuals of the particles with complex composition and morphology observed also in the as-cast condition and partly modified during the solution treatment [46]. Figure 9 also includes some Zr-containing particle with the size of the order of one micron, some of which are also iron-containing. Lastly, the most characteristic particles of secondary phases in the EZ35 alloy were the elongated Zr-Ti-Al particles that can also be seen in Figure 9 (point 11 in maps, with EDS analysis results given in Table 2). These particles, to which the presence of the fine-grain structure of this alloy was attributed [46], clearly appeared in the micrograph taken at lower magnification in Figure 8, where their general intragranular location can be appreciated.

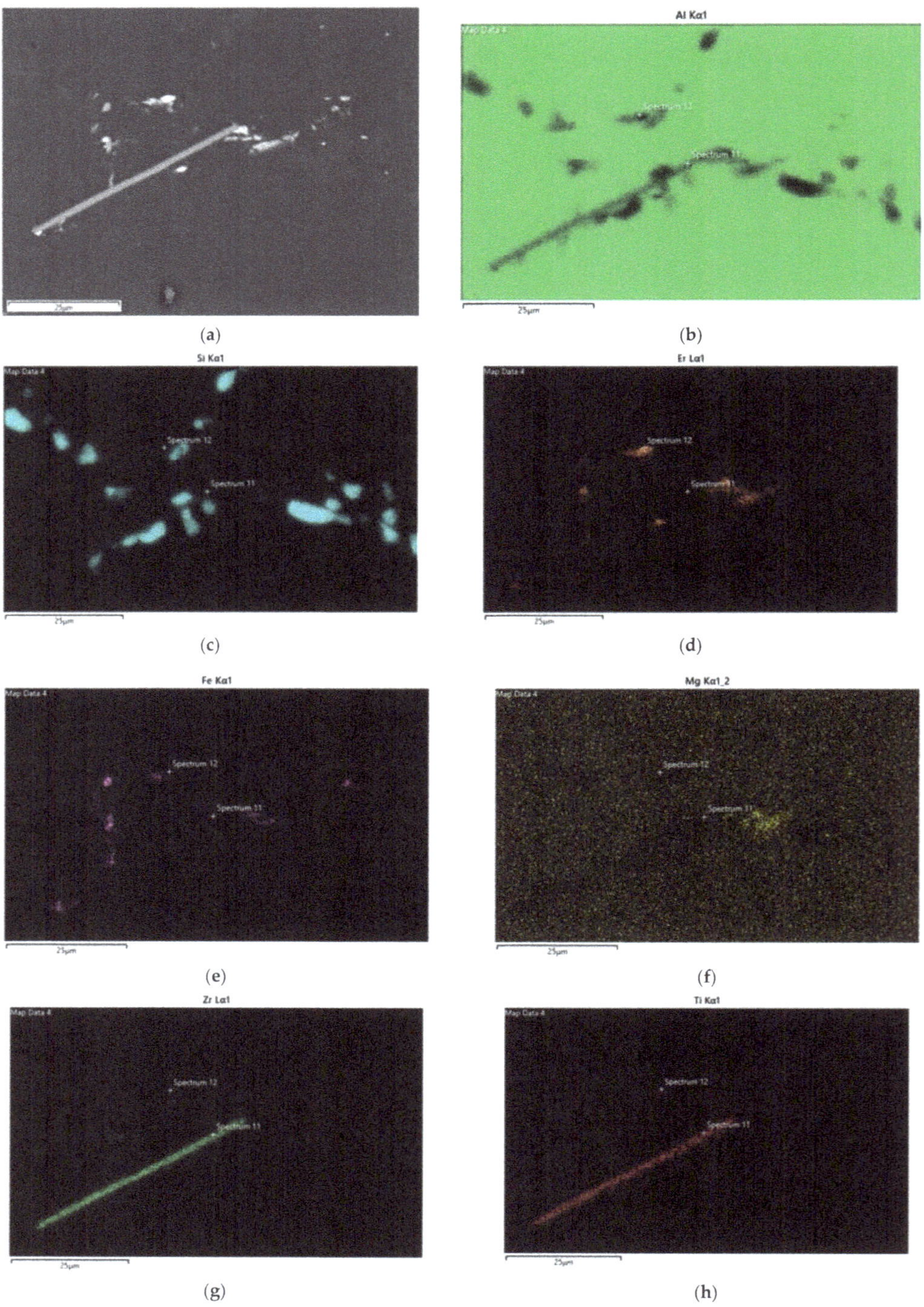

Figure 9. SEM BSE micrograph (**a**) and corresponding EDS map for the alloying elements (Al (**b**), Si (**c**), Er (**d**), Fe (**e**), Mg (**f**), Zr (**g**) and Ti (**h**)) in a sample of EZ353 alloy hold for 100 h at 300 °C.

Table 2. Chemical composition (at%) in points identified as 11 and 12 in maps of Figure 9.

Point#	Al	Si	Ti	V	Fe	Zr	Er	Total
11	75.58	8.19	4.57	0.09	0.47	10.68	0.43	100
12	72.63	11.87	-	-	-	-	15.49	100

4. Discussion

The comparative creep tests performed at 300 °C under nominal stress of 30 MPa suggested that the alloy with the combined addition of Er and Zr was more promising than the other modified alloy, both in terms of an extended lifetime and of a lower minimum creep strain rate.

The creep curves of the three alloys were compared to known literature creep curves of the A356 alloy and of some alloys modified by the addition of elements (different combinations of Cu, Zr, Mo and Mn), tested under the same temperature and stress, but in tension or compression loading and with different initial temper condition [39,40,51,52]. While recently impression creep results have been reported for A356 with different Er and Zr content [53], a direct comparison with them was prevented by the use of lower testing temperature and by the use of impression creep, with no easy correlation between strain rate and applied stresses, these latter reported to be in considerably higher range.

In the case of symmetrical creep behavior, the compressive and tensile creep curves can be compared considering the absolute true stress, strain and strain rate. Further, a power law dependence between the (absolute) true strain rate (de/dt) and the (absolute) true stress σ, the former can be written as:

$$de/dt = f(\sigma_o) \times (1 + \varepsilon)^n \tag{1}$$

where n' is the apparent exponent of the power law, ε and σ_o are the engineering strain and stress, respectively. In the case of tests carried out at the same temperature and the same initial engineering stress, the slope of the curve corresponds to the exponent of the creep strain rate when the true strain rate (de/dt) vs. (1 + ε) is plotted with double logarithmic axes. This kind of plot is adopted in Figure 10a,b to compare different creep curves.

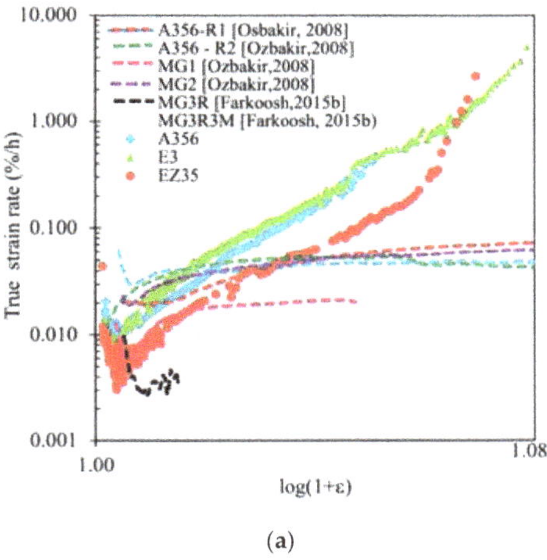

(a)

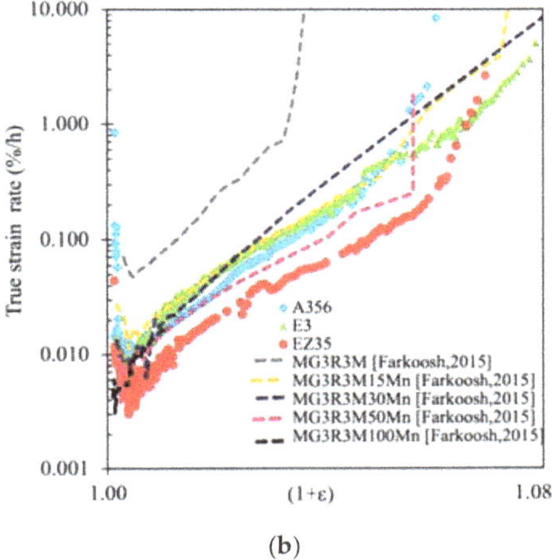

(b)

Figure 10. (a) Comparison between the results of the tensile creep tests performed on A356, E3 and EZ35 alloys and those reported by Ozbakir [52] and Farkoosh [51] for a A356 alloy and modified versions in microstructural conditions mentioned within the text. (b) Comparison between the results of the tensile creep tests performed on A356, E3 and EZ35 alloys and those reported by Farkoosh [39] for modified versions of the A356 alloy in microstructural conditions mentioned within the text.

The results will be discussed both in terms of the minimum strain rate, of strain rate correlation with true stress and of the strain at which creep damage occurs and develops leading to final fracture with high/low ductility indexes.

To the authors' knowledge, a creep curve on the A356 casting alloy at 300 °C and applied stress 30 MPa is available in literature only for the compressive creep. It is reported by E. Ozbakir [52] for an initial slightly overaged condition (SHT, 12 h natural aging, further 5 h aging at 200 °C). Despite the overaged condition, the room temperature tensile properties of the alloy were slightly higher (YS 230–250 MPa and UTS 250–280 MPa) than those previously obtained by the present authors for the A356 alloy [45]. The A356 alloy tested by Ozbakir [52] is referred in Figure 10a as A356-R1, following the author's name. It can be noticed that the slightly overaged A356-R1 alloy reached a minimum true strain roughly double than that of the present alloy.

All the other alloys tested under the same temperature and compression or tension stress level (compared in the same plot) are characterized by the addition of alloying elements with the aim of forming sets of strengthening particles (precipitates or dispersoids) in addition to those of the β-Mg2Si sequence [39,51,52]. For different datasets, the minimum strain rate tended to decrease (and strengthening effect increase) as strengthening alloying elements were added. Thus, the strengthening effect was lower for the present A356 alloy than for E3 and EZ35, for the alloy A356-R1 than A356-R2 (A356 + 0.5%Cu), for these latter than alloys MG1 and MG2 corresponding, respectively, to A356-R1 and A356-R2 with Mn, Cr and Zr additions (0.15 mass% per element). The same holds for creep compression curves under the same testing condition obtained by Farkoosh et al. [51] for Al-7Si-0.5Cu-0.3Mg (referred as MG3R) and for the same alloy with 0.3%Mo (MG3R3M alloy), tested at 300 °C after an optimized aging treatment (SHT + 5 h aging at 200 °C) and 100 h soaking at 300 °C before creep.

A second comparison between experimental and literature data available can be made for tension creep at 300 °C and 30 MPa. While for tension tests a simple comparison on the basis of the times to rupture could be performed, a more detailed analysis including creep resistance, strengthening mechanisms and creep damage and rupture considering the same axes of Figure 10a was here preferred.

The presently investigated alloys in the T6 condition are compared to those by Farkoosh [39,51] for alloys that, after SHT and aging were soaked at 300 °C for 100 h. Specifically, the alloys considered by Farkoosh in [39] were an Al-7Si-0.5Cu-0.3Mg-0.3Mo alloy (referred as MG3R3M) and its modified versions, obtained by adding Mn in 0.15–1 mass% range (the amount of which can be deduced from the number before the final M in the alloy grade name). Concerning the initial condition of different sets of alloys, it can be noticed that, in the alloys analyzed here by Farkoosh [39,51], due to the prolonged soaking anticipating creep, the strengthening effect of the less stable precipitates of the β-Mg2Si sequence, but also those of the quaternary Q-AlSiCuMg sequences, was reasonably lost as experimental data on the aging response recently presented by Pan and Breton show for 300 °C [54]. As a result, most of the creep strengthening for these alloys has to be correlated to the more stable strengthening particles formed with added elements, to atoms in solid solution into the α-Al matrix, and finally to grain size and grain boundary particles, both of them reducing grain boundary sliding and its contribution to the total strain rate. For this reason, the Al-7Si-0.5Cu-0.4Mg alloy referred as MG3R in [51] failed after a couple of hours, while creep life increased in Mo- and Mn-including alloys due to the formation of stable fine dispersoids of Al-(Fe,Mo)-Si or Al-(Fe,Mo,Mn)-Si during SHT. The creep life tended to increase up to 180 h (close to those of the presently investigated alloys) with Mo and Mn additions, at least to certain levels. On the other hand, the minimum creep rate in tension or compression of the alloys reported by Farkoosh [39,51] decreases with the amount of Mn, corresponding to the increased amount of strengthening dispersoids [37,39,51]. The minimum creep strain range of A356, E3 and EZ35 alloys is in the range of the highly alloyed grades considered by Farkoosh. The minimum creep strain rates have thus kept a certain degree of strengthening also in the A356 alloy, to which further effects adds up for the Er- and Zr-containing alloy. The other two elements are responsible for the progressive reduction of grain size, morphology change of eutectic Si and of interdendritic coarse particles and of higher amount of fine intragranular strengthening particles, thus reducing the minimum creep rate. The decrease of minimum creep strain rate with Er- and Zr-additions to the A356 alloy was also observed at the higher stress in impression creep [53].

Both in compression and in tension creep, the correlation between the true minimum strain rate and the term $(1 + \varepsilon)$ (proportional to true stress) displays a range where a regular, almost linear, increase in strain rate occurs. The slope of the curve corresponds to the apparent Norton index (n′ in Equation (1)), which is particularly high. This situation is characteristics of creep strain in particle strengthened alloys and has been recently described also by [55] for an Al-Mg-Zr alloy and also by Spigarelli for a set of Al-alloys strengthened by nanoscale particles [56]. The presence of strengthening particles produces a threshold (or quasi-threshold) effect that in the simplest models' versions, at constant temperature, considering engineering stress and strain can be described by the following equation

$$\mathrm{d}\varepsilon/\mathrm{d}t = (\sigma - \sigma_{th})^n \tag{2}$$

In Equation (2) n is generally referred as the (true) exponent and σ_{th} as the threshold stress, which can be estimated when experimental strain rates ranging of some orders of magnitude are available [39,55,56]. In the presence of this behavior, for alloys of close composition tested under the same level of stress σ, the apparent index (n′, in Equation (1)) is lower for the alloy with lower strengthening effect [39,55]. Further, for the same material, n′ tends to decrease as the stress level increases, reaching the value of the (true) stress exponent n for the matrix at high applied stress.

These effects can be observed, at least qualitatively, in the comparison of creep curves in Figure 10a,b. Figure 10a shows that the slope for the present A356 alloy was slightly higher than that of the A356-R1 alloy, corresponding to the presence of coarser strengthening particles (precipitates). The same figure shows that the alloys characterized by the presence of (Mn + Cr + Zr or Mo) or those containing Cu, all elements forming during heat treatments fine stable intragranular particles (dispersoids or precipitates), were characterized by a relatively steep part of the curves following the minimum creep strain rate.

The same trend to display a faster increase of the strain rate with $(1 + \varepsilon)$ for the curves characterized by the low creep rate (i.e., high creep resistance) can also be observed in Figure 10a for creep tension tests results. For most of the alloys, the part of the curve following the minimum creep rate was almost straight, with a nearly perceptible tendency to decrease before the final steep increase corresponding to creep damage progression. The presently investigated alloys, with a closer minimum strain rate to some of the Farkoosh ones, also had similar slopes for the following part of the curve. For A356, E3 and EZ35 creep curves, the n′ estimated from the above curve was high, even higher than the value close to 30 evaluated by multiple creep tensile tests at lower applied stress by Farkoosh [39] for the MG3R3M15Mg alloy for which the creep curve in Figure 10b is close to those of the E3 and A356 alloy. In the work by Rashno [53] where impression creep at 200–275 °C was performed at the very high stress levels typical of impression creep, the A356 alloys modified with higher amounts of Er and Zr the apparent stress exponent was only 6–7.

As suggested before, the high n′ values observed in the present case could be correlated to the presence of strengthening particles, and to a non-negligible threshold stress value in creep strain rate models, while the role played by grain boundary sliding in the case of fine-grained structures is not straightforward (and in any case of minor importance also for the fine-grained EZ35 alloy). Nevertheless, the extremely high n′ values observed with the present approach based on the analysis of a single specimen could be due to inaccuracy in slope calculation, since the stress range used for the n′ evaluation was small (at 5% strain the true stress increased from 30 to 31.5 MPa). Further, concurrent overaging during the creep test could have contributed to the increase of true strain rate, and thus to increase the calculated n′. When exposed to 300 °C during creep, the precipitates of the β-Mg_2Si sequence are expected to reduce their total volume fraction (a reduction in the range 10–14% is foreseen by thermodynamic calculations performed by the authors using Thermo-Calc software [57]) while particle coarsening is controlled by Mg and Si diffusion (with diffusivities at 300 °C of 3.1×10^{-16} and 2.6×10^{-16} m^2/s, respectively [58]). Modeled evolution of the multiphase precipitates of the Al-Mg-Si alloy showed that the volume fraction of precipitates formed after a temperature increase during aging can have a more complex evolution [59]. The formation of the fine intragranular Si-particle, as typical alloys with excess Si [60] and also observed by the authors during transmission electron microscopy analyses on A356 aged at 200 °C for 168 h [50], can reasonably occur. The changes in precipitates size and volume are in any case thought to leave minor residual strengthening effects within the first tenth of hours. The Mg and Si diffusion rates are not far from that of Er in Al (of the order 4×10^{-19} m^2/s at 300 °C [61]) suggesting a prolonged strengthening effect due to Er-containing fine particles formed during the heat treatment cycles anticipating creep tests in the Er-modified alloy (with bimodal size distribution in T6 condition observed on the alloy of close composition [46]). On the other hand, the 5-orders of magnitude lower diffusivity of Zr in Al suggests that, correspondingly to what observed for isothermal aging of Er- and Zr-containing alloys [23,62], far longer times are needed for coarsening and structural modifications of the fine strengthening phase containing both Er and Zr observed in the alloy in the T6 condition [46].

While the possibility to apply the previous approach to estimate a threshold stress to the case of data from a single creep test on a microstructurally stable material (i.e., no coarsening nor volume change of nanometric strengthening particles) is theoretically possible, in practice this is prevented by the limited strain rate range for single tension creep test. Thus, while the presence of strengthening particles can be considered as a reasonable cause for the high slope of the curves, it cannot be proved.

Focusing now the attention on the last step of the creep curve, the one leading to the final fracture, is worthy of consideration. In compression tests (literature curves in Figure 10a), as the strain level increased above roughly 2–3%, the curves reduced their slope, due to the concurrent barreling effect, which reduced the actual true stress in some portions of the specimen. On the other hand, a different, upward, change of slope is generally noticed in tension creep curves shown in Figure 10b. This occurs at $(1 + \varepsilon)$ corresponding to strain levels of about 4%, 6.5% and 5.5% for A356, E3 and EZ35, respectively. Above these strains, the last stage of the creep tests characterized by creep damage development takes place. In the presently investigated alloys damage is characterized by the formation of the discontinuities presented in Figures 3, 5 and 6. The process of damage onset and development is suggested by the creep curves in Figure 10b. Once again, the results of the present experimental tests can be discussed comparing them to those of several A356-based alloys by Farkoosh [37,39,51]. In these alloys the progressive combined addition of Mo and Mn increased the presence of coarse intermetallics, similarly to what observed for Er and Zr additions in microstructural investigations for E3 and EZ35 alloy. Farkoosh's alloys were further characterized by high aspect ratio of intermetallics. As discussed by these authors [8,39] and noticeable by their creep curves rearranged here in Figure 10b, the alloys with high Mn and Mo contents and characterized by the low creep strain rates are also those for which the high amount of these coarse intermetallics is responsible for the onset of creep damage at low strain, and for the following rapid increase in strain rate.

Analogously, the higher overall amount of coarse intermetallic phases with Zr and Er additions and the morphology changes induced by this latter element on eutectic Si during the alloy solidification process can explain the last stage of the creep curves in the presently investigated alloys.

Thus, the A356 alloy with a softer creep and overaging structure reached both a higher minimum creep rate and a lower level of strain for creep damage onset (approximately 4%, reached at 140 h). In the last stage of the creep test, intergranular damage started at the eutectic Si, still elongated, as revealed by microstructural observations (Figure 5).

The behavior of the E3 alloy was more complex to explain. The alloy in the T6 condition reached a relatively low minimum creep strain rate. The expected coarsening kinetics of fine Er-containing fine strengthening particles, not significantly longer than for modifications of Mg- and Si-containing phases, could be responsible of a relatively fast increase for the strain rate, while a minimal contribution of grain boundary sliding cannot be excluded. The onset of tertiary creep occurs then at an accumulated strain of 6.5% and progresses slowly with strain. As a result, the ductility indexes for E3 were the highest of the investigated alloys. The delay in the onset of damage with respect to the reference alloy and the slow damage progression with strain was reasonably due to the combined presence of globularized Si and of the mostly dissolved and compact coarse intermetallics.

The most stable microstructure, EZ35 alloy, with the higher creep-resistance and the presence of the stable Zr- and Er-containing fine particles [63], after reaching the lowest minimum creep strain rate, slowly accumulates strain. The presence of globularized eutectic Si particles, even if in combination with a high amount and high aspect ratio, but mainly intragranular, intermetallics, delays at about 280 h and 5.5% total strain and the onset of creep damage, which develops intergranularly also in this alloy and which progresses relatively in a relatively fast way.

As a final remark, in the investigated Al-Si-Mg casting alloys the creep behavior was affected by the combined presence of different features induced by minor alloying elements and preliminary heat treatments, such as Si morphology, coarse intermetallics and the fine strengthening particles (overaging or forming, depending on composition and initial alloy temper). When the creep characterization was performed on these alloys in view of specific service conditions, not only the temperature, but also the stress range should be selected, whenever possible, according to foreseen service. Further, when tension stress is considered for a component, the availability of the overall tensile creep strain curve of the alloy could allow one to detect the onset of damage forms. This holds also when multiple stress steps are performed on the same tension creep specimen, since the corresponding strain rate are potentially related to the actual strain and microstructural damage. In addition, the modification of strengthening particle during the multiple stress steps could affect these results. Finally, in situations with particularly long times or high temperatures, changes in coarse precipitates and solid solution composition could also become no more negligible.

5. Conclusions

The analysis of the results of tension creep performed at 300 °C under 30 MPa stress on a A356 alloy, on its E3 and EZ35 modified versions and of the available literature data on differently modified A356 alloys can be summarized in the following points:

- The minimum creep strain rate at 300 °C decreased from the A356 to E3 and EZ35 alloy, corresponding to the greater amount of strengthening particles;
- The analysis of creep strain rate, performed on a true strain rate vs. $(1 + e)$ plot in order to compare creep curves in tension and in compression, shows that the strain rate increased continuously during the constant load test of about one order of magnitude due to the actual increase of true stress and to a high apparent power-law index. Comparisons with literature data confirm this behavior and correlate it to the presence of nanoscale strengthening particles that could be modeled as threshold stress-behavior;
- The above analysis of creep strain rate for a single specimen can be carried out up to the onset of damage (in tension creep) or barreling (in compression creep). Specifically, in tension creep tests the strain level at which the onset of creep damage occurs could reduce significantly as the microstructural features related to it becoming more critical (high aspect ratio or in higher volume fractions);
- The analysis of the overall tensile creep behavior should take into account, in addition to the minimum creep strain rate, also the final damage, causing a final increase of the strain rate up to the final fracture. As a result, the alloy EZ35, characterized by finer dendritic grains, globular eutectic Si and by high amount of stable coarse Zr-rich intermetallics, mostly in the intradendritic position, displays the onset of creep damage at about 5.5% strain and good ductility indexes.

Author Contributions: Conceptualization, E.G. and M.C.; Data curation, E.G.; Formal analysis, E.G.; Investigation, C.C. and M.C.; Methodology, E.G. and C.C.; Resources, E.G.; Visualization, C.C.; Writing—original draft, E.G.; Writing—review and editing, C.C. and M.C. All authors have read and agreed to the published version of the manuscript.

Funding: This research received no external funding.

Institutional Review Board Statement: Not applicable.

Informed Consent Statement: Not applicable.

Data Availability Statement: The raw data required to reproduce these findings cannot be shared at this time as the data also forms part of an ongoing study. The processed data required to reproduce these findings cannot be shared at this time as the data also forms part of an ongoing study.

Acknowledgments: The authors would like to thank Alessandro Morri (Department of Industrial Engineering (DIN), Alma Mater Studiorum, University of Bologna) for the production of alloys, and the master students of Politecnico di Milano, Roberto Fracassi, Hazar Safadi and Mahmoud Aly, who contributed to testing.

Conflicts of Interest: The authors declare no conflict of interest.

References

1. Mishra, S.K.; Roy, H.; Mondal, A.K.; Dutta, K. Damage Assessment of A356 Al Alloy under Ratcheting–Creep Interaction. *Metall. Mater. Trans. A* **2017**, *48*, 2877–2884. [CrossRef]
2. Jeong, C. High temperature mechanical properties of Al–Si–Mg–(Cu) alloys for automotive cylinder heads. *Mater. Trans.* **2013**, *54*, 588–594. [CrossRef]
3. Roy, S.; Allard, L.F.; Rodriguez, A.; Watkins, T.R.; Shyam, A. Comparative Evaluation of Cast Aluminum Alloys for Automotive Cylinder Heads: Part I—Microstructure evolution. *Metall. Mater. Trans. A* **2016**, *48*, 2529–2542. [CrossRef]
4. Caceres, C.H.; Davidson, C.J.; Griffiths, J.R.; Wang, Q.G. The effect of Mg on the microstructure and mechanical behavior of Al-Si-Mg casting alloys. *Metall. Mater. Trans. A* **1999**, *30*, 2611–2618. [CrossRef]
5. Marioara, C.D.; Andersen, S.J.; Zandbergen, H.W.; Holmestad, R. The Influence of Alloy Composition on Precipitatesof the Al-Mg-Si System. *Metall. Mater. Trans. A* **2005**, *36*, 691–702. [CrossRef]
6. Colombo, M.; Gariboldi, E. Prediction of the yield strength and microstructure of a cast Al-Si-Mg alloy by means of physically-based models. *Metall. Ital. Int. J. Ital. Assoc. Metall.* **2017**, *109*, 5–14. Available online: https://pdfs.semanticscholar.org/da8 7/55f77d332b16830fb2f28d91e0ce9c4089f4.pdf?_ga=2.137556403.2019131743.1585306164-1529008527.1567347920 (accessed on 27 April 2021).
7. Gariboldi, E.; Lemke, J.N.; Rovatti, L.; Baer, O.; Timelli, G.; Bonollo, F. High-Temperature Behavior of High-Pressure Diecast Alloys Based on the Al-Si-Cu System: The Role Played by Chemical Composition. *Metals* **2015**, *8*, 348. [CrossRef]
8. Lin, J.; Liu, K.; Chen, X.-G. Improved elevated temperature properties in Al-13%Si piston alloys by Mo addition. *J. Mater. Eng. Perform.* **2020**, *29*, 126–134. [CrossRef]
9. Zhu, S.; Yao, J.-Y.; Sweet, L.; Easton, M.; Taylor, J.; Robinson, P.; Parson, N. Influences of Nickel and Vanadium Impurities on Microstructure of Aluminum Alloys. *JOM* **2013**, *65*, 584–592. [CrossRef]
10. Wang, Q.; Shi, Z.; Li, H.; Lin, Y.; Li, N.; Gong, T.; Zhang, R.; Liu, H. Effects of Holmium Additions on Microstructure and Properties of A356 Aluminum Alloys. *Metals* **2018**, *8*, 849. [CrossRef]
11. Baraderani, B.; Raiszadeh, R. Precipitation hardening of cast Zr-containing A356 aluminium alloy. *Mater. Des.* **2011**, *32*, 935–940. [CrossRef]
12. Pramod, S.L.; Prasada Rao, A.K.; Murty, B.S.; Bakshi, S.R. Effect of Sc addition on the microstructure and wear properties of A356 alloy and A357-TiB2 in situ composite. *Mater. Des.* **2015**, *78*, 85–94. [CrossRef]
13. Shi, Z.M.; Wang, Q.; Zhao, G.; Zhang, R.Y. Effects of erbium modification on the microstructure and mechanical properties of A356 aluminum alloys. *Mater. Sci. Eng. A* **2015**, *626*, 102–107. [CrossRef]
14. Bevilaqua, W.L.; Stadtlander, A.R.; Froehlich, A.R.; Braga Lemos, G.V.; Reguly, A. High-temperature mechanical properties of cast Al-Si-Cu-Mg alloy by combined additions of cerium and zirconium. *Mater. Res. Express* **2020**, *7*, 026532. [CrossRef]
15. Zamani, M.; Morini, L.; Ceschini, L.; Seifeddine, S. The role of transition metal additions on the ambient and elevated temperature properties of Al-Si alloys. *Mater. Sci. Eng. A* **2017**, *693*, 42–50. [CrossRef]
16. Knipling, D.E.; Dunand, D.C.; Seldman, D.N. Criteria for developing castable, creep-resistant aluminium-based alloys—A review. *Z. Metallkunde* **2006**, *97*, 246–265. [CrossRef]
17. Chen, S.; Liu, K.; Chen, X.G. Precipitation behavior of dispersoids and elevated-temperature properties in Al–Si–Mg foundry alloy with Mo addition. *J. Mater. Res.* **2019**, *34*, 3071–3081. [CrossRef]
18. Morri, A.; Ceschini, L.; Messieri, S.; Cerri, E.; Toschi, S. Mo Addition to the A354 (Al–Si–Cu–Mg) Casting Alloy: Effects on Microstructure and Mechanical Properties at Room and High Temperature. *Metals* **2018**, *8*, 393. [CrossRef]

19. Tocci, M.; Donnini, R.; Angella, G.; Gariboldi, E.; Pola, A. Tensile Properties of a Cast Al-Si-Mg Alloy with Reduced Si Content and Cr Addition at High Temperature. *J. Mater. Eng. Perform.* **2019**, *28*, 7097–7108. [CrossRef]
20. Tolley, A.; Radmilovic, V.; Dahmen, U. Segregation in Al3(Sc,Zr) precipitates in Al-Sc-Zr alloys. *Scr. Mater.* **2005**, *52*, 621–625. [CrossRef]
21. Fuller, C.B.; Seidman, D.N. Temporal evolution of the nanostructure of Al(Sc,Zr) alloys: Part II-coarsening of Al3(Sc1-xZrx) precipitates. *Acta Mater.* **2005**, *53*, 5415–5428. [CrossRef]
22. Song, M.; Du, K.; Huang, Z.Y.; Huang, H.; Nie, Z.R.; Ye, H.Q. Deformation-induced dissolution and growth of precipitates in an Al-Mg-Er alloy during high-cycle fatigue. *Acta Mater.* **2014**, *81*, 409–419. [CrossRef]
23. Li, H.; Bin, J.; Liu, J.; Gao, Z.; Lu, X. Precipitation evolution and coarsening resistance at 400 °C of Al microalloyed with Zr and Er. *Scr. Mater.* **2012**, *67*, 73–76. [CrossRef]
24. Wen, S.P.; Gao, K.Y.; Huang, H.; Wang, W.; Nie, Z.R. Precipitation evolution in Al-Er-Zr alloys during aging at elevated temperature. *J. Alloys Compd.* **2013**, *574*, 92–97. [CrossRef]
25. Booth-Morrison, C.; Dunand, D.C.; Seidman, D.N. Coarsening resistance at 400 °C of precipitation-strengthened Al–Zr–Sc–Er alloys. *Acta Mater.* **2011**, *59*, 7029–7042. [CrossRef]
26. Kasprzak, W.; Amirkhiz, B.S.; Niewczas, M. Structure and properties of cast Al-Si Based alloy with Zr-V-Ti additions and its evaluation of high-temperature performance. *J. Alloys Compd.* **2014**, *595*, 67–79. [CrossRef]
27. Hu, X.; Jiang, F.; Ai, F.; Yan, H. Effects of rare earth Er additions on microstructure development and mechanical properties of die-cast ADC12 aluminum alloy. *J. Alloys Compd.* **2012**, *538*, 21–27. [CrossRef]
28. Abdelaziz, M.H.; Doty, H.W.; Valtierra, S.; Samuel, F.H. Mechanical performance of Zr-Containing 354-type Al-Si-Cu-Mg cast alloy: Role of additions and heat treatment. *Adv. Mater. Sci. Eng.* **2018**, *18*, 5715819. [CrossRef]
29. Meng, F.-S.; Zhang, W.-W.; Hu, Y.; Zhang, D.-T.; Yang, C. Effect of zirconium on microstructures and mechanical properties of squeeze-cast Al-5.0Cu-0.4Mn-0.1Ti-0.1RE alloy. *J. Cent. South Univ.* **2017**, *24*, 2231–2237. [CrossRef]
30. Colombo, M.; Gariboldi, E.; Morri, A. Influences of different Zr additions on the microstructure, room and high temperature mechanical properties of an Al-7Si-0.4Mg alloy modified with 0.25%Er. *Mater. Sci. Eng. A* **2018**, *713*, 151–160. [CrossRef]
31. Tsai, Y.-C.; Chou, C.-Y.; Lee, S.-L.; Lin, C.-K.; Lin, J.-C.; Lim, S.W. Effect of trace La addition on the microstructures and mechanical properties of A356 (Al-7Si-0.35Mg) aluminum alloys. *J. Alloys Compd.* **2009**, *487*, 157–162. [CrossRef]
32. Joy-Yii, S.L.; Kurniawan, D. Effect of rare earth addition on microstructure and mechanical properties of Al-Si alloys: An overview. *Adv. Mater. Res.* **2014**, *845*, 27–30. [CrossRef]
33. Mahmoud, M.G.; Samuel, A.M.; Doty, H.W.; Valtierra, S.; Samuel, F.H. Effect of Solidification Rate and Rare Earth Metal Addition on the Microstructural Characteristics and Porosity Formation in A356 Alloy. *Adv. Mater. Sci. Eng.* **2017**, *2017*, 5086418. [CrossRef]
34. Altenbach, H.; Kozhar, S.; Naumenko, K. Modelling creep damage of an aluminium-silicon eutectic alloy. *Int. J. Damage Mech.* **2012**, *22*, 683–698. [CrossRef]
35. Rashno, S.; Ranjbar, K.; Reihanian, M. Impression creep characterization of cast Al-7Si-0.3 Mg alloy. *Mater. Res. Express* **2019**, *6*, 0865e6. [CrossRef]
36. Amiresgandari, B.; Nami, B.; Shahmiri, M.; Abed, A. Effect of Mg and semi solid processing on microstructure and impression creep properties of A356 alloy. *Trans. Nonferrous Met. Soc. China* **2013**, *23*, 2518–2523. [CrossRef]
37. Farkoosh, A.R.; Pekguleryuz, M. The effects of manganese on the T-phase and creep resistance in Al–Si–Cu–Mg–Ni alloys. *Mater. Sci. Eng. A* **2013**, *582*, 248–256. [CrossRef]
38. Wang, Q.; Hess, D.; Yan, X.; Caron, F. Evaluation of a New High Temperature Casting Aluminium for cylinder head applications. In Proceedings of the 122nd Metalcasting Congress, Fort Worth, TX, USA, 3–5 April 2018; pp. 1–7.
39. Farkoosh, A.R.; Grant Chen, X.; Pekguleryuz, M. Interaction between molybdenum and manganese to form effective dispersoids in an Al–Si–Cu–Mg alloy and their influence on creep resistance. *Mater. Sci. Eng. A* **2015**, *627*, 127–138. [CrossRef]
40. Garat, M. Optimization of aluminium cylinder head alloy of the AlSi7Cu3Mg type reinforced by additions of peritectic elements. *Inter. Met.* **2011**, *5*, 17–24. [CrossRef]
41. Roy, S.; Allard, L.F.; Rodriguez, A.; Porter, W.D.; Shyam, A. Comparative Evaluation of Cast Aluminum Alloys for Automotive Cylinder Heads: Part II—Mechanical and Thermal Properties. *Metall. Mater. Trans. A* **2017**, *48*, 2543–2562. [CrossRef]
42. Javidani, M.; Larouche, D. Application of cast Al–Si alloys in internal combustion engine components. *Int. Mater. Rev.* **2014**, *59*, 132–158. [CrossRef]
43. Roy, M.J.; Maijer, D.M.; Dancoine, L. Constitutive behaviour of as-cast A356. *Mater. Sci. Eng. A* **2012**, *548*, 195–205. [CrossRef]
44. Jeong, C.Y.; Kang, C.-S.; Cho, J.-I.; Oh, I.-H.; Kim, Y.-C. Effect of microstructure on mechanical properties for A356 casting alloy. *Int. J. Cast. Met. Res.* **2008**, *21*, 193–197. [CrossRef]
45. Colombo, M.; Gariboldi, E.; Morri, A. Er addition to Al-Si-Mg-based casting alloy: Effects on microstructure, room and high temperature mechanical properties. *J. Alloys Compd.* **2017**, *708*, 1234–1244. [CrossRef]
46. Colombo, M.; Buzolin, R.H.; Gariboldi, E.; Rovatti, L.; Vallant, R.; Sommitsch, C. Effects of Er and Zr Additions on the As-Cast Microstructure and on the Solution-Heat-Treatment Response of Innovative Al-Si-Mg-Based Alloys. *Metall. Mater. Trans. A* **2020**, *51*, 1000–1011. [CrossRef]
47. ISO 6892-2:2018. *Metallic Materials-Tensile Testing—Part 2: Method of Test at Elevated Temperature*; The International Organization for Standardization Publishing: Geneva, Switzerland, 2018.

48. ISO 204:2018. *Metallic Materials—Uniaxial Creep Testing in Tension-Method of Test*; The International Organization for Standardization Publishing: Geneva, Switzerland, 2018.
49. Rasband, W.S. ImageJ, Version 2.0.0-rc-69/1.52i, Distribution Fiji. 2018. Available online: https://imagej.net/Welcome (accessed on 29 March 2021).
50. Colombo, M.; Albu, M.; Gariboldi, E.; Hofer, F. Microstructural changes induced by Er and Zr additions to A356 alloy investigated by thermal analyses and STEM observations. *Mater. Charact.* **2020**, *161*, 11011. [CrossRef]
51. Farkoosh, A.R.; Chen, X.G.; Pekgureryuz, M. Dispersoid strengthening of a high temperature Al-Si-Cu-Mg alloy via Mo addition. *Mater. Sci. Eng. A* **2015**, *620*, 181–189. [CrossRef]
52. Ozbakir, E. Development of Aluminum Alloys for Diesel-Engine Applications. Master's Thesis, McGill University, Montréal, QC, Canada, December 2008.
53. Rashno, S.; Reihanian, M.; Ranjbar, K. Tensile and creep properties of Al-7Si-0.3Mg alloy with Zr and Er addition. *Mater. Sci. Technol.* **2020**, *36*, 1603–1613. [CrossRef]
54. Pan, L.; Breton, F. Evaluation of New AlSiCuMg Cast Aluminium Alloys for Automotive Cylinder Heads. In Proceedings of the 2020 AFS 124th Metalcasting Congress, Cleveland, OH, USA, 21–23 April 2020; Available online: http://afsinc.s3.amazonaws.com/2020%20Proceedings/2020-018_lm.pdf (accessed on 27 April 2021).
55. Griffiths, S.; Croteau, J.R.; Rossell, M.D.; Erni, R.; De Luca, A.; Vo, N.Q.; Dunand, D.C.; Leinenbach, C. Coarsening- and creep resistance of precipitation-strengthened Al–Mg–Zr alloys processed by selective laser melting. *Acta Mater.* **2020**, *188*, 192–202. [CrossRef]
56. Spigarelli, S.; Paoletti, C. A new model for the description of creep behaviour of aluminium-based composites reinforced with nanosized particles. *Compos. Part A Appl. Sci. Manuf.* **2018**, *112*, 345–355. [CrossRef]
57. *Thermo-Calc Software*; Version 2020b with TCAL5.1 Al-Alloys Database; Thermo-Calc Software: Stockholm, Sweden, 2020.
58. Du, Y.; Chang, Y.A.; Huang, B.; Gong, W.; Jin, Z.; Xu, H.; Yuan, Z.; Liu, Y.; He, Y.; Xie, F.-Y. Diffusion coefficients of some solutes in fcc and liquid Al: Critical evaluation and correlation. *Mater. Sci. Eng. A* **2003**, *363*, 140–151. [CrossRef]
59. Du, Q.; Tang, K.; Marioara, C.D.; Andersen, S.J. Modeling over-ageing in Al-Mg-Si alloys by a multi-phase CALPHAD-coupled Kampmann-Wagner Numerical model. *Acta Mater.* **2017**, *122*, 178–186. [CrossRef]
60. Chomsaeng, N.; Haruta, M.; Chairuangsri, T.; Kurata, H.; Isoda, S.; Shiojiri, M. HRTEM and ADF-STEM of precipitates at peak-ageing in cast A356 aluminium alloy. *J. Alloys Compd.* **2010**, *496*, 478–487. [CrossRef]
61. Zhang, Y.; Gao, K.; Wen, S.; Huang, H.; Nie, Z.; Zhou, D. The study on the coarsening process and precipitation strengthening of Al3Er precipitate in Al-Er binary alloy. *J. Alloys Compd.* **2014**, *610*, 27–34. [CrossRef]
62. Huang, H.; Wen, S.P.; Gao, K.Y.; Wang, W.; Nie, Z.R. Age hardening behavior and corresponding microstructure of dilute Al-Er-Zr alloys. *Metall. Mater. Trans. A* **2013**, *44*, 2849–2856. [CrossRef]
63. Manente, A.; Timelli, G. Optimizing the Heat Treatment Process of Cast Aluminium Alloys. In *Recent Trends in Processing and Degradation of Aluminium Alloys*; Ahmad, Z., Ed.; InTech: London, UK, 2011; pp. 197–220. Available online: https://www.intechopen.com/books/recent-trends-in-processing-and-degradation-of-aluminium-alloys/optimizing-the-heat-treatment-process-of-cast-aluminium-alloys (accessed on 27 April 2021).

Article

Effect of Low-Temperature Annealing on Creep Properties of AlSi10Mg Alloy Produced by Additive Manufacturing: Experiments and Modeling

Chiara Paoletti [1,*], Emanuela Cerri [2], Emanuele Ghio [2], Eleonora Santecchia [1], Marcello Cabibbo [1] and Stefano Spigarelli [1]

[1] Dipartimento di Ingegneria Industriale e Scienze Matematiche (DIISM), Università Politecnica delle Marche, Via Brecce Bianche, 60131 Ancona, Italy; e.santecchia@univpm.it (E.S.); m.cabibbo@univpm.it (M.C.); s.spigarelli@univpm.it (S.S.)

[2] Dipartimento di Ingegneria e Architettura (DIA), Università di Parma, V.le G. Usberti 181/A, 43124 Parma, Italy; emanuela.cerri@unipr.it (E.C.); emanuele.ghio@unipr.it (E.G.)

* Correspondence: c.paoletti@pm.univpm.it; Tel.: +39-071-220-4746

Abstract: The effects of postprocessing annealing at 225 °C for 2 h on the creep properties of AlSi10Mg alloy were investigated through constant load experiments carried out at 150 °C, 175 °C and 225 °C. In the range of the experimental conditions here considered, the annealing treatment resulted in an increase in minimum creep rate for a given stress. The reduction in creep strength was higher at the lowest temperature, while the effect progressively vanished as temperature increased and/or applied stress decreased. The minimum creep rate dependence on applied stress was modeled using a physically-based model which took into account the ripening of Si particles at high temperature and which had been previously applied to the as-deposited alloy. The model was successfully validated, since it gave an excellent description of the experimental data.

Keywords: creep; aluminum alloys; additive manufacturing; annealing; modeling

check for **updates**

Citation: Paoletti, C.; Cerri, E.; Ghio, E.; Santecchia, E.; Cabibbo, M.; Spigarelli, S. Effect of Low-Temperature Annealing on Creep Properties of AlSi10Mg Alloy Produced by Additive Manufacturing: Experiments and Modeling. *Metals* **2021**, *11*, 179. https://doi.org/10.3390/met11020179

Academic Editors: Gilbert Henaff and Martin Heilmaier

Received: 1 December 2020
Accepted: 18 January 2021
Published: 20 January 2021

Publisher's Note: MDPI stays neutral with regard to jurisdictional claims in published maps and institutional affiliations.

1. Introduction

1.1. AlSi10Mg Alloy Produced by Additive Manufacturing: Main Structural Features

AlSi10Mg is an aluminum alloy which is extensively used in the production of parts by additive manufacturing (AM). Specifically, laser powder bed fusion (LPBF) is the most widespread AM technique for the production of components made of this alloy [1–3]. Starting from computer aided design (CAD) models, LPBF builds up new parts layer-upon-layer. Each layer is formed by small portions of powder (melt pools), which are selectively melted by the laser. The resulting very high cooling rates (~10^6 °C/s) and the sequence of passes, each of which acts as a thermal treatment on the lower and already solidified layers, produce an extremely fine structure with interesting mechanical properties [4–6]. The complex structure typical of AlSi10Mg AM products can be schematically described as follows:

i. at a macroscopic level (100 μm–1 mm), the most noteworthy features are: (a) the high surface roughness resulting from the presence of melt pools; (b) the possible presence of porosities [7];

ii. at a mesoscopic level (1–100 μm), the melt pool substructure is formed by columnar grains with diameters of several microns, which are subdivided into smaller long cells with diameters of a few hundreds of nm, separated by Si–Mg-rich eutectic regions [7–9];

iii. at a microscopic level (10 nm–1 μm), the long cells are subdivided into equiaxed subcells, again with diameters of a few hundreds of nm, separated by a network of Si-rich eutectic regions. The eutectic regions are richer in Si [10] and contain densely

Metals **2021**, *11*, 179. https://doi.org/10.3390/met11020179

spaced Si particles, the size and distribution of which can vary as a function of process parameters [7,8] or part size [9].

1.2. AlSi10Mg Alloy Produced by Additive Manufacturing: Effect of Stress Relieving/Low Temperature Annealing

As mentioned above, the mechanical properties of as-formed AM parts are indeed peculiar. The extreme thinness of the substructure gives the alloy very high yield and ultimate tensile strengths, which are roughly comparable with those typical of the T6 state, although at the expense of low ductility. Residual stresses and the presence of porosity can further decrease ductility and toughness. For this reason, postprocessing usually includes a stress-relieving or annealing treatment, which increases ductility but significantly reduces the tensile properties of the alloy. An example of the variation in tensile yield stress with annealing temperature for a fixed duration of 2 h is reported in Figure 1 (data from [5,10–13]). The figure includes only the data for the low-temperature treatments, since at higher temperatures (400–550 °C) the meso- and microstructures undergo extensive variations that drastically change their very nature, which progressively approaches that of die-cast alloys.

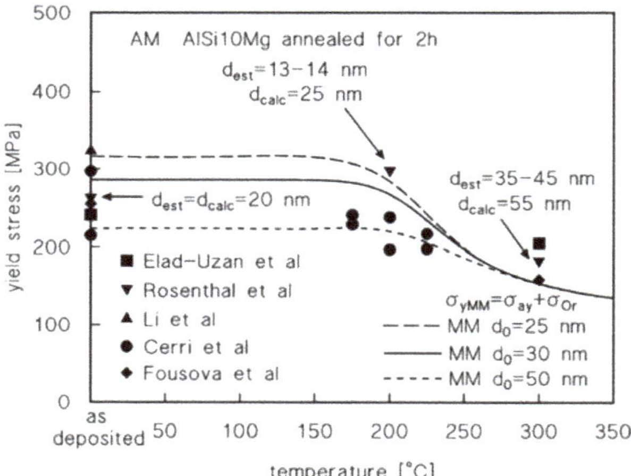

Figure 1. Effect of a 2-h postprocessing heat treatment on yield strength (on the left, the values for the as-deposited alloys) [5,10–13]. The figure also plots the roughly estimated value of the yield stress calculated by taking into account the effects of the Ostwald ripening of Si particles of different initial size (see Appendix A for details on the calculation of the Si particle size d_{calc}, and Table 1 for the meanings of symbols). In addition, data on the observed crystallite size, d_{est}, are reported [5].

The first obvious evidence from Figure 1 is the high variability in the mechanical properties of the alloy in the as-deposited state. This variation can be attributed to the differences in deposition conditions reflected in the different morphologies of the meso- and microstructures.

A point to be addressed is whether the complex trend of the yield strength can be somehow quantitatively described by a physically-based model. A simplified modeling approach [14] assumes that the mechanical properties of the alloy are mainly controlled by its microstructure, i.e., by the interaction between dislocations and obstacles at a submicron scale. The microstructural analysis carried out in [14] confirmed that, in the as-deposited state, the microstructure of the alloy consists of 500 nm-wide cells, surrounded by a network of fine Si particles (Appendix A shows a representative micrograph of the microstructure). The interior of the observed cells, by contrast, contains few remote fine particles. This structure can be modeled as a combination of soft zones surrounded by hard regions. The cell interior can thus be seen as a soft zone which dislocations can easily pass through, while the eutectic region is a hard region which becomes a strong obstacle to dislocation motion. The resulting material model (MM) can be therefore used to attempt a description of the softening phenomena illustrated in Figure 1.

If hard (eutectic) and soft (cells) zones deform under similar strain rates, the strength of the "composite" material can be calculated as

$$\sigma = f_H\,\sigma_H + (1 - f_H)\sigma_S \tag{1}$$

where σ_H and σ_S are the stresses acting on the hard and soft zones, respectively, and f_H is the volume fraction of the hard regions [15–17].

The yield stress of the alloy can then be estimated by assuming that the initial dislocation density is very low, thus showing that dislocation hardening does not significantly contribute to yielding. The stress necessary to start dislocation motion (yielding) in the hard and soft zones can therefore be tentatively calculated as

$$\sigma_{yi} = \sigma_a + \sigma_{Or}^i \tag{2}$$

where σ_a is the yield stress of an alloy containing 0.5% Mg (60 MPa), which includes the solute hardening contribution of this alloying element, and σ_{Or}^i is the Orowan stress in the i-region (soft or hard). The Orowan stress, in turn, can be estimated by its simplest formulation [18], namely

$$\sigma_{Or} = \frac{084mGb}{L} \tag{3}$$

where $m = 3.05$, G is the shear modulus, b is the Burgers vector and L is the surface-to-surface interparticle spacing. Following [14], it can be assumed that each cell has a size of 500 nm [8] and that the volume fraction of the eutectic portions is $f_H = 0.25$. Both cells and eutectic regions contain fine Si particles of initial size d_0. The initial surface-to-surface interparticle distance is assumed to be equivalent to d_0 in the hard zones and to 200 nm in cell interiors.

Once precipitation to the equilibrium volume fraction is complete, particle size evolves during annealing due to coarsening/ripening. Details about the equations used for the calculation of particle size (d_{calc}) and interparticle distance after annealing are presented in Appendix A. The calculated variations in yield stress Equations (2) and (3) obtained for MM with different d_0 values are presented in Figure 1. An analysis of the figure suggests that, in general terms, the combination of MM with the equations describing the Ostwald ripening phenomena of Si particles gives a reasonable description of the yield stress reduction after annealing. Nevertheless, at a closer look, some specific features emerge, namely:

i.   in some cases (see, for example data from [13]), the yield stress monotonically decreases with increasing annealing temperature. The phenomenon is quite correctly described by the model curves;

ii. in other cases (see data from [5] and evidence presented in [10]), precipitation of Si is not completed during the AM process. Thus, in the early stages of low-temperature annealing, an additional precipitation of fine Si particles results in an increase in yield stress. This secondary precipitation is not accounted for by the equations presented in Appendix A, since the model assumes that ripening starts immediately upon annealing. This fact is easily confirmed when comparing the estimated value of the Si crystallites (d_{est}) [5] and the calculated value of Si particle size (Figure 1). The secondary precipitation results in a finer particle size than the one predicted by the ripening equations.

Table 1. List of constitutive model parameters.

Symbol	Meaning
σ	True applied stress (MPa)
$\dot{\varepsilon}_m$	Minimum creep rate (s^{-1})
σ_0	Particle strengthening term (Orowan stress) (MPa)
α	Constant: 0.3
m	Taylor factor: 3.06
R	Gas constant: 8.314 (J mol^{-1}K^{-1})
G	Shear modulus: 30,220–16 T (MPa)
b	Burgers vector: 2.47×10^{-10} (m)
ρ	Dislocation density (m^{-2})
σ_ρ	Dislocation hardening term: $=\alpha m G b \rho^{1/2}$ (MPa)
τ_l	Dislocation line tension: $=0.5 G b^2$ (N)
R_{max}	Maximum strength at the testing temperature [MPa]
k	Boltzmann constant $= 1.38 \times 10^{-23}$ (J K^{-1})
D_{0L}	Pre-exponential factor in the Arrhenius equation describing the temperature dependence of the vacancy diffusion coefficient: 8.3×10^{-6} (m^{2}s^{-1}) [19]
Q_L	Activation energy in the Arrhenius equation describing the temperature dependence of the vacancy diffusion coefficient: 122 (kJ mol^{-1}) [19]
U_{ss}	Energy necessary for Si (and Mg) atoms still in solid solution to jump in and out of the atmospheres that spontaneously form around dislocations; previous calculations gave values close to 10–15 kJ mol^{-1} for Mg [20]. For the sake of simplicity, here U_{ss} is assumed to be 10 (kJ mol^{-1})
$R_{UTS}{}^a$	Room temperature tensile strength of an alloy with the same impurity level, similar content of elements in solid solution and coarse intergranular intermetallics, in the absence of fine Si particles, here roughly estimated to be 115 (MPa)
L	Surface-to-surface interparticle spacing (m)
G_T, G_{RT}	Shear modulus at the testing temperature and at 25 °C, respectively (MPa)
M_{cg}	Temperature dependent dislocation mobility
d_0	Initial dimension of Si particles
d_{est}	Experimental estimate of the size of Si particles at time t
d_{calc}	Calculated value of the size of Si particles at time t
σ_{yi}	Yield stress
σ_a	Yield stress of an alloy containing 0.5% Mg but no Si particles
$\sigma_{Or}{}^i$	Orowan stress in the i-region (i = H,S)

The analysis above indicates that, also during stress relieving/low temperature annealing, AM AlSi10Mg alloy is affected by complex combinations of metallurgical phenomena (secondary precipitation, depletion of elements in solid solution, ripening). These phenomena occur on a microstructure the structural features of which (melt pool size, cell size, particle size, etc.) are strongly dependent upon processing conditions. On the other hand, both the temperature and duration of the postprocessing heat treatment need to be accurately selected in order to reduce brittleness and residual stresses, thus avoiding an excessive softening of the alloy. It is therefore reasonable to select the highest temperature which does not result in a marked decrease in strength (200–300 °C). However, in cases in which Si precipitation only occurs during AM, a marked reduction in strength [10] can be observed for temperatures above 250 °C, presumably due to particle ripening, which is well described by the model curves. On these bases, a temperature of 225 °C/2 h appears to be an interesting candidate for stress relieving/low-temperature annealing treatment.

1.3. AlSi10Mg Alloy Produced by Additive Manufacturing: Creep Response

Although generally exhibiting poor creep resistance, aluminum alloys have attracted considerable attention as case-study materials (see [19,20] for detailed discussions). In addition, Al–Si alloys have been clearly demonstrated to behave as a sort of natural metal matrix composite in which an Al matrix is reinforced by Si particles [21]. In this sense, the creep response of these composites reinforced with nm-sized Si particles is an intriguing matter that deserves detailed investigation.

The creep behavior of AlSi10Mg produced by AM was first investigated in [6,22] and then more extensively addressed in [14], in the latter case by testing the as-deposited material. The experimental data, when described by the well-known phenomenological equation

$$\dot{\varepsilon}_m = A(T)\sigma^n \tag{4}$$

where $A(T)$ is a temperature dependent parameter, gave a stress exponent n that decreased with increasing temperature from 22 to 15. Such high values of the stress exponent are indeed frequently attributed to the strengthening effect of particle–dislocation [17,18]. On these bases, the MM described in Section 1.2 was successfully used in combination with the set of constitutive equations introduced in [19,20], namely:

$$\sigma = \sigma_0 + \sigma_p = \sigma_0 + \alpha m G b \sqrt{\rho} \tag{5}$$

$$\dot{\varepsilon}_m = \frac{2M_{cg}\tau_l bL}{m}\left(\frac{\sigma_\rho}{\alpha m G b}\right)^4 \tag{6}$$

$$M_{cg} \cong \frac{D_{0L}b}{kT}exp\left(\frac{\sigma_\rho b^3}{kT}\right)exp\left\{-\frac{Q_L}{RT}\left[1-\left(\frac{\sigma_\rho}{R_{max}}\right)^2\right]\right\}exp\left(-\frac{U_{ss}}{RT}\right) \tag{7}$$

$$R_{max} = 1.5(R_{UTS}^a + \sigma_{Or})\frac{G_T}{G_{RT}} \tag{8}$$

The meanings of the different symbols are illustrated in Table 1, while the derivation of the model is reported in Appendix B. For a given value of the minimum strain rate, each of Equations (5)–(8) was applied in the hard and soft zones. The particle strengthening term σ_0 was assumed to be equivalent to the Orowan stress and the particle size at the time corresponding to the minimum in creep rate was calculated using the equations in Appendix A, by assuming d_0 = 50 nm. The stress corresponding to the given strain rate was then expressed by Equation (1). The resulting model curves are presented in Figure 2.

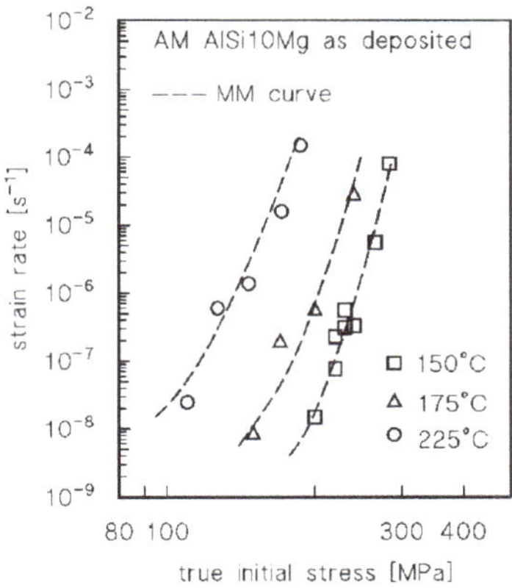

Figure 2. Experimental values of the minimum creep rate in as-deposited additive manufacturing (AM) AlSi10Mg and material model (MM) curves (see Appendix A for the constitutive equations used in combination with the composite model, Equation (1), with d_0 = 50 nm) [14].

The remarkable accuracy of the curves demonstrated that the basic assumption (creep response is mainly controlled by microstructure, i.e., by particle–dislocation interaction at a submicron scale, while, in this context, the mesostructure plays a minor role) was substantially correct.

The study presented here originated from the findings reported in the previous sections. As mentioned above, the creep response of Al10SiMg alloy has been dealt with by only a few studies and the subject is worth of deeper investigation. In addition, it has been recognized that annealing produces a softening of the microstructure but no quantitative models for this effect have been proposed. The aim of the study was thus twofold: i. to investigate the effect of low-temperature annealing (225 °C/2 h) on the creep response of heat-treated AM AlSi10Mg and compare the results with the as-deposited behavior; ii. to validate the quantitative predictions of the constitutive model that had already been successfully used for the as-deposited alloy.

2. Materials and Methods

Dogbone creep samples made of Al-9.6%Si-0.38% Mg alloy with a gauge length of 25 mm and a square section of 3 mm × 3 mm were produced by an SML500 machine (SLM Solutions, Lubecca, Germany) with the following deposition parameters: substrate temperature 150 °C, laser power 350 W, spot size 80 μm, scan speed 1.15 m/s, hatch spacing 170 μm, layer thickness 50 μm. The growth direction was parallel to the sample axis (loading direction). Stress relieving/low-temperature annealing at 225 °C for a duration of 2 h was applied to the as-built samples.

Constant load tensile creep experiments were carried out in air on the as-deposited samples at 150 °C, 175 °C and 225 °C. The samples were mostly strained up to rupture, although longer tests were interrupted after the end of the primary stage. The samples were heated in a three-zone furnace and both elongation, measured by LVDT, and temperature were continuously recorded during the test. Temperature was measured by four K-type thermocouples.

3. Results

After annealing, the microstructure of the alloy remained substantially similar to that shown in Figure A1 due the short annealing time, which, at maximum, could lead to an increase of a few nm in Si particle size.

Figure 3a shows typical examples of the strain vs. time creep curves. Figure 3b shows representative creep strain rate vs. strain curves. The alloy exhibited the conventional three-stage behavior, with a well-defined primary region, a minimum creep rate range and a tertiary stage.

Figure 4 plots the minimum creep rate dependence on applied stress for the annealed alloy. The figure also plots the same data of Figure 2, obtained in [14], by testing the same material in the as-deposited condition. As in the case of the untreated alloy, the annealed material also exhibited high values for the stress exponent (n = 23, 29 and 17 at 150 °C, 175 °C and 225 °C respectively). The creep rates of the heat- treated alloy were higher than those of the as-deposited material, although the difference became smaller as the temperature and/or duration of the test increased. Within the range of experimental conditions investigated, the datasets for the as-deposited and the heat-treated materials align on almost parallel curves, i.e., indicating that the stress exponent was roughly equivalent, only at 150 °C. These findings indicate that, while at low temperatures the two materials behaved quite differently and heat treatment resulted in a marked worsening of the creep response, in the high temperature regime, in particular for long tests, i.e., low stresses, this effect progressively vanished and creep response became more and more similar.

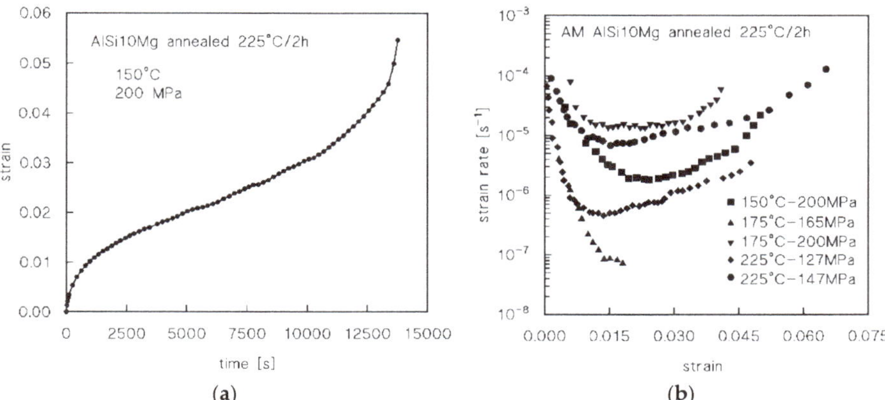

Figure 3. Experimental creep curves: (**a**) example of strain vs. time curve; (**b**) strain rate vs. strain curves at different experimental conditions.

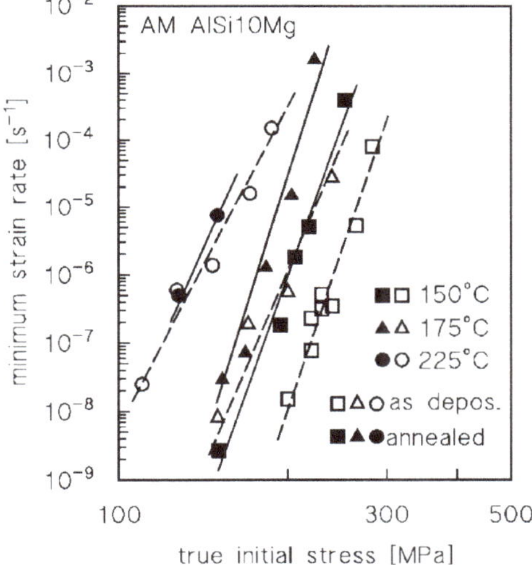

Figure 4. Minimum creep rate dependence on applied stress for the annealed and as-deposited [14] alloy.

Figure 5 plots the time to minimum creep rate (t_m) as a function of the minimum creep rate. The data align on a single line of slope -1. The fact that the experimental data conform to the simple equation

$$\dot{\varepsilon}_m t_m = c_m \tag{9}$$

implies that the same time-dependent mechanisms operate within the range of experimental conditions investigated.

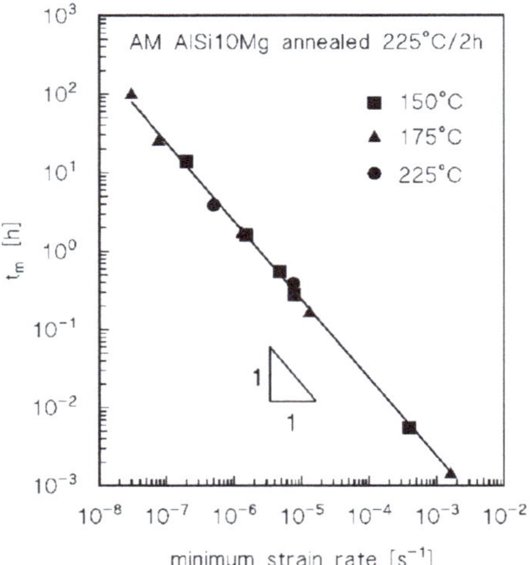

Figure 5. Time to minimum creep rate as a function of minimum creep rate for the annealed alloy.

4. Discussion

4.1. Analysis of the Effect of Low-Temperature Annealing on Creep Response

Section 3 (Figure 4) unambiguously shows that even a low-temperature annealing at 225 °C for 2 h results in a marked decrease in creep properties, at least as long as the testing temperature does not exceed 150 °C. At higher temperatures, the reduction in creep response is more limited, because the as-deposited alloy also undergoes extensive softening during high-temperature exposure. In order to quantify the loss in creep resistance, the data in Figure 4 were used to calculate $\Delta\sigma$, expressed as

$$\Delta\sigma = \sigma^{ad} - \sigma^{ht} \tag{10}$$

where σ^{ad} and σ^{ht} are the stresses applied to obtain a given value of the minimum strain rate in as-deposited and heat-treated alloys, respectively (Figure 6). Figure 6 only includes data for 150 and 175 °C, since at the higher temperature the experimental results for heat-treated and as-deposited alloys substantially overlap.

Figure 6 shows that the as-deposited alloy creep tested at 150 °C maintained a 50 MPa advantage in strength until the minimum creep rate decreased below $10^{-6}\ \mathrm{s}^{-1}$. By contrast, at 175 °C, $\Delta\sigma$ monotonically decreased from 47 to 7 MPa with decreasing minimum creep rate. The softening obviously originates from the microstructural evolution during high-temperature exposure. Postprocessing annealing results in ripening/coarsening of the Si particles, which are the major sources of strengthening at high temperatures. At low testing temperatures, where ripening is a sluggish phenomenon, the advantage in creep strength is maintained until the time of exposure becomes so long that particles start to coarsen even in the as-deposited material. At 175 °C ripening is more rapid and the initial advantage of the as-deposited state, i.e., the presence of smaller particles in the microstructure, is rapidly lost and the two materials progressively become more and more similar. Although reasonable, this qualitative explanation needs to be supported by a more quantitative analysis of the relationship between creep response (the minimum creep rate) and the microstructural features (Si particle size and interparticle distance), which will be dealt with in the following section.

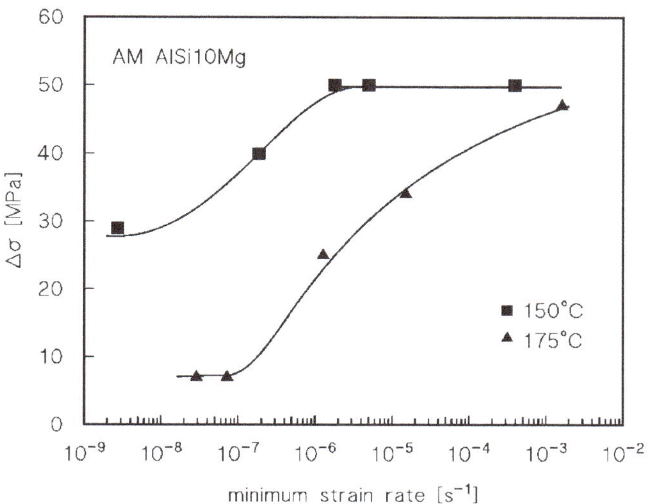

Figure 6. $\Delta\sigma$ as a function of the minimum creep rate for 150 °C and 175 °C.

4.2. Modeling the Effect of Low-Temperature Annealing on Creep Response

Section 4.1 provides useful indications about the effect of annealing on creep response. Nevertheless, the combination of the constitutive equations illustrated in Appendix A with the MM and Equation (1) provides a useful tool for modeling creep response. All the parameters of the different equations are available, since the model was already successfully used to describe the as-deposited creep response (Figure 2). The only missing term is the time corresponding to the minimum creep rate, which, however, can be easily obtained from Figure 5. The annealing time was sufficiently short to not cause major variations in the microstructure, although a barely discernible increase in size was nevertheless expected. Thus, Equation (A6) in Appendix A can be used to estimate particle size after annealing. By taking d_0 = 50 nm, as in [14], after 2 h at 225 °C a particle size close to 59 nm can be obtained. Equation (A6) and Equation (9) can then be combined and used to estimate the size of the particles at t_m for a given value of the minimum strain rate. Since the surface-to-surface interparticle spacing can be expressed as

$$L = d\left(\frac{0.5}{\sqrt{6f/\pi}} - \sqrt{\frac{2}{3}}\right) \tag{11}$$

where f is the volume fraction of the particles (here assumed not to vary during the test, both in hard and soft zones), one obtains

$$L = L_0\frac{d}{d_0} \tag{12}$$

For the particles inside the cells only, particle distance was assumed to increase with time, due to ripening, until it reached half of the cell size (250 nm) and then to remain constant for longer durations. This assumption implies that, even for long times of exposure, at least one single particle remains in the cell interior. Once L has been determined, the equations in Appendix B can be used to calculate the model curves for the heat-treated material shown in Figure 7.

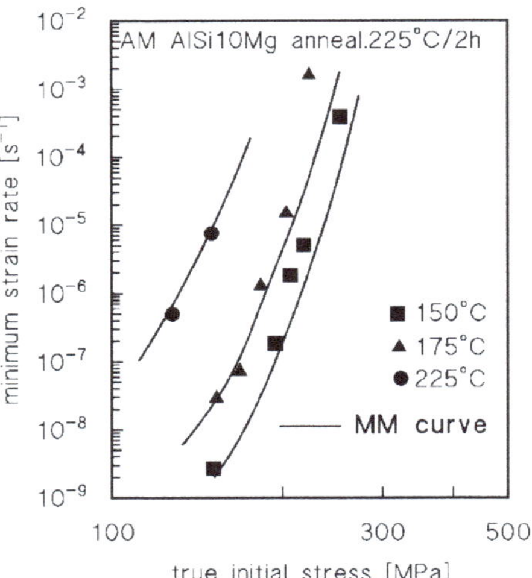

Figure 7. Model curves calculated for the material heat treated at 225 °C for 2 h. The only experimental datum used to calculate the curves is the relationship shown in Figure 4.

If one considers that all the parameters except the c_m constant, which was, in fact, the only experimental value obtained from Figure 4, were calculated for the as-deposited alloy, the accuracy of the description is remarkable. The model significantly overestimates the material strength only in the very high strain rate region.

Figure 8a plots a comparison between the model curves obtained for the as-deposited and the heat-treated alloy; Figure 8b shows the experimental data available at 225 °C for the as-deposited alloy [14] and for the alloy annealed for 2 h at 225 °C (this study) and at 300 °C [22].

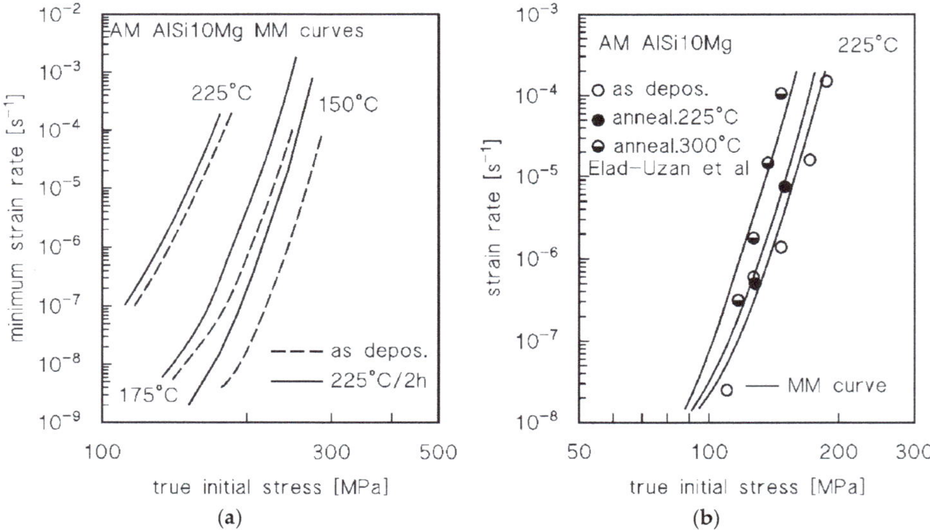

Figure 8. (**a**) Model curves calculated for the as-deposited and the heat-treated (225 °C for 2 h) alloy; (**b**) minimum creep rate obtained at 225 °C for the as-deposited alloy [14] and for the alloy annealed for 2 h at 225 °C (this study) and at 300 °C [22]; (**b**) also includes the model curves calculated by combining the MM and the equations reported in Appendix A.

An interesting feature, which is qualitatively described in the previous subsection and well documented by Figure 8b, is the progressive convergence between the curves obtained for the different initial states as the strain rate decreases, i.e., the time of exposure increases. In addition, the advantage in creep response of the as-deposited alloy decreases as temperature increases (Figure 8a).

Both Figures 7 and 8 validate the prediction of the model proposed in [14], which is sufficiently accurate to constitute a valid tool for the prediction of the creep response of the alloy investigated in different initial states.

The appropriateness of the basic assumption of the model, i.e., that the key factors controlling the creep response are the size and distribution of second-phase particles, has thus been confirmed. Any annealing treatment causes ripening of the Si particles and, as a consequence, a reduction of the creep resistance of the alloy. The accuracy of the description implies that the model is a valid tool for the prediction of the creep response of Al–Si alloys, provided that sufficient data on the microstructural analysis and some information on the shape of the creep curve (the correlation between minimum creep rate and the time at which it is measured) are available. Ingot or rheo/thixo-casting products actually exhibit microstructures which, after solution treatment, largely repeat the same morphology as that shown in Appendix A but at a larger scale, i.e., with much larger "cells" (the grains of Al) surrounded by a network of very coarse Si particles (see, for example, [23,24]). In this regard, the model presented here could in principle also be used for the analysis of the creep response of these conventionally produced materials.

5. Conclusions

The effect of low-temperature annealing at 225 °C on creep response was investigated by constant load experiments carried out at 150 °C, 175 °C and 225 °C. A comparison with the experimental data obtained by testing the as-deposited alloy demonstrated that annealing results in a loss of creep resistance that is more pronounced at lower temperatures. In general, the creep response of the as-deposited and of the heat-treated alloys became more and more similar as temperature increased and/or applied stress decreased. The alloy response was then compared with the prediction obtained from a physically-based model which took into account the effect of Si particle ripening. In order to apply the constitutive equations previously developed to describe pure aluminum and simpler Al alloys, the complex microstructure of AM AlSi10Mg was assimilated to that of a composite formed by soft zones (cell interiors) and hard zones (Si-rich eutectic regions). The constitutive equations were then used on the resulting simplified material model described by the rule of mixture. Since the values of all the parameters had been identified by studying the as-deposited material, only one datum was required for the physically-based model to work, namely the relationship between the time to minimum creep rate and minimum creep rate. The resulting description of the annealed material was excellent, as in the case of the as-deposited alloy, and the model was successfully validated. The straightforward conclusion is that the key factors which determine the creep response of the investigated material are the size and distribution of the second-phase particles. Annealing causes an increase in particle size and a parallel decrease in creep resistance that can be easily quantified by the proposed model.

Author Contributions: Conceptualization, S.S.; methodology, S.S., M.C.; software, C.P.; validation, C.P., S.S.; formal analysis, C.P., S.S.; investigation, C.P., E.S., E.C., E.G., S.S.; resources, E.C., S.S.; data curation, C.P.; writing—original draft preparation, C.P.; writing—review and editing, S.S.; supervision, S.S.; funding acquisition, E.C., S.S. All authors have read and agreed to the published version of the manuscript.

Funding: This research was funded by the Grant of Excellence Departments, MIUR-Italy (ART. 1, 314-337 LEGGE 232/2016).

Institutional Review Board Statement: Not applicable.

Informed Consent Statement: Not applicable.

Data Availability Statement: The data presented in this study are available on request from the corresponding author.

Conflicts of Interest: The authors declare no conflict of interest.

Appendix A

Figure A1 shows the microstructure of the alloy in the as-deposited state, with the cells surrounded by the fine Si particles.

During high-temperature exposure, once the precipitation process ends and the equilibrium volume fraction of the secondary phase (Si) is reached, the particles are thought to evolve by Ostwald ripening, which is described by an equation in the form

$$d^3 = d_0^3 + \frac{64\gamma C_\infty V_m^2}{9RT} D_{eff} t \tag{A1}$$

where d_0 is the initial particle size, d the particle size at time t, γ the interfacial energy, C_∞ the equilibrium concentration of the species that form the particles and V_m the molar volume of the particle.

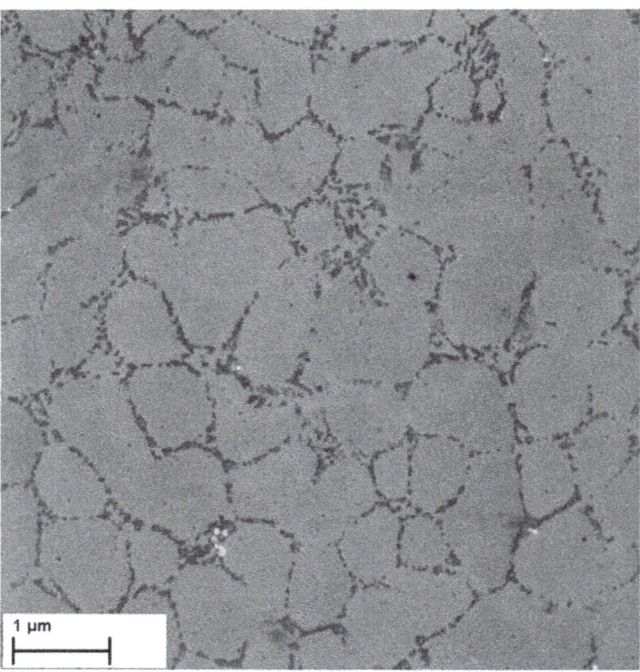

Figure A1. Microstructure of the alloy in the as-deposited state.

The effective diffusion coefficient can be written as

$$D_{eff} \approx D_{Ls} + A_{p2}\rho_m D_{ps} \tag{A2}$$

where $D_{Ls} = D_{0Ls} \exp(-Q_{Ls}/RT)$ is the lattice diffusion coefficient of the element forming the particle, $D_{ps} = D_{0ps} \exp(-Q_{ps}/RT)$ the corresponding pipe diffusion coefficient and ρ_m the mobile dislocation density. A_{p2} is the effective sectional area for diffusion which takes into account the scavenging effect of solute atoms by moving dislocations. The mobile dislocation density can be obtained with the Orowan equation

$$\dot{\varepsilon} = \frac{b}{m}\rho_m v_m \tag{A3}$$

where v_m is the dislocation velocity. Combined with Equation (A1), Equation (A2) gives

$$D_{eff} = D_{Ls} + \frac{A_{p2}m}{bv_m}\dot{\varepsilon}D_{ps} \tag{A4}$$

If A_{p2} is indeed supposed to be proportional to the dislocation velocity v_m, then

$$D_{eff} = D_{Ls}\left(1 + B_{\prime}\dot{\varepsilon}\frac{D_{ps}}{D_{Ls}}\right) = D_{Ls}\left[1 + B\dot{\varepsilon}exp\left(\frac{Q_{Ls} - Q_{ps}}{RT}\right)\right] \tag{A5}$$

where $B = B'D_{0ps}/D_{0Ls}$ is a constant. Equation (A5) is formally analogous to a similar equation proposed by Cohen [25]. The Ostwald ripening equation thus becomes

$$d^3 = d_0^3 + K_g D_{Ls}t + K_g D_{Ls}Bexp\left(\frac{Q_{Ls} - Q_{ps}}{RT}\right)\dot{\varepsilon}t \tag{A6}$$

with

$$K_g = \frac{64\gamma C_\infty V_m^2}{9RT} \tag{A7}$$

Since in the temperature range considered in this study the equilibrium solid solubility of Si in Al can be considered to be almost constant, the weak temperature dependence of K_g was neglected. The activation energies in Equation (A6) were considered to be $Q_{Ls} = 124$ kJ mol^{-1} and $Q_{ps} = 0.6\,Q_{Ls}$; the calculation in [14] gave $D_{0Ls}\,K_g = 2 \times 10^{17}$ nm^3 h^{-1} and $B = 10$ s^{-1}.

In order to verify the accuracy of the model in predicting the effect of high-temperature exposure on particle size, the diameter of the Si particles after 30 h annealing (initial size 50 nm) was estimated and compared with the real microstructure in Figure A2. At first glance, the model overestimates the size of the single particles; however, in the real microstructure, several agglomerates of these smaller Si precipitates can be observed. These agglomerates obviously behave as single larger particles. As a result, one can safely conclude that, although oversimplified, the description of the real microstructure of the alloy provided by the model is accurate.

Equation (A6) was used to calculate the size of the particles in two conditions: i. after annealing, under the assumption that the precipitation process was completed during AM and ripening started just at the beginning of the postprocessing heat treatment; ii. during creep experiments, at the time corresponding to the minimum creep rate. In the latter case, since during primary creep the samples had experienced strain rates much higher than the minimum value, the value $10\,\dot{\varepsilon}_m$ was used for $\dot{\varepsilon}$ in Equation (A6).

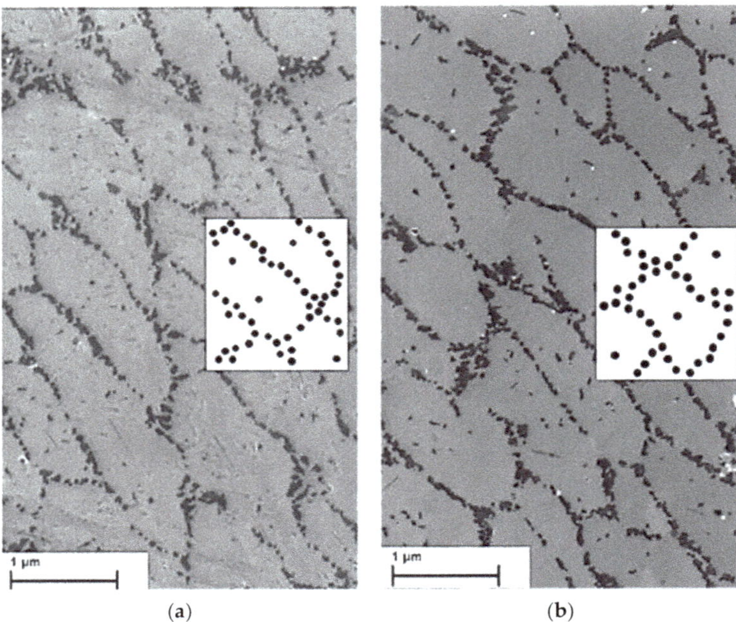

Figure A2. Comparison between the calculated particle size (the black circles in the insert, which qualitatively show the microstructure of the model material) and the real structure after annealing for 30 h. The diameter of the black circles in the inserts corresponds to the calculated particle size at the given magnification. (**a**) 175 °C, calculated size of Si particles = 53 nm; (**b**) 200 °C, calculated size of the particles = 63 nm.

Appendix B

The constitutive model was developed for Cu by Sandström [26–28] and was later modified to describe Al and its alloys [19,20]. The starting point is the well-known Taylor equation, here considered in a simplified form:

$$\sigma = \sigma_0 + \sigma_p = \sigma_0 + \alpha m G b \sqrt{\rho} \tag{A8}$$

where ρ is the dislocation density and $\sigma_\rho = \alpha m G b \rho^{1/2}$ is the dislocation hardening term. The stress σ_0 represents the strengthening contribution due to the interaction between fine particles and dislocations. This formulation of the Taylor equation does not take into account the stress required to move the dislocation in the absence of other dislocations nor the viscous drag stress due to solute atoms, since both these quantities, in a dilute solid solution reinforced by a fine dispersion of particles, are usually negligible when compared to the strengthening term due to particle–dislocation interaction.

The evolution of dislocation density during straining can be expressed as [26–28]

$$\frac{d\rho}{d\varepsilon} = \frac{m}{bL^*} - \omega\rho - \frac{2}{\dot{\varepsilon}} M_{cg}\tau_l\rho^2 \tag{A9}$$

where ω is a constant, τ_l is the dislocation line tension ($\tau_l = 0.5Gb^2$), M_{cg} is the dislocation mobility and L^* is the dislocation mean free path. At high temperatures, the equation can be simplified to

$$\frac{d\rho}{d\varepsilon} = \frac{m}{bL^*} - \frac{2}{\dot{\varepsilon}} M_{cg}\tau_l\rho^2 \tag{A10}$$

The dislocation mean free path represents the distance traveled by a dislocation before it undergoes a reaction. In the presence of fine particles obstructing dislocation mobility, with L interparticle spacing, the dislocation mean free path can be expressed in the form [20]

$$\frac{1}{L^*} = \frac{\sqrt{\rho}}{C_L} + \frac{1}{L} \tag{A11}$$

For alloys with densely spaced particles, Equation (A11) becomes

$$\frac{1}{L^*} \approx \frac{1}{L} \tag{A12}$$

At steady state, the combination of Equations (A8), (A10) and (A11) gives

$$\dot{\varepsilon}_m = \frac{2M_{cg}\tau_l bL}{m} \left(\frac{\sigma_\rho}{\alpha m Gb} \right)^4 \tag{A13}$$

Dislocation mobility in dilute Al–X solid solutions (where X is either Si or Mg) is [26–28]:

$$M_{cg} \cong \frac{D_{0L}b}{kT} exp\left(\frac{\sigma_\rho b^3}{kT} \right) exp\left\{ -\frac{Q_L}{RT} \left[1 - \left(\frac{\sigma_\rho}{R_{max}} \right)^2 \right] \right\} exp\left(-\frac{U_{ss}}{RT} \right) \tag{A14}$$

where R_{max} is the maximum strength of the alloy, tentatively quantified at room temperature as 1.5 times the ultimate tensile strength (R_{UTS}) of the alloy. The U_{ss} term describing the energy necessary for solute atoms to jump in and out of the clouds formed around dislocations has the form [27]

$$U_{ss} = \frac{1}{3\pi} \frac{1+\nu}{1-\nu} \frac{G\Omega\delta R}{k} \tag{A15}$$

where ν is Poisson's ratio (=0.3 in Al), Ω is the average Al atomic volume and δ is the volume atomic misfit.

References

1. Aboulkhair, N.T.; Simonelli, M.; Parry, L.; Ashcroft, I.; Tuck, C.; Hague, R. 3D printing of Aluminium alloys: Additive Manufacturing of Aluminium alloys using selective laser melting. *Prog. Mater. Sci.* **2019**, *106*, 100578. [CrossRef]
2. Herzog, D.; Seyda, V.; Wycisk, E.; Emmelmann, C. Additive manufacturing of metals. *Acta Mater.* **2016**, *117*, 371–392. [CrossRef]
3. Hebert, R.J. Viewpoint: Metallurgical aspects of powder bed metal additive manufacturing. *J. Mater. Sci.* **2016**, *51*, 1165–1175. [CrossRef]
4. DebRoy, T.; Wei, H.L.; Zuback, J.S.; Mukherjee, T.; Elmer, J.W.; Milewski, J.O.; Beese, A.M.; Wilson-Heid, A.; De, A.; Zhang, W. Additive manufacturing of metallic components—Process, structure and properties. *Prog. Mater. Sci.* **2018**, *92*, 112–224. [CrossRef]
5. Rosenthal, I.; Shneck, R.; Stern, A. Heat treatment effect on the mechanical properties and fracture mechanism in AlSi10Mg fabricated by additive manufacturing selective laser melting process. *Mater. Sci. Eng. A* **2018**, *729*, 310–322. [CrossRef]
6. Read, N.; Wang, W.; Essa, K.; Attallah, M.M. Selective laser melting of AlSi10Mg alloy: Process optimisation and mechanical properties development. *Mater. Des.* **2015**, *65*, 417–424. [CrossRef]
7. Trevisan, F.; Calignano, F.; Lorusso, M.; Pakkanen, J.; Aversa, A.; Ambrosio, E.P.; Lombardi, M.; Fino, P.; Manfredi, D. On the selective laser melting (SLM) of the AlSi10Mg alloy: Process, microstructure, and mechanical properties. *Materials* **2017**, *10*, 76. [CrossRef]
8. Wu, J.; Wang, X.Q.; Wang, W.; Attallah, M.M.; Loretto, M.H. Microstructure and strength of selectively laser melted AlSi10Mg. *Acta Mater.* **2016**, *117*, 311–320. [CrossRef]
9. Takata, N.; Kodaira, H.; Suzuki, A.; Kobashi, M. Size dependence of microstructure of AlSi10Mg alloy fabricated by selective laser melting. *Mater. Charact.* **2018**, *143*, 18–26. [CrossRef]
10. Fousová, M.; Dvorský, D.; Michalcová, A.; Vojtěch, D. Changes in the microstructure and mechanical properties of additively manufactured AlSi10Mg alloy after exposure to elevated temperatures. *Mater. Charact.* **2018**, *137*, 119–126. [CrossRef]
11. Li, W.; Li, S.; Liu, J.; Zhang, A.; Zhou, Y.; Wei, Q.; Yan, C.; Shi, Y. Effect of heat treatment on AlSi10Mg alloy fabricated by selective laser melting: Microstructure evolution, mechanical properties and fracture mechanism. *Mater. Sci. Eng. A* **2016**, *663*, 116–125. [CrossRef]
12. Uzan, N.E.; Shneck, R.; Yeheskel, O.; Frage, N. Fatigue of AlSi10Mg specimens fabricated by additive manufacturing selective laser melting (AM-SLM). *Mater. Sci. Eng. A* **2017**, *704*, 229–237. [CrossRef]

13. Cerri, E.; Ghio, E. AlSi10Mg alloy produced by Selective Laser Melting: Relationships between Vickers microhardness, Rockwell hardness and mechanical properties. *Metall. Ital.* **2020**, *7–8*, 5–17.
14. Paoletti, C.; Santecchia, E.; Cabibbo, M.; Cerri, E.; Spigarelli, S. Modelling the creep behavior of an AlSi10Mg alloy produced by additive manufacturing. *Mater. Sci. Eng. A* **2021**, 140138. [CrossRef]
15. Nix, W.D.; Ilschner, B. *Mechanisms Controlling Creep of Single Phase Metals and Alloys*; Haasen, P., Gerold, V., Kostorz, G., Eds.; Pergamon: Aachen, Germany, 1979; Volume 3, pp. 1503–1530. ISBN 978-1-4832-8412-5.
16. Meier, M.; Blum, W. Modelling high temperature creep of academic and industrial materials using the composite model. *Mater. Sci. Eng. A* **1993**, *164*, 290–294. [CrossRef]
17. Spigarelli, S. Constitutive equations in creep of Mg-Al alloys. *Mater. Sci. Eng. A* **2008**, *492*, 153–160. [CrossRef]
18. Orowan, E. *Dislocations in Metals*; Cohen, M., Ed.; The Institute of Metals Division, the American Institute of Mining and Metallurgical Engineers: New York, NY, USA, 1954.
19. Spigarelli, S.; Sandström, R. Basic creep modelling of Aluminium. *Mater. Sci. Eng. A* **2018**, *711*, 343–349. [CrossRef]
20. Paoletti, C.; Regev, M.; Spigarelli, S. Modelling of creep in alloys strengthened by rod-shaped particles: Al-Cu-Mg age-hardenable alloys. *Metals* **2018**, *8*, 930. [CrossRef]
21. Spigarelli, S.; Evangelista, E.; Cucchieri, S. Analysis of the creep response of an Al-17Si-4Cu-0.55Mg alloy. *Mater. Sci. Eng. A* **2004**, *387–389*, 702–705. [CrossRef]
22. Uzan, N.E.; Shneck, R.; Yeheskel, O.; Frage, N. High-temperature mechanical properties of AlSi10Mg specimens fabricated by additive manufacturing using selective laser melting technologies (AM-SLM). *Addit. Manuf.* **2018**, *24*, 257–263. [CrossRef]
23. Toschi, S. Optimization of a354 Al-Si-Cu-Mg alloy heat treatment: Effect on microstructure, hardness, and tensile properties of peak aged and overaged alloy. *Metals* **2018**, *8*, 961. [CrossRef]
24. Salleh, M.S.; Omar, M.Z.; Syarif, J.; Alhawari, K.S.; Mohammed, M.N. Microstructure and mechanical properties of thixoformed A319 aluminium alloy. *Mater. Des.* **2014**, *64*, 142. [CrossRef]
25. Buffington, F.S.; Cohen, M. Self-Diffusion in Alpha Iron Under Uniaxial Compressive Stress. *JOM* **1952**, *4*, 859–860. [CrossRef]
26. Sandström, R. Basic model for primary and secondary creep in copper. *Acta Mater.* **2012**, *60*, 314–322. [CrossRef]
27. Sandström, R. Influence of phosphorus on the tensile stress strain curves in copper. *J. Nucl. Mater.* **2016**, *470*, 290–296. [CrossRef]
28. Sandström, R. The role of cell structure during creep of cold worked copper. *Mater. Sci. Eng. A* **2016**, *674*, 318–327. [CrossRef]

Article

High Strain Rate Superplasticity of WE54 Mg Alloy after Severe Friction Stir Processing

Marta Álvarez-Leal [1,†], **Fernando Carreño** [1,*], **Alberto Orozco-Caballero** [1,‡], **Pilar Rey** [2] and **Oscar A. Ruano** [1]

[1] Physical Metallurgy Department, CENIM, CSIC, Av. Gregorio del Amo 8, 28040 Madrid, Spain; m.alvarez@cetemet.es (M.Á.-L.); alberto.orozco.caballero@upm.es (A.O.-C.); ruano@cenim.csic.es (O.A.R.)
[2] Technological Centre AIMEN, Relva 27A-Torneiros, O'Porriño, 36410 Pontevedra, Spain; prey@aimen.es
* Correspondence: carreno@cenim.csic.es
† Now at Department of R&D and Programmes, Technology Centre of Metal-Mechanical and Transport (CETEMET), Avda. 1° de Mayo, s/n. Linares, 23700 Jaén, Spain.
‡ Now at Department of Mechanical Engineering, Chemistry and Industrial Design, Polytechnic University of Madrid, Ronda de Valencia 3, 28012 Madrid, Spain.

Received: 2 November 2020; Accepted: 23 November 2020; Published: 25 November 2020

Abstract: Friction stir processing (FSP) was used on coarse-grained WE54 magnesium alloy plates of as-received material. These were subjected to FSP under two different cooling conditions, refrigerated and non-refrigerated, and different severe processing conditions characterized by low rotation rate and high traverse speed. After FSP, ultrafine equiaxed grains and refinement of the coarse precipitates were observed. The processed materials exhibited high resistance at room temperature and excellent superplasticity at the high strain rate of 10^{-2} s^{-1} and temperatures between 300 and 400 °C. Maximum tensile superplastic elongation of 726% was achieved at 400 °C. Beyond 400 °C, a noticeable loss of superplastic response occurred due to a loss of thermal stability of the grain size. Grain boundary sliding is the operative deformation mechanism that can explain the high-temperature flow behavior of the ultrafine grained FSP-WE54 alloy, showing increasing superplasticity with increasing processing severity.

Keywords: friction stir processing; WE54 magnesium alloy; superplasticity; grain boundary sliding; processing severity

---

1. Introduction

Magnesium, as the lightest structural metallic material, is being used in different engineering applications and may replace aluminum once its strength and ductility are improved with the help of intensive research on this material [1,2]. Such investigations will decrease the weight of structures, especially in the automotive and aerospace industries, increasing fuel-efficiency and minimizing its CO_2 footprint.

The strength of magnesium can be increased by alloying with rare earth elements, which usually present high solubility in the magnesium lattice and are effective for precipitation and age hardening [3,4]. This addition can also significantly improve the creep resistance of magnesium alloys, as proven with the addition of yttrium and gadolinium, heavy rare earth (RE) elements that usually improve high temperature strength [3–5].

An additional issue for the widespread use of magnesium comes from its poor workability at room temperature, which makes the fabrication of complex-shaped components a difficult task. This is due to the limited number of slip planes available and the activation of twinning. Such a problem is partially solved by extensive grain refining that may increase ductility at room temperature [6–8].

Among numerous techniques used for achieving ultrafine grain (UFG) microstructures, severe plastic deformation (SPD) processing techniques are nowadays the most promising route [9–11]. The large stresses and huge deformation values imposed during SPD generate high dislocation densities in the materials, which evolve into ultrafine grain structures through recovery and recrystallization processes.

Among the SPD techniques, friction stir processing (FSP) is rapidly developing because it allows the attainment of large products with UFG microstructures and can be easily implemented in the industry. FSP is a versatile solid-state processing technique that can tailor the microstructure and, therefore, the mechanical properties by controlling the processing parameters and the cooling rate. In this work, magnesium samples were friction stir processed under two different cooling conditions, non-refrigerated and cryogenic (liquid nitrogen) cooling, and at various tool rotation and traverse speeds seeking a high severity in the introduction of plastic deformation.

A direct consequence of microstructure refinement is the possibility of deforming in a superplastic way. In such a case, elongations to failure are much larger than for coarse-grained materials, usually reaching values larger than 400%, or 200% for high strain rates ($>10^{-2}$ s^{-1}) [12]. This behavior is needed for the use of superplastic forming as a feasible production process of geometrically complex parts. In order to make superplastic forming a profitable method at an industrial scale, strain rates equal to or higher than 10^{-2} s^{-1} are required. The underlying deformation mechanism in superplasticity is grain boundary sliding (GBS), which requires a fine grain size, less than 20 µm, and a high fraction of high angle grain boundaries. The grain sliding during GBS requires a diffusion-controlled mechanism that acts as an accommodation process. Therefore, the temperature should be relatively high, which may lead to grain growth as a counter-productive side effect. These two factors should be closely controlled to attain successfully large elongations. In addition, the finer the grain size, the higher the superplastic strain rate, the lower the superplastic temperature, and the slower the grain growth. Therefore, it is interesting to process the material with the highest severity in order to obtain the smallest grain size and benefit from these advantages in the superplastic forming of high added-value structural parts.

2. Material and Experimental Procedure

Extruded WE54 plates of dimensions $300 \times 80 \times 5$ mm^3 were received from Magnesium Elektron (Manchester, UK). The plates were solution-heat treated at 525 °C for 8 h, followed by hot water quenching (60 °C) and final aging at 250 °C for 12 h, which consist of the so-called T6 temper [13]. The composition of the alloy was the following (in wt.%): 5% Y–1.5% Nd–1.5% RE–0.45%Zr–balance Mg.

The microstructural characterization was carried out by means of electron back-scattered diffraction (EBSD). EBSD samples were prepared by mechanical polishing up to a final 1 µm diamond particles step, followed by chemical polishing for 5 to 10 s using a solution of 4.2 g picric acid, 10 mL acetic acid, 10 mL H$_2$O, and 70 mL ethanol.

The as-received plates were subjected to FSP under different processing and cooling conditions. The tool to perform FSP was made of an MP159® nickel-cobalt base alloy and it comprises a scrolled shoulder 9.5 mm in diameter and a concentric threaded conical pin with flutes 4.7–4.1 mm in diameter and 1.8 mm in length. The as-received WE54 alloy in the T6 condition was processed by FSP with a special emphasis on high processing severity to obtain the finest possible grain size. In order to achieve the above-mentioned high severity, the material was processed using low rotation rate (ω) and high traverse speed (v) to minimize the heat input (HI), which can be estimated as HI $\propto \omega^2/v$. Such an approach is convenient for industrial applications, since we decrease the processing time. Furthermore, the sheets were firmly mounted on two different backing plates, one made of steel and the other one made of copper. The latter contains a series of cavities where liquid nitrogen flows to increase the cooling rate. The liquid nitrogen applied flow rate was such that the upper surface of the copper backing anvil reached −60 °C, facilitating a high heat extraction during FSP and restricting any possible grain growth. The temperature of the steel backing plate was room temperature. The different processing conditions, characterized by the traverse speed, v, and the rotational speed, r or ω, and its nomenclature are given in Table 1. The specific cooling conditions are named "refrigerated" and "non-refrigerated" for the copper and steel backing plates, preceded by the letter R or N, respectively.

Table 1. Nomenclatures and values of the friction stir processing (FSP) conditions.

ω (rpm)	v (mm/min)	HI (rpm^2/(mm/min))	Nomenclature
1400	500	3920	N or R14r05v
1000	500	2000	N or R10r05v
1000	1000	1000	N or R10r10v

The six sets of processing conditions, selected following a logarithmic increase in the severity, were named N14r05v, N10r05v, N10r10v, and R14r05v, R10r05v, and R10r10v.

Constant crosshead speed tensile tests (CCST) initially at 10^{-2} s^{-1} were used to characterize the mechanical behavior of the alloy. In addition, strain rate change tests (SRCT) in tension ranging from 10^{-1} to 10^{-5} s^{-1} were performed to characterize the high temperature (300–450 °C) behavior. The tests were performed in air using an Instron 1362 universal testing machine equipped with a four-lamp ellipsoidal furnace. Planar dog-bone tensile samples with 6.5 × 2 × 1.7 mm gage area dimensions were electro-discharge machined so that their longitudinal axis was parallel to the extrusion or FSP direction, as shown in Figure 1. The tensile samples were mirror-polished in their upper surfaces to a final thickness of 1.6 mm to monitor their changes after tensile testing at various temperatures. The true strain, ε, is calculated as $\varepsilon = \ln(1 + e)$, where $e = (l - l_o)/l_o$, l_o is the initial gage length and l is the instantaneous length. The true stress is defined as $\sigma = F/A_o (1 + e)$, where A_o is the initial section of the sample and F is the supported load during tensile testing.

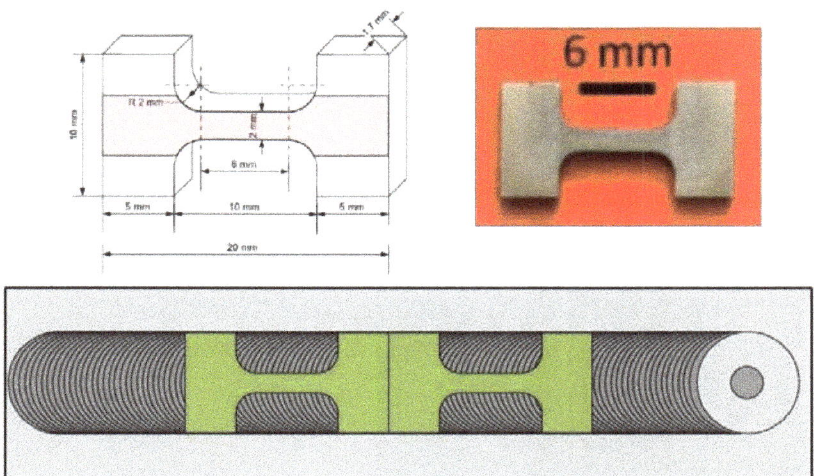

Figure 1. Geometry and orientation of machined tensile samples of the FSPed WE54 Mg alloy.

3. Results

3.1. Microstructures

Figure 2 shows an orientation-imaging map of the as-received WE54 Mg alloy. The micrograph, taken in the L plane, reveals equiaxed grains of diameters in the range 50 to 250 μm. A description of similar microstructures is given elsewhere [14–16].

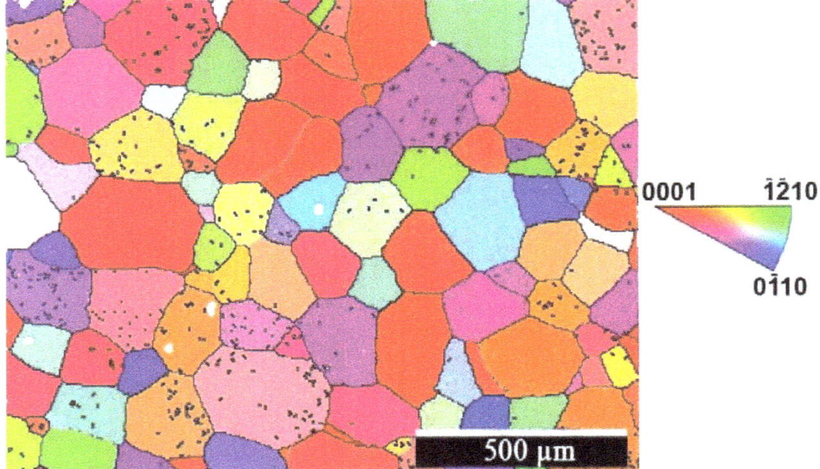

Figure 2. Orientation-imaging map (OIM) obtained by electron back-scattered diffraction (EBSD) of the as-received WE54 Mg alloy, T6 condition.

Figure 3 shows orientation-imaging maps of the WE54 alloy (EBSD inverse pole figure maps) after FSP of the 10r10v material, (a) non-refrigerated and (b) refrigerated.

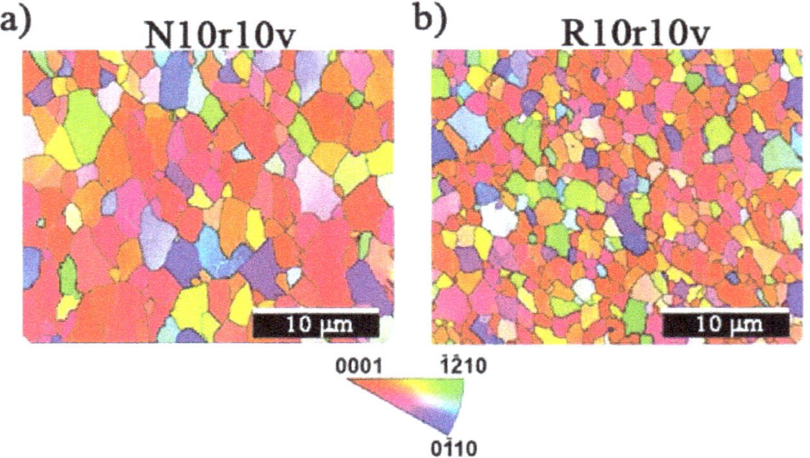

Figure 3. Orientation-imaging maps of the WE54 alloy (EBSD inverse pole figure maps) after FSP in the conditions (**a**) N10r10v and (**b**) R10r10v.

The grain size values of the R10r10v and N10r10v materials, which correspond to the most severe FSP condition, measured by Feret diameters, are 0.9 and 1.3 μm, respectively. As is evident from comparing Figures 2 and 3 (please note the different scale, 500 vs. 10 μm), a huge refinement has been attained by severe processing of this magnesium alloy.

3.2. Tensile Tests to Rupture of WE54 Alloy Processed by FSP

Figure 4 shows stress–strain curves at initial strain rates of $10^{-2}\ \mathrm{s}^{-1}$ and various temperatures of the WE54 alloy processed by FSP under three processing conditions and two cooling conditions, refrigerated and non-refrigerated. Two different behaviors can be distinguished, one at low (high stresses and low ductility) and another one at high (low stresses and high ductility) temperatures, with a transition temperature around 250–300 °C. The low temperature regime (25–200 °C) shows very similar low tensile ductility values < 25% for all samples and high stress values, which decrease slightly with increasing temperature in this interval. On the contrary, the high temperature interval (350–450 °C) shows very large tensile elongations (as high as 726%), and very low stress values. This contrast between low and high temperature behavior increases with increasing processing severity and with the use of refrigeration. The transition temperatures (250–300 °C) are also affected by refrigeration and processing severity. For the non-refrigerated conditions, at 300 °C the ductility values increase noticeably with increasing severity, while for the refrigerated conditions, the increasing ductility is observed from 250 °C for the two most severe processing conditions. Additionally, the largest tensile ductility value is observed for all conditions at 400 °C, accompanied by very low stress values. At 450 °C, the stress values increase and the ductility values drop, which indicates changes in the deformation mechanism.

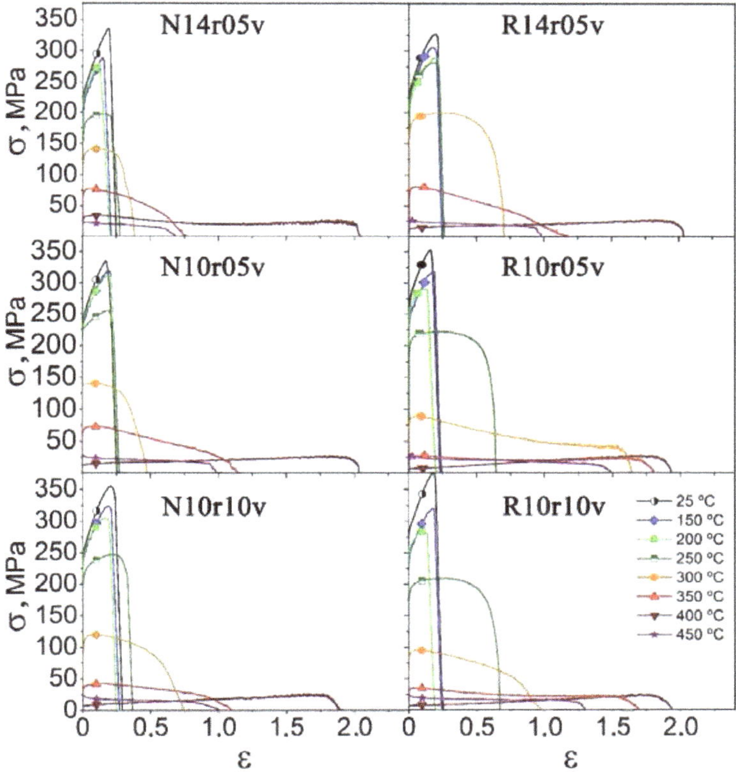

Figure 4. Stress–strain curves of materials processed by FSP and tested at an initial strain rate of $10^{-2}\,\mathrm{s}^{-1}$ at temperatures from 25 to 450 °C, non-refrigerated (N, left) and refrigerated (R, right).

Table 2 gives the mechanical parameters of all the curves presented in Figure 4. It is remarkable the high elongation values reached at 400 °C for all materials under all conditions at this high strain rate. A maximum elongation to failure of 726% is observed for the R14r05v material. Nonetheless, very high values of 562, 593, 599, 664, and 666% are also obtained for the rest of the conditions at 400 °C. It is also interesting to note that very high elongations are also obtained at 350 °C and in some conditions at 300 and 450 °C. However, at 450 °C, there is a clear loss of ductility with respect to 400 °C, which is attributed to massive grain growth, and will be discussed in depth in the next section.

Table 2. Yield stress ($\sigma_{0.2}$), flow stress (σ), uniform elongation (e_u), and elongation to failure (e_F) at different test temperatures of the alloy WE54 processed by FSP and tensile tested at 10^{-2} s^{-1}.

Condition	T (°C)	$\sigma_{0.2}$ (MPa)	σ (MPa)	e_u (%)	e_F (%)
	20	195	335	21	22
	150	165	288	15	17
	200	205	281	13	14
	250	146	197	18	24
N14r05v	300	111	141	12	46
	350	63	78	7	107
	400	26	33	17	666
	450	21	22	2	98
	20	220	336	19	24
	150	213	318	22	24
	200	218	312	21	24
	250	224	255	21	26
N10r05v	300	138	140	8	61
	350	60	73	8	210
	400	12	17	69	664
	450	26	28	1	166
	20	230	355	23	27
	150	223	324	21	24
	200	219	305	18	22
	250	194	247	23	39
N10r10v	300	90	120	7	107
	350	32	43	14	198
	400	6	14	76	562
	450	22	24	1	169
	20	203	327	22	24
	150	194	306	20	23
	200	179	290	21	25
	250	195	282	20	25
R14r05v	300	148	200	26	101
	350	60	81	6	226
	400	16	23	10	726
	450	20	23	3	222
	20	248	354	17	20
	150	213	282	13	14
	200	231	292	13	15
	250	179	223	27	89
R10r05v	300	71	90	7	420
	350	23	28	7	512
	400	6	15	96	599
	450	29	30	0.4	345
	20	270	375	20	22
	150	209	319	19	21
	200	230	283	11	15
	250	159	209	25	94
R10r10v	300	75	94	8	166
	350	26	35	7	440
	400	6	12	97	593
	450	22	23	0.6	265

Figure 5 shows curves of the elongation to failure at 10^{-2} s^{-1} as a function of temperature for the WE54 alloy after FSP under various conditions. For comparison, data from the as-received material in the T6 condition (not subjected to FSP) are included in the figure [17]. The ductility continuously increases with temperature and a strong peak is observed at 400 °C. At this temperature, the ductility values for all samples are around 700% with a maximum of 726% observed in the R14r05v material. It can also be observed that at 250–350 °C, the tensile ductility values of the refrigerated conditions (continuous lines) are higher than those corresponding to the non-refrigerated ones (dashed lines). At 450 °C, the elongation to failure drastically drops, pointing to a change in the deformation mechanism as a result of grain growth. This peak and the subsequent drop do not exist in the non-processed WE54 material, since the ductility increases continuously up to the highest temperature of 450 °C [17].

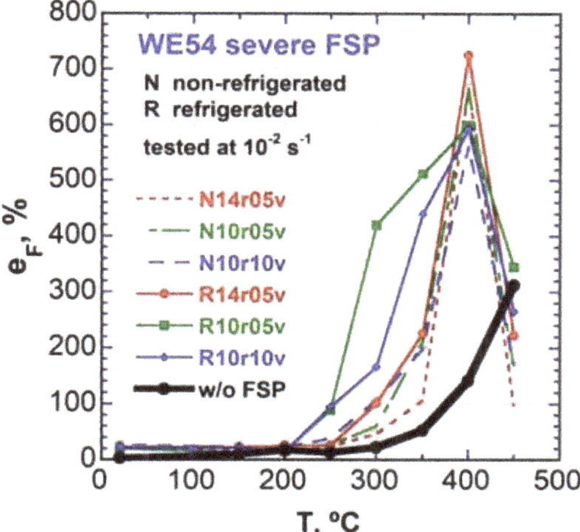

Figure 5. The elongation to failure as a function of temperature for the FSP WE54 alloy tested at 10^{-2} s^{-1}.

Very low stress values are also observed at high temperatures in most of the tensile tests in the range 350–450 °C. The flow stress, σ, as a function of temperature at 10^{-2} s^{-1}, is represented in Figure 6. Again, for comparison, data from the as-received material in the T6 condition (not subjected to FSP) are included in the figure [17]. The figure shows a moderate decrease in the stress with temperature up to 200 °C. Above this temperature, the stress strongly drops. The same behavior is observed for all FSP and cooling conditions. The highest stress value at room temperature is given for the material processed under the most severe condition, and refrigerated, R10r10v. On the contrary, at high temperatures, this material shows the lowest stress values. It should be noted the slight stress increase at 450 °C, which is attributed to rapid grain coarsening of all our WE54 processed materials at such a high temperature. This behavior can be compared with that of the material without FSP [17], for which the increase at 450 °C does not exist and the curve continuously decreases from 250 °C.

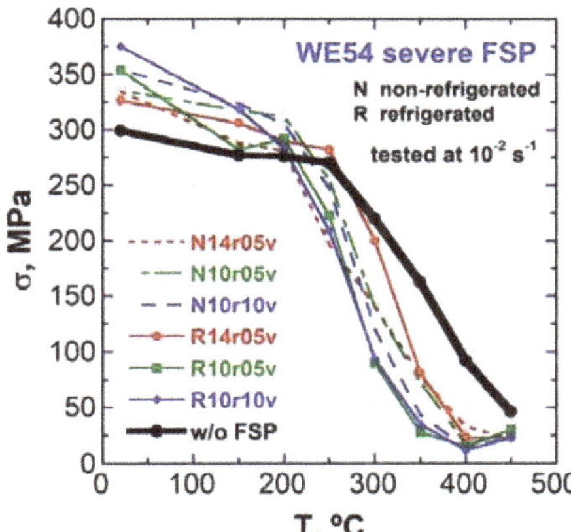

Figure 6. The flow stress as a function of temperature for the FSP WE54 alloy tested at 10^{-2} s^{-1}.

3.3. Surface Morphology of Samples after the Constant Strain Rate Tests

The surface morphology of the tested samples gives a clear hint of the underlying microstructures, especially under superplastic conditions for which sliding of the grains is clearly visible. Figure 7 presents those SEM micrographs after testing at 10^{-2} s^{-1} for the 14r05v materials under the two cooling conditions. Similar microstructures are obtained for the other two processing conditions, 10r05v and 10r10v. Fine grains are depicted in all cases except at 450 °C. Before testing, the Feret diameters were about 0.9 and 1.3 μm for the refrigerated and non-refrigerated materials, respectively. These grain sizes are similar to those tested at 250 °C, for which deformation is small. Larger grains are observed in samples deformed at higher temperatures attesting grain growth during testing. At 350–400 °C, clear hints of grain boundary sliding are observed with the grains sliding against each other, as shown in Figure 7. This fine microstructure is able to attain very large elongations, 726% at 400 °C, at high strain rate, as shown in Figure 5. At 450 °C, the microstructure is unstable, tending to grow quickly so that most of the grains become too large to withstand superplastic deformation for most of the FSPed microstructures.

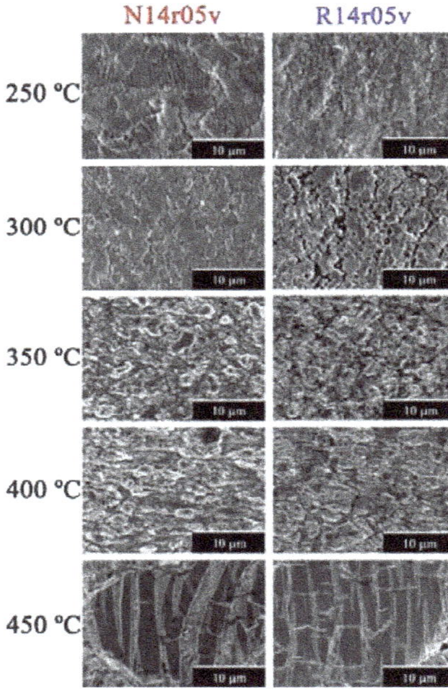

Figure 7. SEM micrographs of the topography of samples tested at 10^{-2} s^{-1} at various temperatures for the processing condition 14r05v, non-refrigerated (N, **left**) and refrigerated (R, **right**).

3.4. Strain Rate Change Tests

Figure 8 shows data from the strain rate change tests performed in the range 300 to 450 °C at strain rates from 10^{-5} to 10^{-1} s^{-1} for all processing conditions and two cooling conditions. The slope of the curves corresponds to the apparent stress exponent, n. A low n value of about 2 is usually observed at low strain rates and high temperature, which is typical of superplastic materials where grain boundary sliding controls deformation. The increase to higher n values at high strain rates and lower temperatures in these materials is usually attributed to a change into slip creep as the controlling deformation mechanism. As observed in Figure 8, there is a gradual decrease in stress values (from as high as 200 MPa at 300 °C to as low as 1–2 MPa at 400 °C) and n values (towards 2 at 400 °C) with increasing testing temperature, for all samples. At 350–400 °C, there is a wide interval of high strain rates for which the FSPed alloys show low n values close to two, which can be associated to GBS of the ultrafine grains. Although the high temperature behavior of the different samples appears to be quite similar, some hints show that there is an influence of the processing severity and refrigeration on the superplastic response, which will be highlighted in the discussion section. The anomalous behavior at 450 °C is noticeable, for which their stress values should be lower than those of 400 °C; however, most of them are even higher.

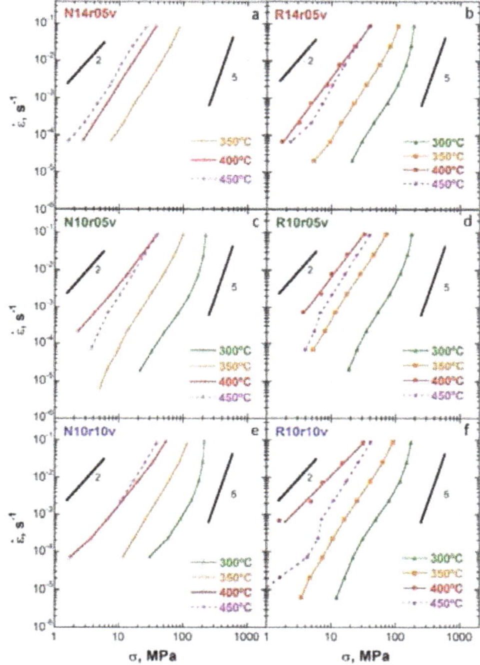

Figure 8. (a–f) Strain rate as a function of stress obtained by strain rate change tests for the six FSP processing conditions, non-refrigerated (N, left) and refrigerated (R, right).

Regarding the activation energy, only an estimation can be provided since there is no constant value of stress exponent and the microstructure is constantly coarsening. For instance, a rough evaluation has been made in the R14r05v material between 300 and 400 °C at 40 MPa, in the range of low n values, that gives a value of about 180 kJ/mol, somewhat higher than the activation energy value associated with lattice diffusion, Q_L. However, negatives values (without any physical meaning) are found when trying to obtain an activation energy value between 400 and 450 °C, associated with a change in the deformation mechanism.

4. Discussion

The present study makes evident the need to understand the influence of FSP conditions and the resulting microstructure on the wide range of deformation behaviors of the materials processed at different rates and cooling conditions. Figure 6 shows the high resistance at low temperatures, which is a result of the severe grain refinement due to the strategy of imposing high severity processing conditions. This is reached by a combination of low rotation speed, minimizing the frictional heat generation, and by increasing the traverse speed to ease a quick heat dissipation, minimizing the process temperature and reducing grain growth just after the tool pass [18–20]. The Hall–Petch law predicts an increase in the stress with decreasing grain size, at low temperature, as occurs for the processed material, but other factors such as the grain size distribution, refinement of the second phases, and the presence of elements in solid solution may also influence this behavior [21]. On the other hand, the figure also shows the rapid decrease in the stress above 250 °C for all materials, and especially, in the processed materials. This rapid stress drop with increasing temperature is attributed to the start of a creep behavior that is mainly controlled by diffusional processes. These diffusional processes generally involve the ability of dislocations to bypass the obstacles. These obstacles are usually overcome through a climb mechanism, but, in the presence of solutes, an interaction among dislocations and solutes may be controlling as well. Additionally, when the material presents small, equiaxed, and highly misoriented grains, another diffusionally accommodated mechanism may take control of the creep behavior under certain strain rate and temperature conditions, which is GBS in the so-called superplastic window. The three possible mechanisms differ in their microstructural evolution, and stress and ductility values at high temperature. The solid solution creep mechanism brings higher stresses due to the reinforcing solute–dislocation interactions, and showing elongated grains parallel to the imposed deformation. On the contrary, the GBS mechanism yields easy deformation at much lower stresses than those of dislocation-based creep mechanisms and the morphology of grains does not change even after very large deformations, remaining equiaxed and highly misoriented. The ductility obtained by GBS is usually much higher than for any other mechanism, in parallel to the much lower creep stresses measured during high temperature deformation. Additionally, the smaller the grain size, the larger the ductility, the lower the creep stress, and the higher the strain rate sustaining GBS. This is the reason underlying our strategy to process the material severely, in order to obtain the finest microstructure as possible [18–20].

As shown clearly in Figure 6, the six processed materials present much lower stress values than those of the as-received WE54 magnesium alloy, at high temperature, up to a minimum value at 400 °C. The stress values can be as low as a factor of four, which is very useful for the industrial forming of complex parts. This behavior corresponds to a different deformation mechanism than that operating in the non-processed as-received alloy.

Another feature important to note in Figure 6 is the increase in the stress at 450 °C in the FSP materials at which the microstructure coarsens significantly (Figure 7). Therefore, the mechanism controlling deformation at high temperature is one depending on grain size. In contrast, the material not subjected to FSP does not show such behavior which is an indication that, at high temperature, the controlling mechanism is different from that of the processed materials. Indeed, the behavior of the as-received material was attributed to a solute-drag slip creep mechanism [17,22], as a rate-controlling process which is grain size-independent. Correspondingly, Figure 5 clearly shows the rapid increase in the elongation to failure from 20 to 400 °C and their high values, much higher than those corresponding to the as-received material, up to about 700%, typical of a grain boundary sliding mechanism for which the ductility is sensibly dependent on grain size [23–25].

In addition, a homogeneous equiaxed microstructure is present after deformation, up to 400 °C, as shown in Figure 7, and a low stress exponent, close to 2, is obtained, as shown in Figure 8. This figure also shows very low stress values, which corroborate the strong decrease in stresses above 300 °C shown in Figures 6 and 8. This evidence points clearly to the grain boundary sliding mechanism as the one controlling deformation. The high strain rates where n is still close to 2 at temperatures above 300 °C are remarkable. This wide range of strain rates is associated to the fine grain size obtained by severe friction stir processing of the WE54 Mg alloy, around 1 μm at the start of testing. At 300 °C and above 10^{-2} s^{-1}, a typical power law region is observed with n values higher than 5, pointing to a dislocation creep mechanism at lower temperatures. At 450 °C, for which the microstructure strongly coarsens, as revealed in Figure 7, a change back to a dislocation creep mechanism is taking place.

The various graphs of Figure 8 from strain rate change tests corresponding to the six processed materials show similar behavior. A relatively large range of strain rates and temperatures is found, where n is about 2 and, therefore, superplastic properties can be expected in such a large window, even at strain rates as high as 10^{-1} s^{-1} at the highest temperatures. This behavior corresponds to a

GBS mechanism, as the processed alloys possess ultrafine, equiaxed, and highly misoriented grains, the stresses are low, and the ductility values are very high. In the range of low temperature and high strain rates, a change towards dislocation creep is observed, increasing the values of the stress exponent, as expected, and corroborated by Figure 6, as well. Another interesting feature revealed in these graphs is the behavior at 450 °C. As grain coarsening is taking place rapidly during testing at this temperature in the processed alloys, the stresses are higher than expected, and they even cross the lines corresponding to the data at 400 °C, showing higher stress exponents. In fact, as shown in Figure 8c,d,f, there is an abrupt change in slope at 450 °C at low strain rates because increasing coarsening is leading the behavior of the ultrafine processed alloy towards the behavior corresponding to a coarse-grained alloy. Effectively, the solute-drag slip creep mechanism, characterized by a stress exponent of 3, was demonstrated to occur in the same WE54 material that was not severely plastically deformed by FSP and had a coarse grain size [17]. Since the sample tested at 450 °C showed large grain growth, it is reasonable to assume that the sample is in the process of changing mechanisms from GBS to solute-drag creep. This is especially important during deformation at the lowest strain rates, where the sample remains at this temperature for a long time. Due to the continuous grain coarsening at high temperature, even extreme at 450 °C, leading to a change in deformation mechanism, it is difficult to obtain a true activation energy value associated with the high temperature creep of the processed alloys. In the range 300–400 °C, an approximate value of 180 kJ/mol is obtained for the activation energy for creep in the R14r05v material, although in the range 400–450 °C, the value becomes negative, without physical meaning, due to the change in deformation mechanism. Temperature-dependent activation energy values were also found for a similar fine-grained Mg–Y–Nd alloy [26]. However, the value obtained, although slightly high, is not far from that corresponding to lattice self-diffusion of magnesium, 135 kJ/mol [27,28], and may be regarded as typical of reinforced magnesium alloys. In summary, between 300 and 400 °C, the materials deform by a GBS mechanism and at 450 °C, the ultrafine-grained alloys coarsen their microstructures rapidly and change their superplastic behavior towards a solute drag creep mechanism, as the initial, unprocessed coarse WE54 alloy [17].

Figure 9 gives a comparison of the curves at 300 and 400 °C between the two cooling conditions, non-refrigerated and refrigerated, for the N10r10v and R10r10v materials. It is observed that the refrigerated material, the most severely processed one, is less resistant at both temperatures, which is a result of the finest grain size obtained. This means that at 300 °C, there is already an influence of grain size on stress values, corroborated by Figure 6. Additionally, the ductility values are also much higher for the processed alloys than for the unprocessed initial WE54, as observed in Figure 5, at 300 °C, thus showing that it is at about this temperature from which the start of deformation controlled by grain boundary sliding can be attributed.

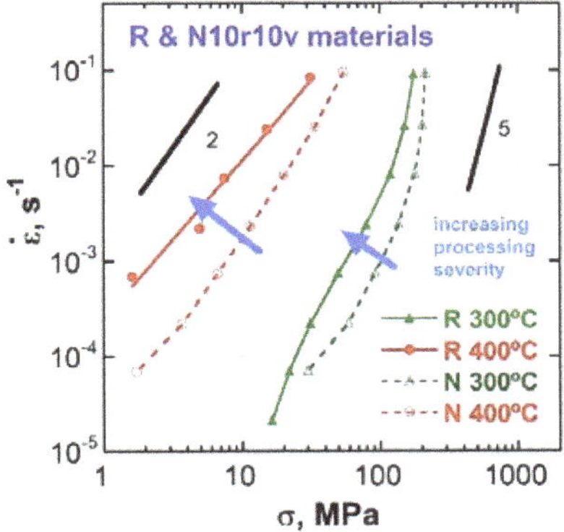

Figure 9. Comparison of FSP WE54 10r10v materials, refrigerated (R) and non-refrigerated (N).

An important aspect that influences the creep behavior of FSP materials is the severity of the processing. Figure 10 shows strain rate versus stress curves at 400 °C for the refrigerated materials with different processing conditions. As mentioned before, the severity of the processing increases following the order 14r05v < 10r05v < 10r10v. The figure shows that the stress decreases at an imposed strain rate, or conversely, the strain rate at a given stress increases in the same order, according to the applied processing severity. This favors the forming process, which is a consequence of the microstructure refining with increasing severity. Therefore, it can be concluded that processing severity is an important factor that influences the creep behavior of the WE54 magnesium alloy by achieving ultrafine grain sizes which allow for superplasticity at higher strain rates. Similarly, a low heat input applied to a Mg-9Li-1Zn alloy has also been successful to obtain ultrafine grains [29]. Furthermore, the ultrafine grain sizes obtained also improve the resistance of the WE54 alloy at room temperature.

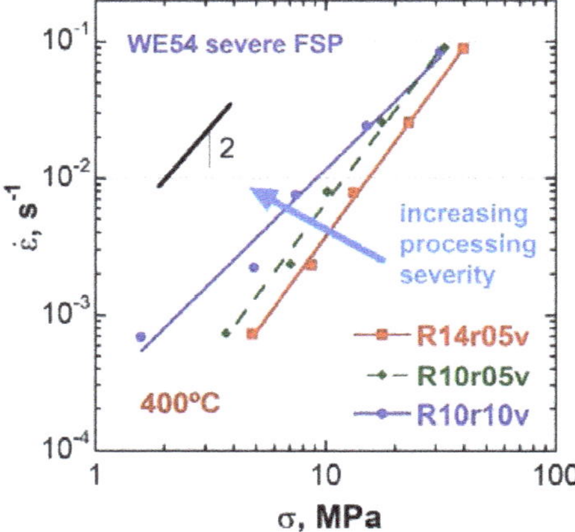

Figure 10. The strain rate as a function of stress at 400 °C for the refrigerated materials with different processing conditions to show the influence of the processing severity.

5. Conclusions

1. Friction stir processing under highly severe conditions was successfully applied to the coarse WE54 alloy, so that refinement of the coarse precipitates and fine equiaxed grains was obtained.
2. The processed materials showed fine grains of the order of 1 µm and exhibited excellent high strain rate superplasticity at temperatures between 300 and 400 °C, and strain rates even higher than 10^{-2} s^{-1}. This behavior contrasts with the higher resistance and lower ductility of the unprocessed WE54 Mg alloy.
3. Large elongations, with a maximum of 726%, were achieved at 400 °C and 10^{-2} s^{-1}. Additionally, low stress and stress exponent values associated with grain boundary sliding are obtained, which were lower the higher the processing severity. Beyond 400 °C, the elongations rapidly decreased as a consequence of microstructure coarsening.
4. At 450 °C, a transition of mechanisms occurs from grain boundary sliding to solute drag creep as the microstructure coarsens. This last mechanism is the operative mechanism of the coarse unprocessed alloy at high temperatures.
5. Severe friction stir processing has been proven to be a method of obtaining ultrafine grains in Mg alloys and, thus, producing materials of high resistance at low temperatures, and easy to form superplastically at high temperatures and high strain rates.

Author Contributions: Conceptualization and methodology, F.C. and O.A.R.; investigation, M.Á.-L., F.C. and O.A.R.; data curation and visualization, M.Á.-L., F.C. and O.A.R.; supervision: A.O.-C., F.C. and O.A.R.; sample manufacturing by FSP, M.Á.-L. and P.R.; material characterization, M.Á.-L.; writing—original draft preparation, O.A.R. and F.C.; writing—review and editing, O.A.R., M.Á.-L., A.O.-C. and F.C.; resources, funding acquisition, and project administration, F.C. and P.R. All authors have read and agreed to the published version of the manuscript.

Funding: Project MAT2015-68919-R MINECO/FEDER, Spain, and FPI fellowship number BES2013-063963 (MINECO/FEDER/ESF).

Acknowledgments: Financial support from MINECO (Spain), Project MAT2015-68919-R (MINECO/FEDER) is gratefully acknowledged. A.O.-C. also thanks CENIM, CSIC, for a contract funded by the aforementioned project. M.Á.-L. thanks MINECO for a FPI fellowship, number BES2013-063963 (MINECO/FEDER/ESF).

Conflicts of Interest: The authors declare no conflict of interest.

References

1. Mordike, B.L.; Ebert, T. Magnesium: Properties—Applications—Potential. *Mater. Sci. Eng. A* **2001**, *302*, 37–45. [CrossRef]

2. Luo, A.A. Recent Magnesium Alloy Development for Elevated Temperature Applications. *Int. Mater. Rev.* **2004**, *49*, 13–30. [CrossRef]

3. Bettles, C.; Gibson, M.; Zhu, S. Microstructure and Mechanical Behaviour of an Elevated Temperature Mg-Rare Earth Based Alloy. *Mater. Sci. Eng. A* **2009**, *505*, 6–12. [CrossRef]

4. Tekumala, S.; Seetharaman, S.; Almajud, A.; Gupta, M. Mechanical Properties of Magnesium-Rare Earth Alloy Systems: A Review. *Metals* **2015**, *5*, 1–39. [CrossRef]

5. Maruyama, K.; Suzuki, M.; Sato, H. Creep Strength of Magnesium-Based Alloys. *Metall. Mater. Trans.* **2002**, *33*, 875–882. [CrossRef]

6. Kaibyshev, O.A.; Kazachkov, I.V.; Salikhov, S.Y.A. The Influence of Texture on Superplasticity of the Zn-22% Al Alloy. *Acta Metall.* **1978**, *26*, 1887–1894. [CrossRef]

7. McFadden, S.X.; Mishra, R.S.; Valiev, R.Z.; Zhilyaev, A.P.; Mukherjee, A.K. Low-Temperature Superplasticity in Nanostructured Nickel and Metal Alloys. *Nature* **1999**, *398*, 684–686. [CrossRef]

8. Fukusumi, M.; Somekawa, H.; Mukai, T. Texture and Mechanical Properties of a Superplastically Deformed Mg-Al-Zn Alloy Sheet. *Scripta Mater.* **2009**, *61*, 883–889.

9. Kawasaki, M.; Langdon, T.G. Review: Achieving Superplastic Properties in Ultrafine-Grained Materials at High Temperatures. *J. Mater. Sci.* **2015**, *51*, 19–32. [CrossRef]

10. Alizadeh, R.; Mahmudi, R.; Ngan, A.H.W.; Pereira, P.H.R.; Huang, Y.; Langdon, T.G. Microstructure, Texture, and Superplasticity of a Fine-Grained Mg-Gd-Zr Alloy Processed by Equal-Channel Angular Pressing. *Metall. Mater. Trans. A* **2016**, *47*, 6056–6069. [CrossRef]

11. Kim, Y.S.; Kim, W.J. Microstructure and Superplasticity of the as-Cast Mg-9Al-1Zn Magnesium Alloy after High-Ratio Differential Speed Rolling. *Mater. Sci. Eng. A* **2016**, *677*, 332–339. [CrossRef]

12. *Glossary of Terms Used in Metallic Superplastic Materials*; JIS H 7007; Japanese Standards Association: Tokyo, Japan, 1995.

13. Nie, J.F.; Muddle, B.C. Characterisation of Strengthening Precipitate Phases in a Mg-Y-Nd Alloy. *Acta Mater.* **2000**, *48*, 1691–1703. [CrossRef]

14. Antion, C.; Donnadieu, P.; Perrard, F.; Deschamps, A.; Tassin, C.; Pisch, A. Hardening Precipitation in a Mg-4Y-3RE Alloy. *Acta Mater.* **2003**, *51*, 5335–5348. [CrossRef]

15. Kiełbus, A. The Influence of Ageing on Structure and Mechanical Properties of WE54 Alloy. *J. Achiev. Mater. Manuf. Eng.* **2007**, *23*, 27–30.

16. Liu, H.; Zhu, Y.M.; Wilson, N.C.; Nie, J.F. On the Structure and Role of β'_F in β_1 Precipitation in Mg-Nd Alloys. *Acta Mater.* **2017**, *133*, 408–426. [CrossRef]

17. Ruano, O.A.; Álvarez-Leal, M.; Orozco-Caballero, A.; Carreño, F. Large Elongations in WE54 Magnesium Alloy by Solute-Drag Creep Controlling the Deformation Behavior. *Mater. Sci. Eng. A* **2020**, *791*, 139757. [CrossRef]

18. Orozco-Caballero, A.; Cepeda-Jiménez, C.M.; Hidalgo-Manrique, P.; Rey, P.; Gesto, D.; Verdera, D.; Ruano, O.A.; Carreño, F. Lowering the Temperature for High Strain Rate Superplasticity in an Al-Mg-Zn-Cu Alloy via Cooled Friction Stir Processing. *Mater. Chem. Phys.* **2013**, *142*, 182–185. [CrossRef]

19. Orozco-Caballero, A.; Hidalgo-Manrique, P.; Cepeda-Jiménez, C.M.; Rey, P.; Verdera, D.; Ruano, O.A.; Carreño, F. Strategy for Severe Friction Stir Processing to Obtain Acute Grain Refinement of an Al-Zn-Mg-Cu Alloy in Three Initial Precipitation States. *Mater. Charact.* **2016**, *112*, 197–205. [CrossRef]

20. Orozco-Caballero, A.; Ruano, O.A.; Rauch, E.F.; Carreño, F. Severe Friction stir Processing of an Al-Zn-Mg-Cu Alloy: Misorientation and Its Influence on Superplasticity. *Mater. Design* **2018**, *137*, 128–139. [CrossRef]

21. Chiang, C.R. The Grain Size Effect on the Flow Stress of Polycrystals. *Scripta Metall.* **1985**, *19*, 1281–1283. [CrossRef]

22. Taleff, E.M.; Henshall, G.A.; Nieh, T.G.; Lesuer, D.R.; Wadsworth, J. Warm-Temperature Tensile Ductility in Al-Mg Alloys. *Metall. Mater. Trans. A* **1998**, *29*, 1081–1091.

23. Ruano, O.A.; Miller, A.K.; Sherby, O.D. The Influence of Pipe Diffusion on the Creep of Fine-Grained Materials. *Mater. Sci. Eng.* **1981**, *51*, 9–16. [CrossRef]

24. Langdon, T.G. The Mechanical Properties of Superplastic Materials. *Mater. Trans. A* **1982**, *13*, 689–701. [CrossRef]

25. Ruano, O.A.; Sherby, O.D. On Constitutive Equations for Various Diffusion-Controlled Creep Mechanisms. *Revue Phys. Appl.* **1988**, *23*, 625–637. [CrossRef]

26. Vavra, T.; Minarik, P.; Vesely, J.; Kral, R. Excellent Superplastic Properties Achieved in Mg-4Y-3RE Alloy in High Strain Rate Regime. *Mater. Sci. Eng. A* **2020**, *784*, 139314. [CrossRef]

27. Shewmon, P.G.; Rhines, F.N. Rate of Self-Diffusion in Polycrystalline Magnesium. *JOM* **1954**, *6*, 1021–1025. [CrossRef]

28. Frost, H.J.; Ashby, M.F. *Deformation-Mechanism Maps*, 1st ed.; Pergamon Press: Oxford, UK, 1982; p. 44.

29. Zhou, M.; Morisada, Y.; Fujii, H.; Wang, J.-Y. Pronounced Low-Temperature Superplasticity of Friction Stir Processed Mg-9Li-1Zn Alloy. *Mater. Sci. Eng. A* **2020**, *780*, 139071. [CrossRef]

Publisher's Note: MDPI stays neutral with regard to jurisdictional claims in published maps and institutional affiliations.

Review

Review on Dynamic Recrystallization of Martensitic Stainless Steels during Hot Deformation: Part I—Experimental Study

Hamed Aghajani Derazkola [1],*, Eduardo García Gil [1], Alberto Murillo-Marrodán [1] and Damien Méresse [2]

[1] Department of Mechanics, Design and Industrial Management, University of Deusto, 48007 Bilbao, Spain;
e.garcia@deusto.es (E.G.G.); alberto.murillo@deusto.es (A.M.-M.)

[2] LAMIH UMR CNRS 8201, Université Polytechnique Hauts-de-France, CEDEX 9,
F-59313 Valenciennes, France; dmeresse@uphf.fr

* Correspondence: h.aghajani@deusto.es

Abstract: The evolution of the microstructure changes during hot deformation of high-chromium content of stainless steels (martensitic stainless steels) is reviewed. The microstructural changes taking place under high-temperature conditions and the associated mechanical behaviors are presented. During the continuous dynamic recrystallization (cDRX), the new grains nucleate and growth in materials with high stacking fault energies (SFE). On the other hand, new ultrafine grains could be produced in stainless steel material irrespective of the SFE employing high deformation and temperatures. The gradual transformation results from the dislocation of sub-boundaries created at low strains into ultrafine grains with high angle boundaries at large strains. There is limited information about flow stress and monitoring microstructure changes during the hot forming of martensitic stainless steels. For this reason, continuous dynamic recrystallization (cDRX) is still not entirely understood for these types of metals. Recent studies of the deformation behavior of martensitic stainless steels under thermomechanical conditions investigated the relationship between the microstructural changes and mechanical properties. In this review, grain formation under thermomechanical conditions and dynamic recrystallization behavior of this type of steel during the deformation phase is discussed.

Keywords: martensitic stainless steels; microstructure changes; hot deformation

Citation: Derazkola, H.A.; García Gil, E.; Murillo-Marrodán, A.; Méresse, D. Review on Dynamic Recrystallization of Martensitic Stainless Steels during Hot Deformation: Part I—Experimental Study. *Metals* **2021**, *11*, 572. https://doi.org/10.3390/met11040572

Academic Editor: Stefano Spigarelli

Received: 22 February 2021
Accepted: 25 March 2021
Published: 1 April 2021

Publisher's Note: MDPI stays neutral with regard to jurisdictional claims in published maps and institutional affiliations.

1. Introduction

Martensitic stainless steels (MSS) are a family of alloys composed mainly of iron, chromium (Cr), and carbon (c) as main elements [1]. MSS have a martensitic crystal structure in the hardened condition [2]. They are ferromagnetic, treatable and have better corrosion resistance comparing to other types of stainless steels (SS). Chromium (Cr) in these steels is usually 10.5–18% weight percentage (wt %) with a higher percentage of carbon (C) in comparison to ferritic steels [3–5]. The specific amount of C and Cr is used to ensure the formation of a martensitic structure after a complete cycle of heat treatment [6]. In MSS, an adequate Cr amount is needed to provide corrosion resistance, which is achieved by the formation of a chromium oxide film on the surface [7,8]. The steel requires at least 11 wt % Cr to have stainless characteristics. On the other hand, the austenite phase cannot be stable in steels with more than 10.5% Cr. For this reason, carbon, nitrogen, nickel, and manganese are added to MSS as austenite stabilizers. In addition, some other elements like molybdenum (Mo) and tungsten (W) are added to carbide for strengthening the MSS [9]. In such cases, Mo_2C and W_2C can improve the strength of MSS. Elements such as titanium and aluminum are added to the low carbon MSS to promote the precipitation of intermetallics (IMC), such as NiTi and NiAl [10].

In addition, the high carbon content favors the formation of some carbides with other elements like niobium, silicon, tungsten, and vanadium, which increase the wear resistance of MSS. In some cases, small amounts of nickel are added to MSS to improve corrosion

113

resistance and toughness [9]. During MSS production, the temperatures in the furnace are high enough to transform the steel structure into austenite. After cooling treatment in air, the microstructure of MSS will transform allotropically into the martensite phase [11]. With different heat treatments, such as quenching and tempering, it is possible to achieve a wide range of strength (between 275–1900 MPa), wear and corrosion resistance [12]. In 1913, Harry Brearley invented the first class of MSS that called 410 grades. The 410 is the primary grade of MSS commercialized and standardized in the 1930s and 1940s [13]. MSS are used in various applications like surgical instruments, springs, valves, shafts, bearings, turbine blades, and petrochemical tools [14]. The 410 stainless steel (SS) has evolved through the years by adding chemical elements to enhance its use in specific applications [15]. Figure 1 illustrates a schematic view of MSS grades with additional elements compared with 410.

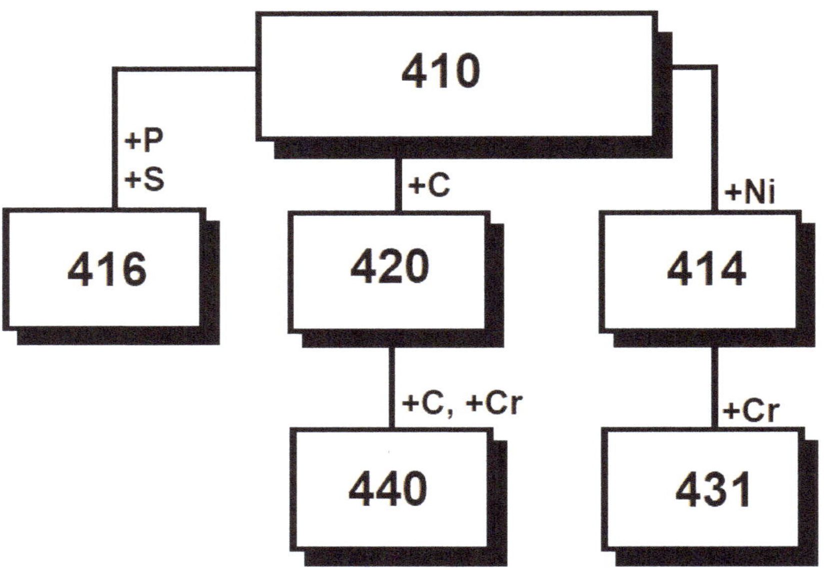

Figure 1. Development of other grades of martensitic stainless steels (MSS) from the 410 [11].

In addition, the chemical composition of carbide strengthened MSS is presented in Table 1. The steels in Table 1 are divided into three groups. The first group is the low to medium C (0.10 ± 0.30 wt %) alloys with approximately 12 wt % of Cr. They are used in creep resistance applications and also in steel components that require ductile fracture resistance at high temperatures. The second group is those containing higher levels of C (0.60–1.2 wt %) and rather large amounts of Cr (16–18 wt %). Higher content of C and Cr is added in this group to achieve higher hardness and promotion of wear resistance. The third group is the high chromium cold work die steels (D2 and D7). The high Cr content and cold working on these steels improve wear and corrosion resistance simultaneously. As mentioned before, the MSS is usually used in components where corrosion and oxidation resistance with high strength are required [16]. From a metallurgical point of view, three types of MSS exist, and their behavior as traditional (first and second types), new or uncategorized (third type) MSS could be studied [17]. The first type contains specific wt % carbon (as reported in Table 1) and is strengthened by iron carbide precipitation when tempered at low temperatures or by alloy carbide precipitation when tempered at higher temperatures (secondary hardening), as discussed before. The second type contains a lower amount of carbon and is strengthened by the precipitation of particles of copper or intermetallic on tempering [18]. Another type is the precipitation-strengthen MSS. Precipitation strengthened MSS are low-C steels, strengthened by the precipitation of second-phase particles, other than alloy carbides, during thermal treatment. Pure copper particles, NiAl, NiTi, Ni_3Ti and Ni_3Be, are well-known precipitates in this MSS type [18]. Table 2 presents some grades of precipitation strengthened MSS.

Table 1. Chemical composition of various grades of MSS.

Grade	C	Mn	Si	Cr	Mo	W	P	V	S	Ref.
410	0.15	1.0	0.5	11.5–13	-	-	0.04	-	0.03	[19]
416	0.15	1.25	1.0	12.0–14.0	0.6	-	0.04	-	0.15	[20]
420	0.15–0.4	1.0	1.0	12.0–14.0	-	-	0.04	-	0.03	[21,22]
422	0.23	1.0	0.75	12.0	1.0	1.0	0.04	0.22	-	[23]
431	0.2	1.0	1.0	15.0–17.0	1.25–2.00	-	0.04	-	0.03	[24]
440A	0.60–0.75	1.0	1.0	16.0–18.0	0.75	-	0.04	-	0.03	[25]
440B	0.75–0.95	1.0	1.0	16.0–18.0	0.75	-	0.04	-	0.03	[26]
440C	0.95–1.20	1.0	1.0	16.0–18.0	0.75	-	0.04	-	0.03	[27]
D2	1.50	0.3	0.25	12	1.0	-	-	1.0	-	[28]
D7	2.35	0.4	0.4	12	1.0	-	-	4.0	-	[29]
8Cr13MoV	0.775	0.458	0.333	14.68	0.213	-	0.031	0.182	0.004	[30]

Table 2. Chemical composition of precipitation MSS grades.

Grade	C	Co	Cu	Cr	Si	Mo	Ni	Al	other	Ref.
PH15-5	0.04	-	3.0	15	-	-	4.7	-	0.20 Nb	[29,31]
PH7-17	0.04	-	-	17.06	0.385	-	7.2	1.06	0.674 Mn	[32]
Custom 450	0.04	-	1.5	11.5	-	-	8.5	-	0.7 Nb	[33]
PH17-4	0.07	-	3.5	16.5	-	-	4.0	-	0.3 Nb + Ta	[34,35]
PH13-8	0.03	-	-	12.6	-	1.7	7.9	1.0	-	[36]
Custom 465	0.02	-	-	11.8	-	1.0	11.0	-	1.7	[37]
Custom 455	0.05	-	2.0	11.5	-	0.5	8.5	-	1.1	[38]
SM2Mo	0.02	-	-	12.59	0.42	1.90	5.01		0.0062	[36,39,40]
SM2MoNb	0.022	-	-	12.91	0.41	2.05	5.16		0.0043	[41–43]
Pyromet X-15	0.01	20	-	15	-	2.9	-	-	-	[11]
Pyromet X-23	0.03	10	-	10	-	5.5	7.0	-	-	[11]
2Cr13	0.18	-	-	12.86	0.39	-	0.12	-	0.53Mn + 0.002S	[44]
00Cr13Ni5Mo2	0.013	-	-	12.97	0.18	2.04	4.92	-	0.59 Mn + 0.001S	[44]
18Cr-5Ni-4Cu-N	0.06	-	4.13	17.63	0.86	2.11	5.55	-	0.08 Nb + 0.13 V + 0.41 N + 0.025 P	[45]

The third type corresponds to those MSS, which are strengthened by the precipitation of both alloy carbides and intermetallic. The chemical composition of this group (3rd type) involves the combined precipitation of NiAl and alloy carbides (secondary hardening). This type has standard high chromium content and alloying combinations that allow quenching from the austenitizing temperature to obtain an almost completely martensitic structure with a small amount of retained austenite [18]. Table 3 presents the chemical composition of 3rd group samples.

Table 3. Chemical composition of intermetallic-carbide-strengthened MSS grades.

Grade	C	Cr	Ni	Mo	Co	V	Nb	Ref.
AFC77	0.15	14.5	-	5	13.5	-	-	[46]
AFC260	0.08	15.5	2.0	4.3	13.0	-	-	[13]
HSL 180	0.20	12.5	1.0	2.0	15.5	-	-	[11]
CSS-42L	0.13	13.8	2.1	4.7	12.5	0.60	0.04	[11]

Metal forming (metalworking) consists of deformation processes in which a metal billet or blank is shaped by the action of tools or dies. The design and control of such processes or final products depend on the base material's characteristics, the conditions at the tool/workpiece interface, the mechanics of plastic deformation (metal flow), the equipment used, and the finished product requirements [47]. Among various metals and alloys, stainless steel is favorable for industries due to its excellent properties and reasonable price. Hot metalworking is one of the major technologies used to produce stainless steel products. In this process, the stainless steel as raw material is in billet, rod, or slab form, and the surface-to-volume ratio in the formed part increases considerably under the action of primarily compressive loading [48,49]. Rolling, skew mills, ring rolling, forging, draw benches for tube and rod, and extrusion are samples of manufacturing processes that form stainless steels in hot condition. Among metallurgical changes of metallic materials, dynamic recrystallization (DRX) is very crucial to the microstructural evolution during hot deformation of alloys with low, medium, and high stacking fault energy (SFE) [50,51]. This phenomenon changes the microstructure substantially through nucleation and growth of new strain-free grains at the expense of pre-existing ones [52,53]. The DRX is of critical importance from the industrial viewpoint because it reduces the stored strain energy, and the corresponding softening effect decreases the required force to shape a workpiece. Taking advantage of DRX to decrease the required energy for shaping a product and avoiding plastic instabilities or premature fracture is quite a practical way in different industrial manufacturing processes [6]. In addition to this, DRX often contributes to improving microstructure, thereby refining the grain structure and improving the final product's mechanical properties. For a careful design of the hot working process, having in-depth information about a material's dynamic recrystallization behavior is essential [54]. There is sufficient literature about DRX phenomena in stainless steel during hot forming, but there is not comprehensive literature or review about DRX in martensitic stainless steels during how working. Based on the mentioned underlying phenomena, this review article's aim is a deep understanding of DRX in MSS from an experimental point of view.

2. Hot Formability of MSS

Generally, hot working refers to the temperature of mechanical processing that is above half of the melting temperature (Tm) of base metal. Being above 0.5 Tm decreases yield strength and hardness and increases the ductility of metallic materials [55]. From a metallurgical point of view, the plastic deformation above the metal's recrystallization temperature is called hot working (forming) [56].

From the manufacturing processes point of view, metal forming is divided into two main groups: sheet metal forming and bulk metal forming. One of the main parameters to find specific material that the specific process can form is workability or formability test. There are several kinds of formability tests in sheet metal forming. Uniaxial and multiaxial tensile tests in various temperatures are a well-known approach to find the formability of sheet metals. The formability limitation, mechanical properties, and prediction of fracture during sheet metal forming are accessible by formability test results.

On the other hand, in bulk metal forming, like forging, extrusion, coining, rolling, and stretching, another type of workability test is implemented—this type of test is called a hot compression test. The dimension of raw materials and applied stress concept informing process are the main results that caused researchers to use hot compression test in the laboratory. The compression stress for the flow of materials is the primary stress used in bulk metals forming like hydrostatic forging, orbital forging, skew rolling, tube forming, extrusion, and even solid-state additive manufacturing hot conditions. Some research has been done to develop new workability tests for materials in bulk scale, but so far, the primary test for formability (workability) of metallic materials, especially steel, is the hot compression test.

Various forming processes of metals, like forging, rolling, drawing, and extrusion, can be done in hot conditions [57]. In order to understand the behavior in hot conditions of metallic materials, such as MSS, hot compression tests are usually carried out to characterize hot workability (formability). This test consists of isothermal tests in a wide range of temperature and strain rates. This experiment is pervasive and well-known in laboratories and industry [58]. The sample is heated up to the desired temperature (near the actual temperature of considered hot forming). It remains at the temperature for a specific period, and after this, the compression test is carried out on the sample while the high temperature is maintained. At the final stage, the piece is quenched to ambient temperature as a real situation in the industry [59].

The temperature, duration time of heating, and heat treatment could vary depending on both the chemical composition of the steel and the forming process (various strain rates) studied. For the test, a cylindrical specimen with a specific section and height is machined from the as-received bar. First, the sample is heated using a constant heating rate until the highest temperature of the test campaign is reached. This temperature is then maintained for a certain time to ensure homogeneity and, after this period, it is cooled down to the test temperature. Then the isothermal compression test is carried out at various temperatures and strain rates. There are standards for testing procedures like ASTM-E209 (standard practice for compression tests of metallic materials) that can be used [60]. The result is the flow stress behavior of the material under different plastic deformation conditions. Figure 2 shows an example of a hot compression test setup that was employed for various MSS. As seen, the heating rate, heating period, and deformation time (strain rate) are different.

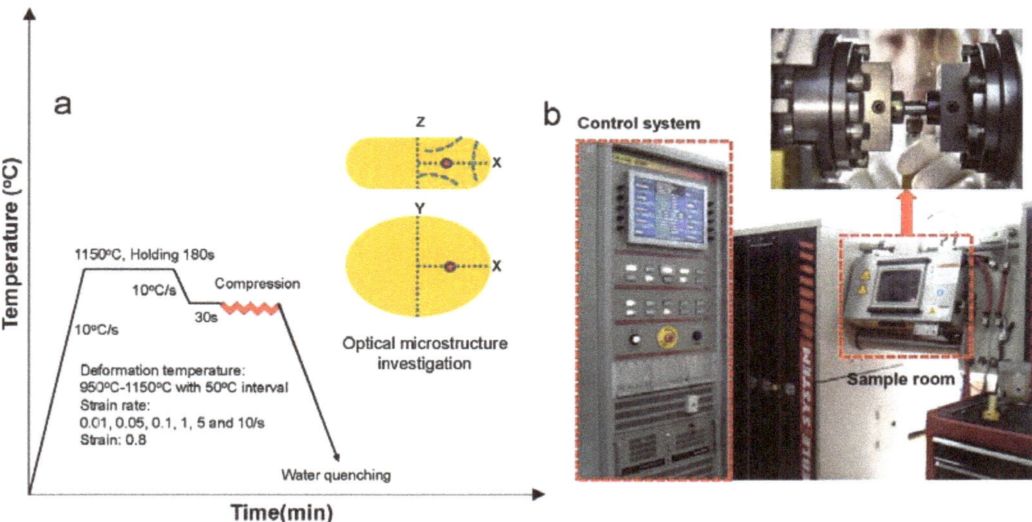

Figure 2. (**a**) Schematic representation of hot compression test and (**b**) illustration of the testing machine [61].

3. Dynamic Recrystallization of MSS under Hot Deformation

The word recrystallization usually refers to the replacement of a deformation microstructure by new grains [62,63]. The process of recrystallization during heat treatment is called static recrystallization (SRX) [64,65]. SRX (primary recrystallization) is one of the most studied phenomena, which refers to the nucleation and growth of new grains during heat treatment [40,66]. The schematic view of this process is depicted in Figure 3.

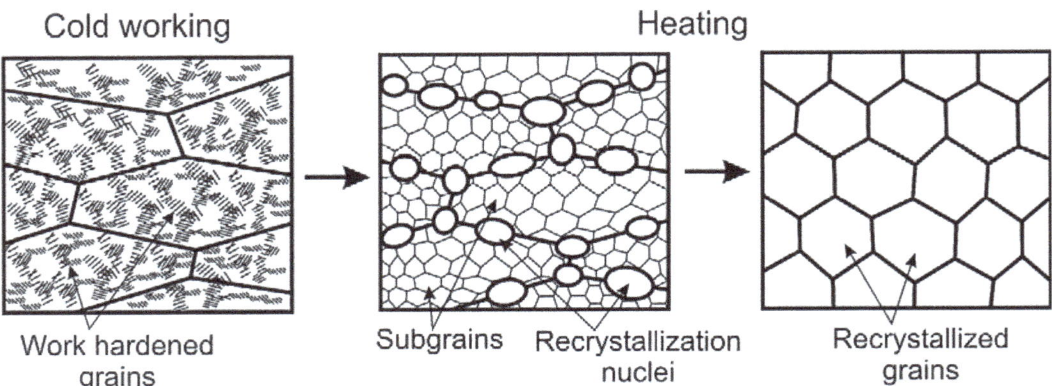

Figure 3. Schematic representation of the discontinuous static recrystallization (dSRX) taking place during the annealing of strain hardened materials [67].

During the early stages of heat treatment, static recovery (SRV) develops recrystallization nuclei as fine dislocation-free crystallites. In the early stages of heat treatment, the development of recrystallization nuclei development is started. In this stage, the recrystallization nuclei are fine dislocation-free crystallites. The nuclei grow by the distant migration of the boundaries, which consume the strain-hardened microstructure. In the early-stages of SRX, the nuclei are outlined by low angle boundaries (LAGBs), the misorientations, of which gradually increase until they attain values typical of high angle boundaries (HABs) [68,69]. During heat treatment, the microstructure is a mixture of recrystallized and strain hardened grains. This type of microstructure is known as discontinuous static recrystallization (dSRX) [70]. Alongside, the sub-boundary misorientation (θ) increases gradually till all the low-angle boundaries (LABs) are transformed into HABs, and this type of microstructure remains homogeneous all over. This structure is referred to as in situ or continuous static recrystallization (cSRX). SRX occurs to strain hardened metals heated above approximately half of their melting point (~0.5 Tm). During hot forming processes, the material is heated up before the process, and after its deformation, it tolerates another thermal cycle as heat treatment. For this reason, a billet could have partial SRX in its microstructure before starting the hot process [70].

At high strains and temperatures above about 0.5 Tm (as a hot deformation requirement), the new grains appear at the nucleation strain and then completely replace the initial microstructure. This mechanism is called dynamic recrystallization (DRX). DRX phenomena are acknowledged as an important feature to restore the ductility of those materials, which are being work hardened during the deformation process. DRX allows large deformation in the material without crack or damage appearance [71,72].

Typically, DRX is favored by high temperatures and low strain rates and is also acknowledged as a grain refinement mechanism under most forming conditions. A new grain structure appears after DRX by the formation and migration of high-angle grain boundaries, promoting the grain refinement of the deformed alloy [67,68]. The DRX has three main subgroups that consist of discontinuous dynamic recrystallization (dDRX), continuous dynamic recrystallization (cDRX) and geometric dynamic recrystallization (gDRX) [54,73–76]. The dDRX has two main phases, namely, nucleation and growth stages [77]. The dDRX occurs by means of bulging mechanisms. Increasing deformation leads to fluctuations in the grain boundary, leading to the formation of serrations and bulges that eventually transform into new strain-free grains [78–80]. It consists of grain bulging, dislocation rearrangement, and boundary movement [81].

During plastic deformation, the dislocations glide and climb to form sub-grains by dynamic recovery (DRV) [82]. Dislocation motions prepare new grains nucleation. In this phase, the bulges without or with a few dislocations are surrounded by accumulated dislocations. After that, the bulges separate to generate nuclei in the early stages of dDRX [83]. Nuclei are high-energy sites that pin the movement of the dislocations. In the final stages of deformation, the nuclei grow until original grains are replaced by finer recrystallized grains. There is a slow transformation of the sub-grains mainly formed in the vicinity of the boundaries into nuclei delineated by HABs and referred to as discontinuous dynamic recrystallization (dDRX) [84]. The dDRX is rarely reported in high chromium MSS during hot deformation. This type of recrystallization (dDRX) is common in low and medium stacking-fault energy (SFE) steels, and MSS are in the category of high SFE materials [85]. The schematic view of dDRX is depicted in Figure 4.

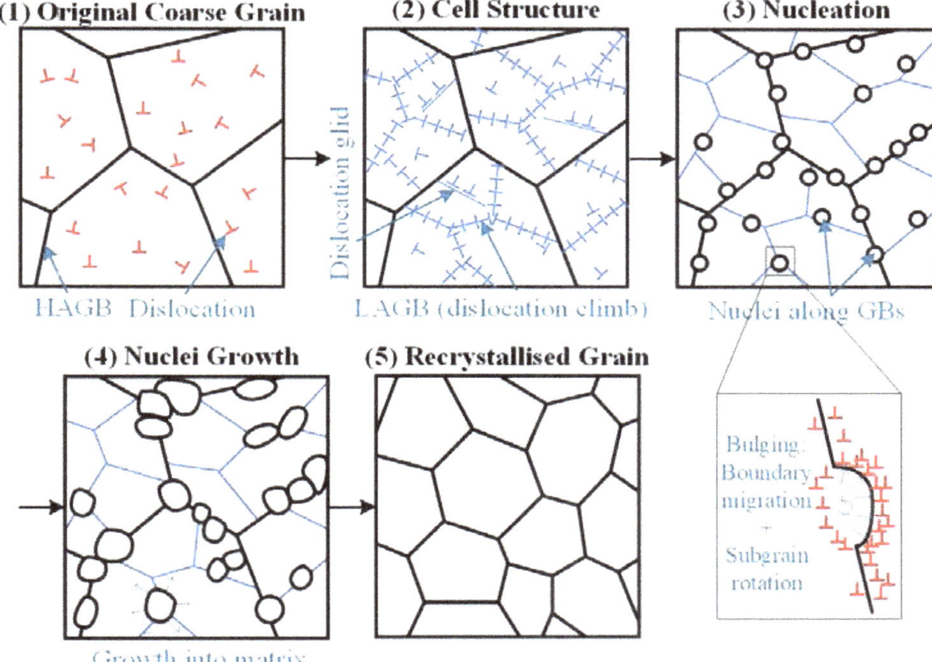

Figure 4. Schematic illustration from the development of new grains during dDRX [86].

The cDRX directly changes the orientation of sub-grains and forms new grains [87,88]. The cDRX steps consist of cell structure formation, sub-grain rotation and increasing of the misorientation, and conversion of grains from LAGB to HAGB [89]. During cDRX, dislocations are generated and rearranged to form LAGB sub-grains boundaries and newborn grains surrounded by HAGB. The cDRX is very common in high stacking-fault energy (SFE) materials like MSS.

In gDRX, during hot deformation with large deformation, substantial grain refinement occurs via grain elongation and thinning. Grain elongation and thinning increase the grain boundary area dramatically [90–93]. The grain boundaries become serrated as a result of LAGB (sub-grain) formation [94]. This behavior compresses grains serrations on opposite sides that lead the grains to collapse each other, causing grain fragmentation [95]. Both gDRX and the cDRX have the same features, which are continuous growth of HAGB area and absence of nuclei [96]. At the same time, gDRX usually occurs due to large deformations when the grains are extremely elongated and thinned by grain migration [97,98]. The schematic view of cDRX and gDRX is shown in Figure 5a,b, respectively.

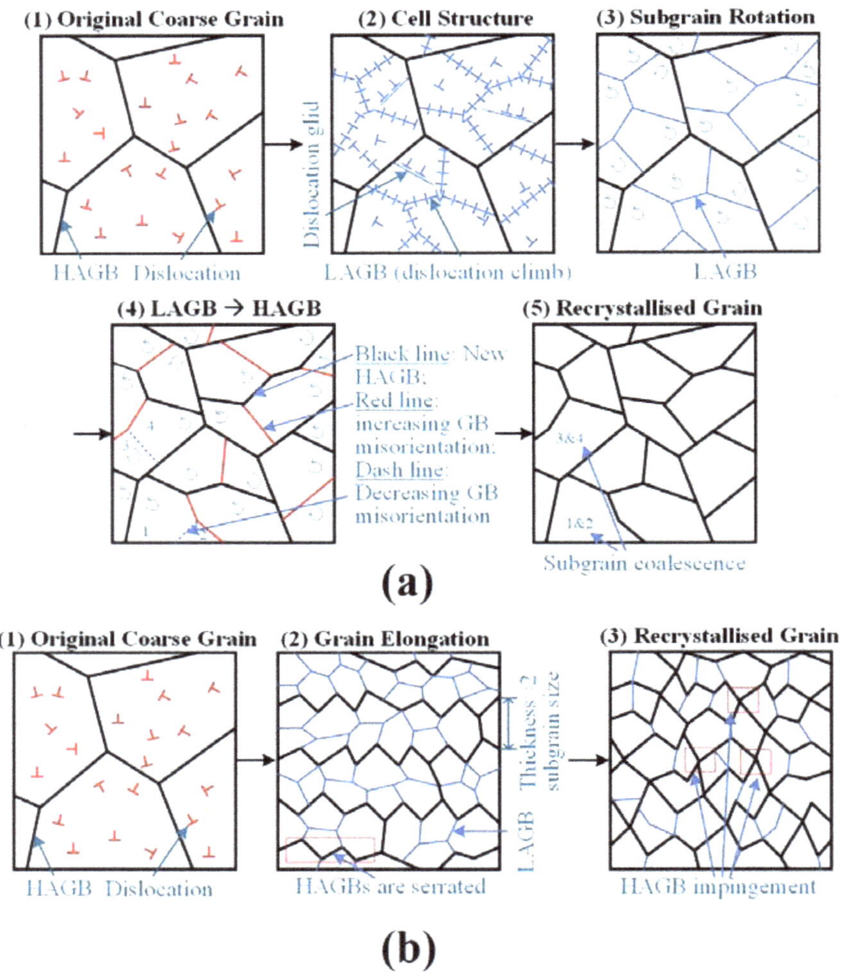

Figure 5. Schematic illustration from the development of new grains during (**a**) cDRX and (**b**) geometric dynamic recrystallization (gDRX) [86].

From a metallurgical aspect, the hot formability of an MSS is generally limited by the generation of various deformation defects [99]. At the same time, Dynamic recovery (DRV) and dynamic recrystallization (DRX) impede the accumulation of defects. Thus, the DRX and DRV are the softening mechanisms during hot forming that allow the formability of MSS and prevent defect formation [4,67]. Only a few small dynamic recrystallization grains are distributed along the grain boundary, whereas most deformed microstructures usually are composed of dynamically recovered grains. DRV is a unique flow softening mechanism that plays a softening role in counter-balance of the work hardening. As mentioned, the microstructure behavior is related to the temperature and strain rate. At high-temperature deformation of MSS, the necklace structure is formed by fine grains originated in the recrystallization at the elongated grain boundaries. It is reported that, at low-temperature and high-strain rate (in adiabatic deformation situation of MSS), different shear forces are created in diverse parts of the sample [100]. In such case, the dislocations are stuck on the interface of grain boundaries, and inhomogeneous grain distributions are developed [101]. Inhomogeneous grain distributions lead to internal shear force, which causes a flow localization band during hot deformation. The flow localization is an induced instability, which is undesirable for mechanical properties [102–104]. Figure 6 shows the different microstructure of 14% Cr at various temperatures and strain rates in adiabatic hot deformation situations. The microstructure evolution mechanism of 14% Cr MSS at different strain rates and temperatures during hot compression deformation can be divided into six regions. Partial DRX represents the region in which the DRX occurs partially, and full DRX represents the region in which the DRX occurs entirely. Flow localization occurs in the unstable region, and the grains are elongated in the DRV region [105]. During the hot compression test, when the strain rate decreases and the temperature increases, the MSS is placed in an unstable region. From a formability point of view, the undesirable microstructure forms at unstable regions due to the plastic flow localization. The plastic flow localization occurs when during hot compression of MSS, the adiabatic temperature rises at low-temperature and strain rate increases. This type of temperature strain rising in the unstable area causes different shear forces in diverse parts of the specimen. In this case, When the deformation temperature rises, the average grains become coarse after hot compression. It indicates that the power dissipation of microstructure evolution is dominated by recrystallization grains during hot deformation.

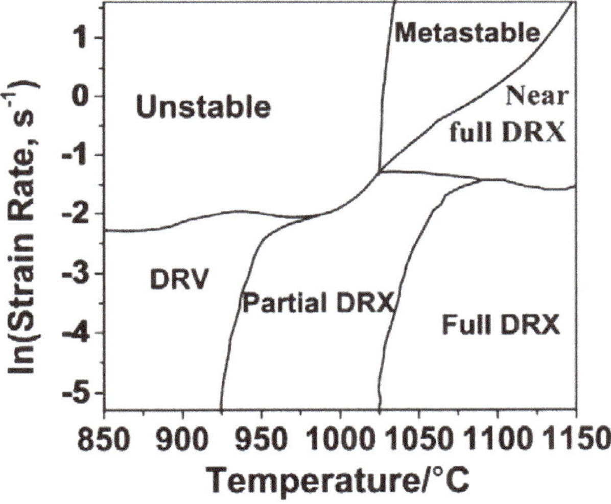

Figure 6. The microstructure evolution mechanism of 14% Cr MSS during deformation conditions [105].

Some research has been done to monitor cDRX with imaging of microstructure at various strains [32]. It is shown that the original microstructure of the 17-7PH MSS is austenite matrix (γ) with the coaxial ferrite phase and δ ferrite. Figure 7 shows the microstructure of deformed 17-7PH MSS samples at 950 °C and strain rate of 0.01 s⁻¹ at strains of 0.15, 0.3, 0.45, and 0.6. It is shown that the γ grains and δ islands have almost remained equiaxed at low strains (0.15). In the middle of this sample, work hardening region as the serrated grain boundaries detected that indicate the progress of DRV in the substructure. The appearance of serrated grain boundaries at low strains is a sign of fast DRV in the 17–7PH MSS. Some small grains, presumably due to DRX, are appeared around the original grain boundaries and the necklace structure in the early stages of recrystallization not observed.

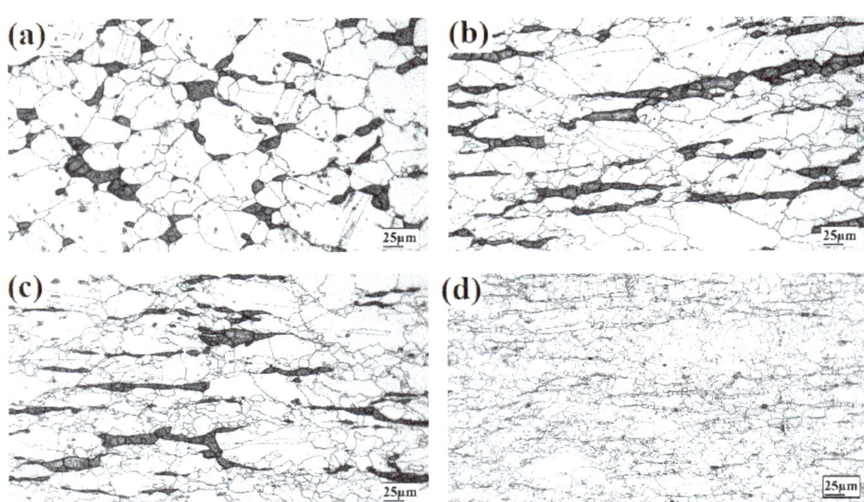

Figure 7. Microstructure of deformed samples at 950 °C and strain rate of 0.01 s^{-1} to strain of (**a**) 0.15, (**b**) 0.3, (**c**) 0.45 and (**d**) 0.6 [106].

The absence of necklace structure indicates that the probability of dDRX, which occurs through the nucleation and growth of new grains almost at prior grain boundaries, is very weak in this steel [106]. The number of new cDRX grain has increased as deformation continues, and the matric and δ islands are elongated perpendicular to the compression direction. The new grains are observed inside the elongated original grains and along the grain boundaries. With increasing strain, decomposition and combination of the subgrain boundaries could lead to the formation of HABs, which are the result of the rapid growth of the sub-grains during hot plastic deformation [107,108]. As the deformation continues and the strain increases to the number of recrystallized grains increases again. At high strain, the dissociation of original grains by the cDRX grains has progressed, and the fraction recrystallization increased. In this alloy, some islands of original deformed grains were detected that revealed that the microstructure of a PH7-17 has not been fully recrystallized even at high strain during the hot compression test. In the same case, if the strain rate increased, the progress of DRV decreased at low strain, which is responsible for the formation of serrations on the grain boundaries. The reduction of deformation time due to the increase in the strain rate has probably led to a limitation of diffusion and the difficulty of dislocation's motion. The weakness of DRV at a higher strain rate possibly retards the evolution of sub-grains and, therefore, postpones cDRX [106].

Figure 8 shows the microstructure of 12% Cr MSS after 0%, 25% and 50% hot compression [57]. The original austenite in base metal has disappeared, while there is a high number of small sub-grains formed with an increase of deformation. The gDRX has been observed at elevated temperatures where sub-grain boundary formation has been observed. The gDRX is detected during hot deformation at a very high strain [81,95,109]. After gDRX, the initial grains of the sample are fragmented into the new smaller grains with HABs. With gDRX, the HAB area can dramatically increase at high stacking fault energy (SFE) metals, but not in the same way as cDRX [110]. The cDRX includes the sub-grains boundaries' misorientation. The misorientation of sub-grains by cDRX is just a few degrees and much lower than HABs. The sub-grain boundaries are mobile and annihilate with other boundaries decreasing the sub-grain size throughout large strain hot deformation [111–113]. In cDRX, sub-grains transform into new grains within the deformed original grains. When the dislocations accumulate progressively in LAGBs, the formation of HAGBs proliferates, leading, consequently, to an increase of misorientation [112]. Deformation conditions like temperature and strain rate determine the number of sub-grains and stored energy after hot forming.

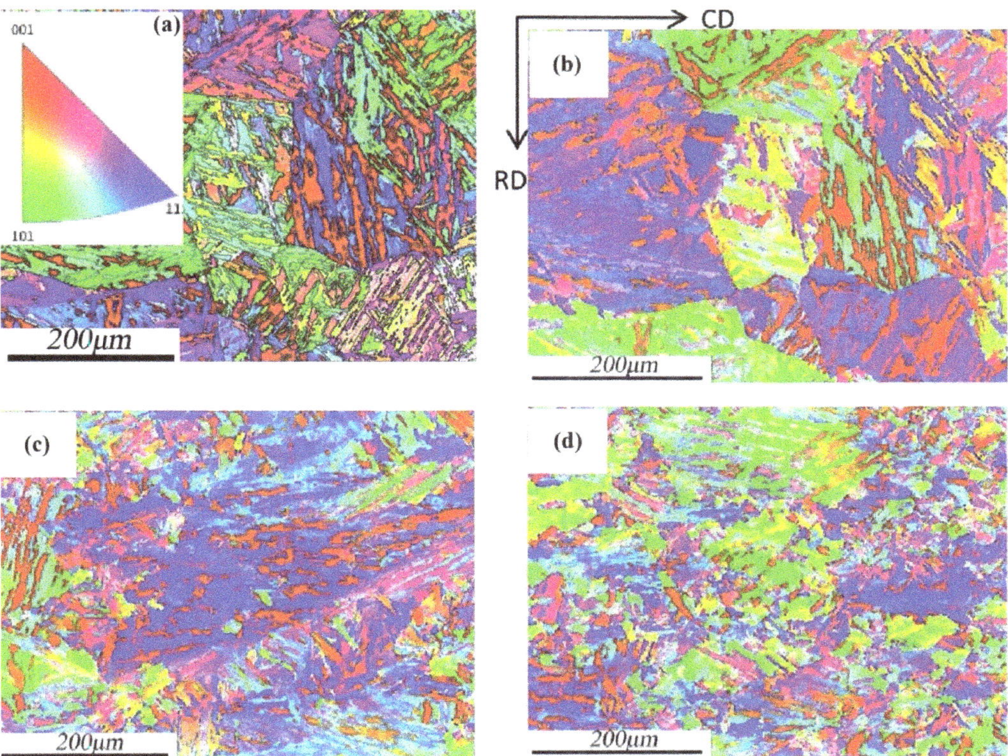

Figure 8. (**a**) Initial microstructure of 12Cr USC rotor steel, microstructure of rotor steel at 1050 °C/0.01 s^{-1} with different deformations (**b**) 0%, (**c**) 26%, (**d**) 50% [57].

In the hot forming of MSS at a low strain rate, DRV is the main softening phenomenon because there is enough time for dislocation to climb and cross-slip [114]. Low-angle grain boundaries are formed by the migration of dislocations and have a lower energy state [115]. Generally, during hot deformation of MSS at low-temperature and high strain rate, only a few DRX grains form along the boundaries, whereas most deformed microstructures are DRV grains. The local bulging of grain boundaries is frequently observed as the initial step to the nucleation of DRX, which is often termed as strain-induced grain boundary migration (SIBM) [116]. From an industrial point of view, DRX is very important since it is a common microstructural phenomenon during the hot forming processes of MSS [117]. For the industrial hot forming process, determining the required strains for the initiation and completion of DRX is necessary. It is almost proved that during hot forming, the build-up of dislocation density reaches a critical value from which grain boundaries start to bulge through SIBM [117]. Quite close to the peak strain (critical strain) of the DRX flow curve, SIBM happens [118]. Locally bulged boundaries sweep away dislocations and cause a drop in the work hardening rate under the control of dynamic recovery [119]. After the critical point, the local bulges in boundaries grow to occupy the prior boundaries and form the necklace structure at the peak [120]. The kinetics of DRX beyond the peak determines the strain required to reach the steady-state flow. It is, therefore, evident that the critical peak and steady-state strains are essential in characterizing the DRX behavior of a particular alloy.

In most MSS, hot deformation is performed in the stable region of the austenite phase, where DRX is the dominant microstructural modifying mechanism [120]. Therefore, it can be a desirable process to take advantage of DRX to refine the microstructure of austenite and to prevent deformation defects prior to the transformation into martensite [121]. The softening due to DRX results in a flow stress decrement after work hardening, leading to a steady-state regime associated with the dynamic balance between work hardening and flow softening [122]. The steady-state flow dominated by the occurrence of DRX prevents the formation of plastic instabilities and results in microstructural modification [106,123,124]. In MSS, the distance between the partial dislocations is short, and for this reason, during hot forming, the cross-slipping of dislocations happens easily. The dislocation migration prompts the annihilation and rearrangement of dislocations, which can decrease the dislocation density, lower the amount of stored energy and eventually retard DRX [115]. This leads to DRV forms during hot forming. Also, at different strains, elongated and recovered grains can be seen in MSS microstructures. Increasing the strain during hot forming is advantageous to drop the size of the average substructure and refine the MSS microstructure. This emanation results in an increase in dislocation density and in the number of LAGBs, which is beneficial to reduce the average substructure size and to refine the microstructure. On the other hand, the dislocation migration can be promoted by increasing the temperature during the deformation process [125]. The microstructure changes during the hot compression test of AISI 420 steel is depicted in Figure 9.

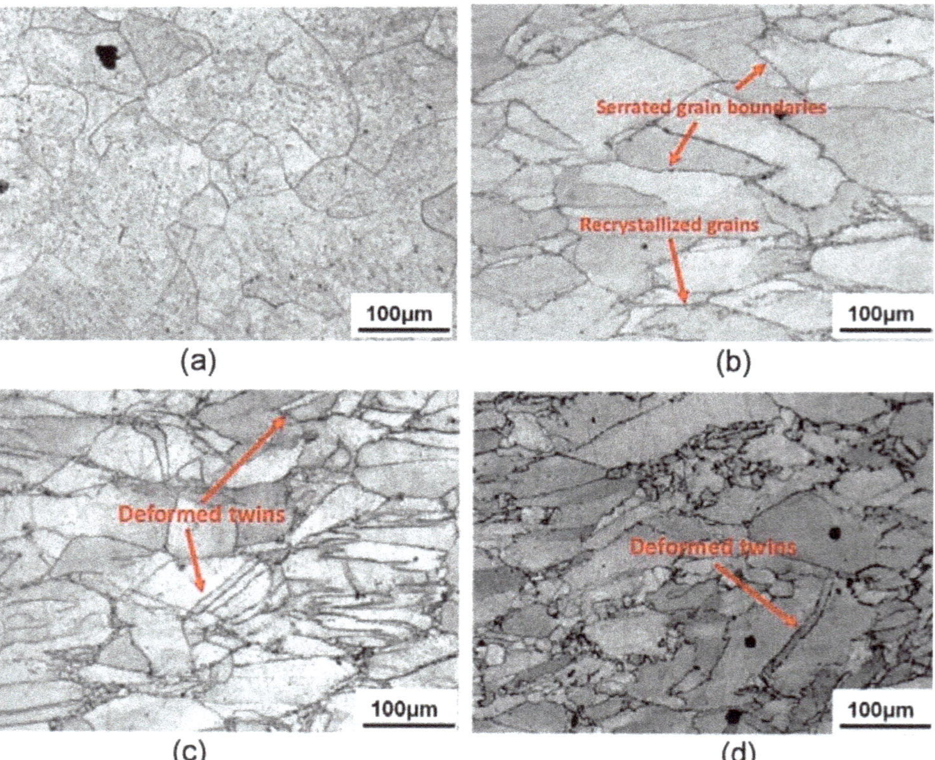

Figure 9. (**a**) Initial microstructure and typical microstructures of AISI 420 martensitic stainless steel deformed at a true strain of 0.7 under different conditions: (**b**) 1223 K, 0.1 s^{-1}, (**c**) 1273 K, 10 s^{-1} and (**d**) 1373 K, 10 s^{-1} [61].

The micrograph of base metal revealed coarse austenite grains with 158 μm size. At the deformation temperature of 1223 K and strain rate of 0.1 s^{-1}, the original coarse austenite grains are elongated, and some sub-grains are formed along boundaries (Figure 9b). Microstructure investigation revealed that the DRX is insufficient and incomplete in high strain rate (10^{-1}), and partial DRX or DRV was detected as the primary deformation mechanism. Some serrated grain boundaries, together with local bulges, were observed on the newly formed boundaries. The reason may be that the grain boundary (or deformation band) has a larger driving force and more nucleation sites for DRX. It seems that the larger density of dislocation in the corresponding regions is the root of this phenomenon. Figures 8d and 9c show the typical microstructures of the deformation twin phenomenon, deformed at the strain rate of 10 s^{-1} and the deformation temperature 1273 K, 1373 K, respectively. The deformation twin phenomenon may be that the grain deformation and grain boundary sliding are inhibited by the low-level dynamic recovery [61].

The obtained microstructure at different thermo-mechanical regimes of stable deformation during hot forming of 410 MSS shows an equiaxed appearance of original austenite grains extracted from the final martensitic structure, which implies the occurrence of DRX. It is very well documented that grain size remains almost unchanged at strains over the onset of steady-state flow. The steady-state grain size, also known as DRX grain size, increases with temperature, decreasing with strain rate. This demonstrates that DRX grain size is almost independent of the initial grain size. Therefore, different models are recommended to express the dependence of DRX grain size to processing parameters incorporated in the Zener–Hollomon parameter (Z). Thus, concluded that DRX grain size depends on the Z parameter and is almost independent of the initial grain size so that the DRX grain size decreases with an increase in the Z [126].

Kishor et al. studied the characteristics of 13Cr-4Ni microstructure during hot deformation [62]. The optical microstructure of ASR 13Cr-4Ni martensitic stainless steels before the deformation is shown in Figure 10a, consisting of a full-lath martensite structure. According to the achieved results, after the hot deformation, all the microstructures predominantly exhibited a martensitic structure for all the deformation conditions. Microstructural evolution after hot deformation was characterized to validate the occurrence of DRX and to confirm the stable and unstable regions. Uniform grain growth and small equiaxed grains along the grain boundaries were observed at the deformation conditions of 1000 °C, 0.001 s^{-1} and 1050 °C, 0.1 s^{-1}, as can be seen in Figures 9c and 10b, respectively. These features demonstrate the occurrence of DRX that was achieved by optimum hot working domains. For 950 °C, 0.1 s^{-1} deformation condition (Figure 10d), the microstructure revealed partially recrystallized features (seen as equiaxed grains) and remaining recovered structure (seen as inhomogeneous distribution of grains). This is representative of the incomplete softening mechanism during 950 °C, 0.1 s^{-1} deformation condition. Microstructure in Figure 10e corresponding to the instability region 900 °C, 10 s^{-1} revealed localized flow structure and shear bands (marked by arrow) due to the absence of steady-state behavior. It also shows refined grains formed due to recrystallization within the bands. According to the processing map, this condition is located in the instability region.

Microalloying of MSS can change the DRX during hot deformation. The standard and well-known element in MSS is niobium (Nb) that has an intense effect on DRX. The microstructure changes of MSS micro-alloy during the hot compression test are shown in Figure 11. The initial microstructure was a completely coarse equiaxed austenitic grain structure. At a low strain rate, the austenite grains were elongated, and DRX was not detected. Many Nb precipitates with lamellar shape appeared alongside the austenite grain boundaries. With increasing temperature, the precipitates along grain boundaries decreased, and a small amount of DRX with fine grains formed in hot deformed areas. It is proved that a high density of dislocations could be produced during the martensitic transformation and distribution of a large number of nanosized precipitates [127].

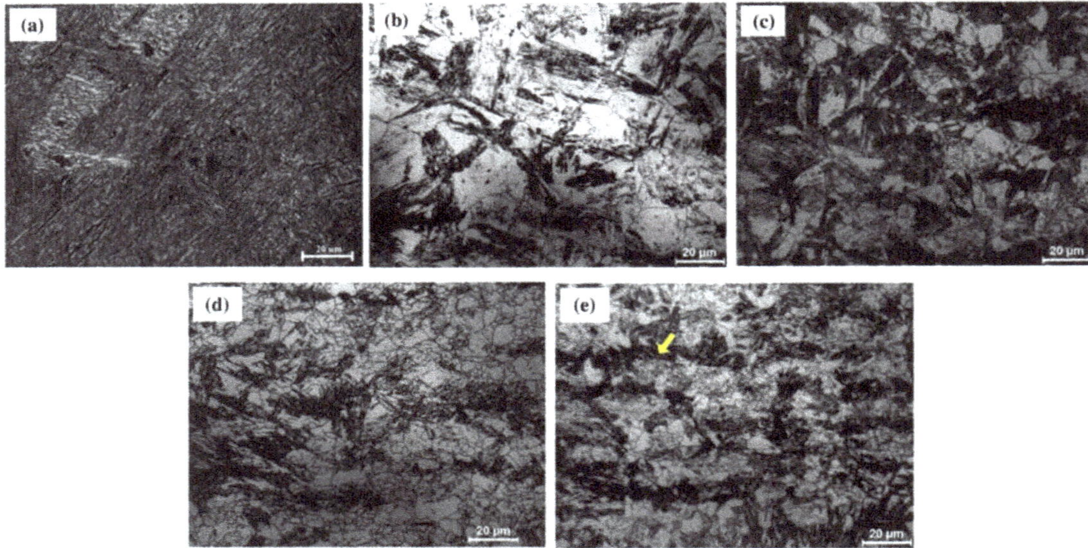

Figure 10. Optical micrographs of (**a**) as-received MSS, and hot deformed specimens exhibiting dynamic recrystallization (DRX) at the deformation conditions of (**b**) 1000 °C, 0.001s^{-1}, (**c**) 1050 °C, 0.1s^{-1}. Incomplete restoration characteristics are observed for deformation condition of (**d**) 950 °C, 0.1s^{-1}. Instability features are observed in (**e**) 900 °C, 10 s^{-1} specimen [62].

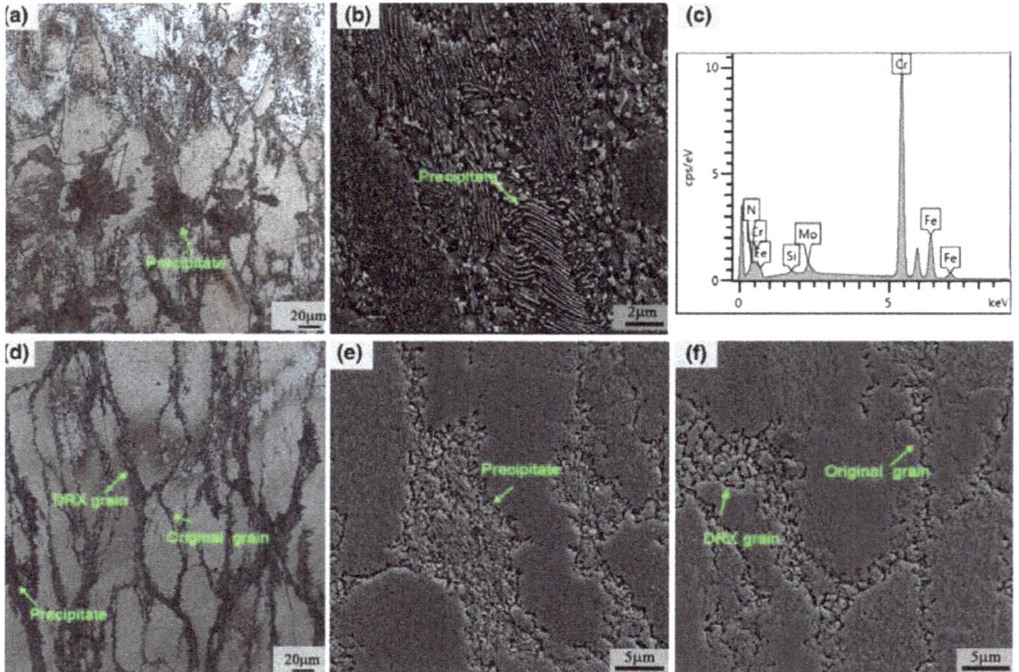

Figure 11. Microstructures of 15% Cr MSS deformed with a strain rate of 0.005 s^{-1} at temperatures of (**a–c**) 1223 K, (**d–f**) 1273 K [127].

Restoration mechanisms in metals during hot work have been one of the main issues in recent decades. In most cases, DRV has been reported as the only restoration mechanism for alloys with high stacking fault energy [128,129]. Besides, in metals and alloys with high SFE, cDRX occurs instead of dDRX. cDRX plays a prominent role in producing finer grains through hot deformation [3–5]. This mechanism mostly occurs in low carbon ferritic steels and even to some extent in austenitic stainless steels as well as MSS [130]. Generally, ferritic microstructures prefer the work softening by an extended DRV, resulting in cDRX at higher applied strains [88,131]. Dynamic restoration leads to an increase in the density of dislocations at LAGBs, thereby transforming them into HAGBs [132]. It accepted that sub-grains might rotate to the point that the adjacent sub-grains reach a similar orientation due to boundary diffusion processes [133]. In this case, already existing LAGBs will be deleted [11]. These sub-grains will combine and convert to larger subgrains. The driving force necessary for this process will be obtained by decreasing the surface area of LAGBs in unit volume [134].

4. Analysis of MSS Flow Stress under Hot Deformation

Metals flow stress graph relates strain, strain rate and temperature. Many mechanical or metallurgical parameters could change the results of flow stress. The addition of alloying elements in MSS could increase the flow stress value and reduce the hot workability. The properties of stress–strain graph are determined by the deformation temperature, strain rate, microstructural, and the final crystallized grain size. Generally, the flow stress decreases with increasing temperature and increases with increasing strain rate at a constant parameter during compression test [77].

During hot deformation of MSS, three types of flow stress graphs have been reported that indicate that there are cDRX and DRV [135]. In the first case, during high-temperature plastic forming, the flow stress graph shows a continuous increase until reaching saturation stress. This behavior, depicted in Figure 11a, indicates a steady state of flow, which is the result of strain hardening decrease during deformation. In these cases, DRV is the main restoration process during hot deformation [136]. This type of graph is usually seen during hot deformation of MSS with low strain rates ($\varepsilon < 1$), and it is justified by the fact that the softening produced by DRV is able to balance the strain hardening rate. In the second case, which is referred to as the dDRX phenomenon, the flow stress graph in high-temperature compression shows a gradual increase until it reaches a stress peak (Figure 12). Then with increasing strain, the flow stress decreases until a steady-state trend. In exceptional cases, some decreasing peaks of heights form before the steady-state behavior. During straining (from starting point and before reaching peak stress), the graph's increase is the result of new grains appearance, which leads to softening due to decreasing the rate of work hardening. After the peak stress, the steady flow stress is the result of the dynamic equilibrium existing between strain softening and strain to harden. This dynamic equilibrium is the result of new grains formation with grain boundary migration. The third type of graph is obtained at high strain rate (5^{-1}) experiments. In these cases, DRV occurs in the substructures at the early stage of straining and sub-grains misorientations till they attain HAB values [84,87,112,135]. During the hot forming of MSS, the relation between the DRV rate and the grain boundaries migration velocity determines the formation of dDRX and cDRX. With increasing temperature, the critical strain for the dDRX decreases, which leads to a reduction in metal grain size [135,137]. The addition of elements in MSS during production can affect the flow stress during hot deformation. For example, Niobium (Nb) precipitates effectively postpone or inhibit DRX in MSS. During hot deformation, Nb atoms impose solute dragging force on the moving dislocations and grain boundaries.

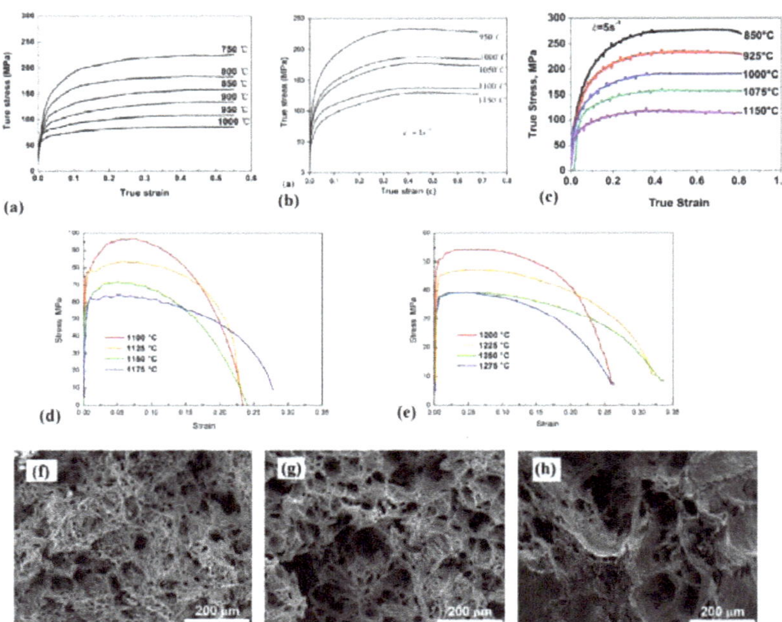

Figure 12. Obtained stress–strain curves after the hot compression test of MSS that shows (**a**) DRV in 17% Cr MSS, (**b**) dDRX in 12% Cr MSS and (**c**) cDRX in 14% Cr MSS [105]. (**d,e**) hot tension curves of 13CrMoNbV steel at temperatures in the range of 1100 °C–1275 °C (dr 1). Microstructure of the tensile samples' fractured surface after deformation at (**f**) 1100 °C, (**g**) 1150 °C and (**h**) 1200 °C [138].

In another type of hot testing, a hot tensile test has been done on 13CrMoNbV steel. The engineering stress–strain curves of the steel at various temperatures are depicted in Figure 12d,f [138]. The increasing temperature decreased the engineering stress during the tensile test, similar to the hot compression behavior. On the other hand, the ultimate tensile strength values increase with decreasing temperature. The 13CrMoNbV steel shows high values of plasticity in the range of 1100 °C–1275 °C [138]. The fracture surface of tensile samples consists of traces of intensive deformation before failures, such as micro-voids, dimples, and sharp crests. The fracture surface becomes more heterogeneous with a temperature increase from 1100 °C to 1200 °C. The dimples' size increases with increasing deformation temperature due to a more intensive plastic deformation and the coalescence of micro-voids during a longer deformation process at high deformation temperatures [139].

5. Precipitates and Dislocations

Generally, the martensitic matrix of MSS contains precipitates after thermo-mechanical processing. Depending on the strain rate and temperature, the size of precipitates and also morphologies of obtained martensite after transformation could be different [140]. During thermo-mechanical processing of MSS, the probability of nanosized precipitates formation and density of dislocations increase. During the hot working, the martensitic matrix changes to martensitic blocks, and the blocks are further divided into several parallel laths [141–145]. TEM images of martensitic lath and precipitates in 403Nb steel after hot working are shown in Figure 13. The small-block and martensitic lath are detectable in the TEM images. The large carbides (thick black arrows), nanosized MX carbides (thin black arrows), and undissolved precipitates (white arrows) arisen during solid solution treatment are detected [146]. The big carbides that mostly had $M_{23}C_6$ chemical composition precipitated along with the martensite lath and blocks. This carbide type is regularly detected in conventional 9–12% Cr steel [147–149].

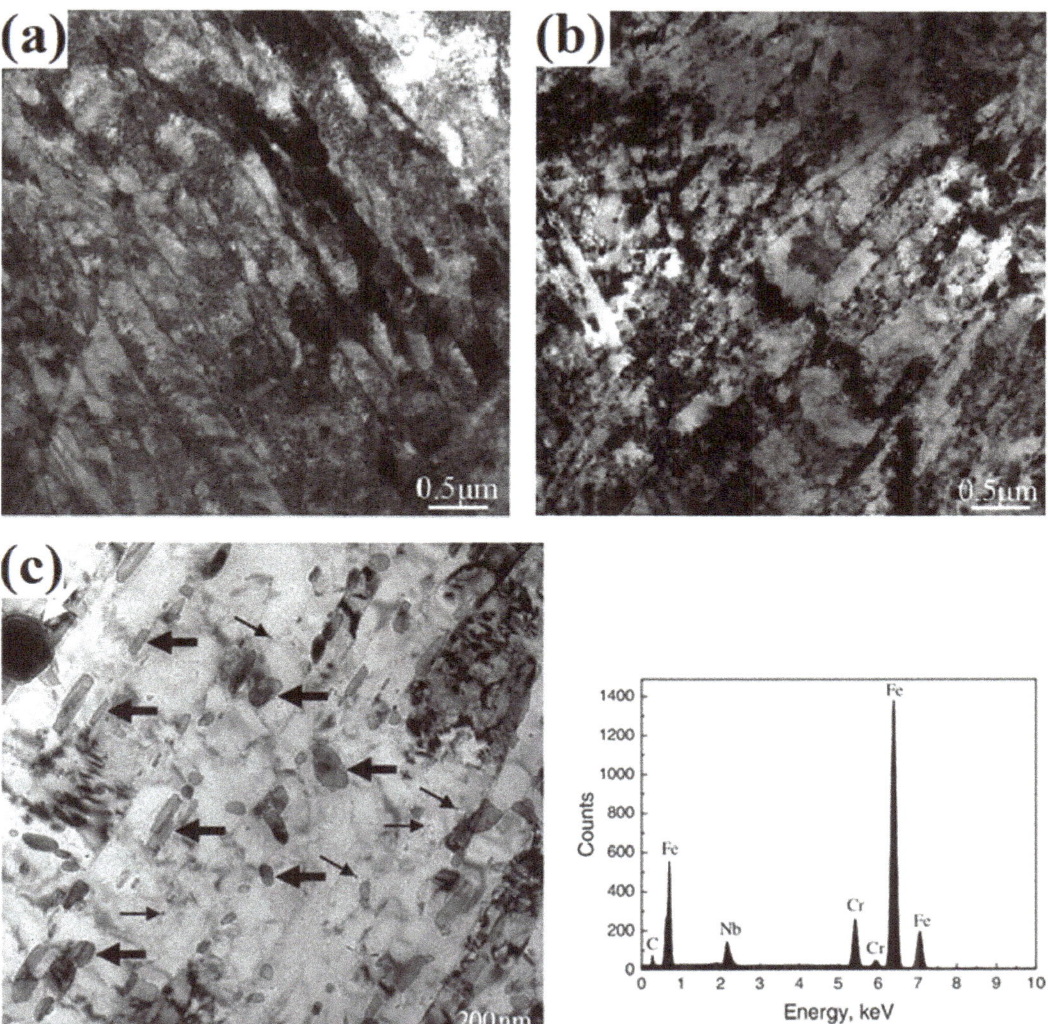

Figure 13. (**a**,**b**) TEM images showing the martensitic lath of 403Nb steel after thermomechanical treatment, (**c**) TEM images showing the MX carbides precipitate of 403Nb steel [141].

NbC, Cr2N, M23C6 carbides are detected in precipitation MSS grades [45]. In this type of MSS, the increase in strain rate rapidly increases the deformation energy in unit time, which leads to quick dislocation movement, accumulation, entanglement, and plugging, which makes the kinetic recovery and recrystallization incomplete [150]. Because of the solid solution in steel, nitrogen retards the growth of the detrimental intermetallic phase, and the precipitation of nitride and M23C6 causes hardening of the grain boundary [151]. With increasing hot compression test temperature, the atomic kinetic energy increases, and the thermal deformation is more easily activated. Simultaneously, dynamic recrystallization nucleation becomes easier, and the high-temperature dynamic recrystallization process is accelerated. These two factors greatly accelerate the softening process and reduce flow stress. In Cu-bearing MSS, the degree of dynamic recrystallization is tremendous, and for starting DRX lower strain rate is needed. In this type of MSS, the CrN and Cr23C6 phases were detected during hot compression, strengthening the grain boundary and significantly affecting the strain limit of the material [152].

The fine MX carbides are distributed mainly within that lath and along boundaries. In some cases, deformation-induced phase transformation caused an increase of dislocation density in the martensite and induced a concentration of dislocations at the boundaries [153]. During the hot forming, the piling up of dislocations on the boundaries creates internal stress concentration [154]. Furthermore, the number of dislocations in martensite lath boundaries increased. Therefore, the density of dislocations near the transformed lath is much higher than in the initial martensite matrix [155,156].

During the hot compression test of MSS, the metal atoms can diffuse sufficiently at a low strain rate, which is advantageous for the occurrence of dynamic recrystallization [157]. At low strain rates, the longer DRX times make it easier for recrystallized grains to grow [158]. Further, a lower strain rate results in less coarse energy obtained by metal deformation. Consequently, the recrystallization driving force decreases accordingly, and the area where recrystallization can occur also decreases simultaneously [159]. In precipitation MSS grades, carbides at grain boundaries have hindered the movement of dislocations, grain boundary slip, and metal recrystallization growth. During hot compression test, at low-temperature and low strain rate, these carbide remains in the microstructure of MSS, but at a higher temperature and high strain rate, the carbides at grain boundaries gradually dissolved into the matrix [160]. On the other hand, at the high-temperature compression test, it is shown that during the high strain rate hot compression test of MSS, the number of carbides at grain boundaries was reduced [161]. This phenomenon indicated that strain is greater than the time effect on dissolving of carbides. This issue can be used in the processing efficiency of MSS [162]. The carbide precipitates dissolved at the grain boundaries, thereby weakening the pinning effect on grain boundaries, and consequently, DRX was easier to occur in precipitation MSS grades at a higher temperature [163].

6. Summary and Further Investigation

The main contributions of the experimental characterization of dynamic recrystallization of martensitic stainless steels during hot deformation are analyzed in this paper, focusing on its applicability to bulk forming operations.

The experimental technique mostly used for the characterization of MSS workability in hot forming conditions is the hot compression test, as for most metals. Flow stress behavior of the material under different plastic deformation conditions are obtained, as well as relevant information of grain evolution that has been used for understanding DRX phenomena. There are three types of DRX: discontinuous, continuous and geometrical. In the dDRX, dislocation motion and accumulation favor new grain nucleation at grain boundaries leading to a new finer cell structure. It is rarely reported in high chromium MSS during hot deformation since it is related to low and medium stacking-fault energy (SFE) steels, and MSS are in the category of high SFE materials. The cDRX directly changes the orientation of sub-grains and forms new grains. During cDRX, dislocations are generated and rearranged to form LAGB sub-grains boundaries. The cDRX is very common in high stacking-fault energy (SFE) materials like MSS. Finally, gDRX is a mechanism of grain refinement by grain elongation, thinning, and fragmentation. gDRX and the cDRX have the same features, which are continuous growth of HAGB area and absence of nuclei. These types of grain refinement added to the analysis of stability have been recurrently studied in the literature reviewed for different grades of MSS.

Regarding the analysis of MSS flow stress during hot deformation (at temperatures above 0.5 T_m), it has been found that DRV and DRX phenomena counteract strain hardening and lead to strain softening at large strains. In MSS, as high SFE materials, deformation is controlled mainly by DRV, especially at low strain rates ($\dot{\varepsilon} < 1\,\mathrm{s}^{-1}$). During hot forming, the redisposition and annihilation of dislocations readily happen over DRV, leading to the formation of subgrains. DRV and cDRX cause strain-softening of MSS during hot deformation. In cDRX, a critical strain is needed to nucleate new grains and HABs are formed during high strain deformation conditions. These new grains contain high densities of dislocations. The grains structure after cDRX is nearly homogeneous.

In addition, the presence of precipitates, main carbides of different compositions (NbC, Cr_2N, $M_{23}C_6$) and Cr compounds (CrN and $Cr_{23}C_6$) cause hardening of the grain boundary, facilitates the activation of thermal deformation and ease the nucleation of the DRX process. As a result, an acceleration of the softening process and a reduction of flow stress are usually observed in MSS under high-temperature forming conditions due to precipitations. Despite the significant effort made in studying the microstructural structures of MSS produced by DRV and DRX reviewed here, some notable features remain to be explained. First, the application of cDRX to microstructure control in MSS grades does not consider allotropic transformations. Second, the effects of the continuous multiphase hot bulk metal forming on the DRX of MSS have not been considered. In hot bulk metal forming, the sub-DRX and microstructural changes could be interesting. Third, the deformation twinning role in producing cRDX grain refinement is also an interesting contribution for the future.

Author Contributions: Conceptualization, H.A.D., A.M.-M., D.M., and E.G.G.; software, H.A.D.; validation, E.G.G., A.M.-M. and D.M.; formal analysis, H.A.D., A.M.-M., D.M., and E.G.G.; investigation, H.A.D., A.M.-M., D.M., and E.G.G.; resources, H.A.D.; data curation, H.A.D., A.M.-M., D.M., and E.G.G.; writing—original draft preparation, H.A.D.; writing—review and editing, H.A.D., A.M.-M., D.M., and E.G.G.; supervision, E.G.G.; and D.M.; project administration, E.G.G.; All authors have read and agreed to the published version of the manuscript.

Funding: This research received no external funding.

Institutional Review Board Statement: Not applicable.

Informed Consent Statement: Not applicable.

Data Availability Statement: MDPI Research Data Policies.

Conflicts of Interest: The authors declare no conflict of interest.

References

1. Stichel, W. ASM speciality handbook: Stainless steels. Hrsg. von J. R. Davis, 577 S., ASM International, Materials Park, Ohio, USA, 1994, £ 136.00 (ASM Member £ 102.00) ISBN 0-87170-503-6. In Europa zu beziehen durch: American Technical Publishers Ltd, 27–29 Knowl Pi. *Mater. Corros.* **1995**, *46*, 499. [CrossRef]
2. Hara, T.; Semba, H.; Amaya, H. *Pipe and Tube Steels for Oil and Gas Industry and Thermal Power Plant*; Elsevier: Amsterdam, The Netherlands, 2020; ISBN 978-0-12-803581-8.
3. Iannuzzi, M. Environmentally assisted cracking (EAC) in oil and gas production. In *Woodhead Publishing Series in Metals and Surface Engineering*; Raja, V.S., Shoji, T.B.T.-S.C.C., Eds.; Woodhead Publishing: Sawston, UK; Cambridge, UK, 2011; pp. 570–607. ISBN 978-1-84569-673-3.
4. Kim, M.-S.; Park, K.-S.; Kim, D.-I.; Suh, J.-Y.; Shim, J.-H.; Hong, K.T.; Choi, S.-H. Heterogeneities in the microstructure and mechanical properties of high-Cr martensitic stainless steel produced by repetitive hot roll bonding. *Mater. Sci. Eng. A* **2021**, *801*, 140416. [CrossRef]
5. Kassis, K.; Masmoudi, J.; Kolsi, A.W.; Dubois, B. Effect of phosphorus on the formation of titanium carbide in 17% Cr ferritic steels stabilized with titanium. *Rev. Met. Paris* **2004**, *101*, 685–693. [CrossRef]
6. Singh, R. *Welding Corrosion Resistant Alloys—Stainless Steel*; Singh, R.B.T.-A.W.E., Ed.; Butterworth-Heinemann: Boston, MA, USA, 2012; Chapter 6; pp. 191–214. ISBN 978-0-12-391916-8.
7. Darvell, B.W. Steel and Cermet. In *Woodhead Publishing Series in Biomaterials*; Darvell, B.W.B.T.-M.S., Tenth, E., Eds.; Woodhead Publishing: Sawston, UK; Cambridge, UK, 2018; Chapter 21; pp. 540–554. ISBN 978-0-08-101035-8.
8. Dong, J.; He, Y.; Kim, M.; Shin, K. Effect of Creep Stress on the Microstructure of 9–12% Cr Steel for Rotor Materials. *Microsc. Microanal.* **2013**, *19*, 95–98. [CrossRef] [PubMed]
9. Manilova, E. Examination of Minor Phases in Martensitic 12% Cr-Mo-W-V Steel. *Microsc. Microanal.* **2006**, *12*, 1612–1613. [CrossRef]
10. Tanaka, T. *Fatigue Strength: Improving by Surface Treatment*; Buschow, K.H.J., Cahn, R.W., Flemings, M.C., Ilschner, B., Kramer, E.J., Mahajan, S., Veyssière, P.B.T.-E., Eds.; Elsevier: Oxford, UK, 2001; pp. 2990–2994. ISBN 978-0-08-043152-9.
11. Garrison, W.M. *Stainless Steels: Martensitic*; Buschow, K.H.J., Cahn, R.W., Flemings, M.C., Ilschner, B., Kramer, E.J., Mahajan, S., Veyssière, P.B.T.-E., Eds.; Elsevier: Oxford, UK, 2001; pp. 8804–8810. ISBN 978-0-08-043152-9.

12. Somers, M.A.J.; Christiansen, T.L. *Gaseous Processes for Low Temperature Surface Hardening of Stainless Steel*; Mittemeijer, E.J., Somers, M.A.J.B.T.-T.S.E., Eds.; Woodhead Publishing: Oxford, UK, 2015; pp. 581–614. ISBN 978-0-85709-592-3.

13. Webster, D. Development of a high strength stainless steel with improved toughness and ductility. *Metall. Mater. Trans. B* **1971**, *2*, 2097–2104. [CrossRef]

14. Villalobos, J.C.; Del-Pozo, A.; Campillo, B.; Mayen, J.; Serna, S. Microalloyed Steels through History until 2018: Review of Chemical Composition, Processing and Hydrogen Service. *Metals* **2018**, *8*, 351. [CrossRef]

15. Krauss, G. *Physical Metallurgy of Steels: An Overview**Every Effort has been Made to Trace Copyright Holders and to Obtain their Permission for the Use of Copyright Material. The Publisher Apologizes for Any Errors or Omissions in the Acknowledgements*; Rana, R., Singh, S.B.B.T.-A.S., Eds.; Woodhead Publishing: Sawston, UK; Cambridge, UK, 2017; pp. 95–111. ISBN 978-0-08-100638-2.

16. Yang, Q.; Zhou, Y.; Li, Z.; Mao, D. Effect of Hot Deformation Process Parameters on Microstructure and Corrosion Behavior of 35CrMoV Steel. *Materials* **2019**, *12*, 1455. [CrossRef]

17. Magnee, A.; Drapier, J.M.; Coutsouradis, D.; Habrakan, L.; Dumont, J. *Cobalt-Containing High-Strength Steels*; IAEA Report: Brussels, Belgium, 1974.

18. Pickering, F.B. *Physical Metallurgy and the Design of Steels/F. B. Pickering*; Materials Science Series; Applied Science Publishers: London, UK, 1978; ISBN 0853347522.

19. Zhang, H.; Wei, Z.; Xie, F.; Sun, B. Assessment of the Properties of AISI 410 Martensitic Stainless Steel by an Eddy Current Method. *Materials* **2019**, *12*, 1290. [CrossRef]

20. Harsha, A.-P.; Limaye, P.-K.; Tyagi, R.; Gupta, A. Effect of Temperature on Galling Behavior of SS 316, 316 L and 416 Under Self-Mated Condition. *J. Mater. Eng. Perform.* **2016**, *25*, 4980–4987. [CrossRef]

21. Ben Lenda, O.; Tara, A.; Lazar, F.; Jbara, O.; Hadjadj, A.; Saad, E. Structural and Mechanical Characteristics of AISI 420 Stainless Steel After Annealing. *Strength Mater.* **2020**, *52*, 71–80. [CrossRef]

22. Giordana, M.F.; Alvarez-Armas, I.; Armas, A. On the Cyclic Softening Mechanisms of Reduced Activity Ferritic/Martensitic Steels. *Steel Res. Int.* **2012**, *83*, 594–599. [CrossRef]

23. Landon, P.R.; Caligiuri, R.D.; Duletsky, P.S. The influence of the M23 (C,N)6 compound on the mechanical properties of type 422 stainless steel. *Metall. Trans. A* **1983**, *14*, 1395–1408. [CrossRef]

24. Hemmati, I.; Ocelík, V.; De Hosson, J.T.M. Microstructural characterization of AISI 431 martensitic stainless steel laser-deposited coatings. *J. Mater. Sci.* **2011**, *46*, 3405–3414. [CrossRef]

25. Huang, C.A.; Tu, G.C.; Yao, H.T.; Kuo, H.H. Characteristics of the rough-cut surface of quenched and tempered martensitic stainless steel using wire electrical discharge machining. *Metall. Mater. Trans. A* **2004**, *35*, 1351–1357. [CrossRef]

26. Venske, A.F.; de Castro, V.V.; da Costa, E.M.; dos Santos, C.A. Sliding Wear Behavior of an AISI 440B Martensitic Stainless Steel Lubricated with Biodiesel and Diesel–Biodiesel Blends. *J. Mater. Eng. Perform.* **2018**, *27*, 5427–5437. [CrossRef]

27. Hasan, S.; Thamizhmanii, S. Tool flank wear analyses on AISI 440 C martensitic stainless steel by turning. *Int. J. Mater. Form.* **2010**, *3*, 427–430. [CrossRef]

28. Aghaie-Khafri, M.; Adhami, F. Hot deformation of 15-5 PH stainless steel. *Mater. Sci. Eng. A* **2010**, *527*, 1052–1057. [CrossRef]

29. Zukerman, I.; Raveh, A.; Kalman, H.; Klemberg-Sapieha, J.E.; Martinu, L. Thermal stability and wear resistance of hard TiN/TiCN coatings on plasma nitrided PH15-5 steel. *Wear* **2007**, *263*, 1249–1252. [CrossRef]

30. Yu, W.-T.; Li, J.; Shi, C.-B.; Zhu, Q.-T. Effect of Spheroidizing Annealing on Microstructure and Mechanical Properties of High-Carbon Martensitic Stainless Steel 8Cr13MoV. *J. Mater. Eng. Perform.* **2017**, *26*, 478–487. [CrossRef]

31. Rafi, H.K.; Starr, T.L.; Stucker, B.E. A comparison of the tensile, fatigue, and fracture behavior of Ti–6Al–4V and 15-5 PH stainless steel parts made by selective laser melting. *Int. J. Adv. Manuf. Technol.* **2013**, *69*, 1299–1309. [CrossRef]

32. Mortezaei, S.; Arabi, H.; Seyedein, H.; Momeny, A.; Soltanalinezhad, M. Investigation on Microstructure Evolution of a Semi-Austenitic Stainless Steel Through Hot Deformation TT. *IUST* **2020**, *17*, 60–69. [CrossRef]

33. Azizpour, A.; Hahn, R.; Klimashin, F.F.; Wojcik, T.; Poursaeidi, E.; Mayrhofer, P.H. Deformation and Cracking Mechanism in CrN/TiN Multilayer Coatings. *Coatings* **2019**, *9*, 363. [CrossRef]

34. Murayama, M.; Hono, K.; Katayama, Y. Microstructural evolution in a 17-4 PH stainless steel after aging at 400 °C. *Metall. Mater. Trans. A* **1999**, *30*, 345–353. [CrossRef]

35. Reddy, V.V.; Kumar, A.; Valli, P.M.; Reddy, C.S. Influence of surfactant and graphite powder concentration on electrical discharge machining of PH17-4 stainless steel. *J. Braz. Soc. Mech. Sci. Eng.* **2015**, *37*, 641–655. [CrossRef]

36. He, J.; Chen, L.; Tao, X.; Antonov, S.; Zhong, Y.; Su, Y. Hydrogen embrittlement behavior of 13Cr-5Ni-2Mo supermartensitic stainless steel. *Corros. Sci.* **2020**, *176*, 109046. [CrossRef]

37. An, G.; Liu, R.; Yin, G. Fatigue Limit of Custom 465 with Surface Strengthening Treatment. *Materials* **2020**, *13*, 238. [CrossRef] [PubMed]

38. Dennies, D.P. Proposed Theory for the Hydrogen Embrittlement Resistance of Martensitic Precipitation Age-Hardening Stainless Steels such as Custom 455. *J. Fail. Anal. Prev.* **2013**, *13*, 433–436. [CrossRef]

39. Ma, X.; Wang, L.; Subramanian, S.V.; Liu, C. Studies on Nb Microalloying of 13Cr Super Martensitic Stainless Steel. *Metall. Mater. Trans. A* **2012**, *43*, 4475–4486. [CrossRef]

40. Zhang, Y.; Yin, Y.; Li, D.; Ma, P.; Liu, Q.; Yuan, X.; Li, S. Temperature Dependent Phase Transformation Kinetics of Reverted Austenite during Tempering in 13Cr Supermartensitic Stainless Steel. *Materials* **2019**, *9*, 1203. [CrossRef]

41. De Oliveira, M.P.; Calderón-Hernández, J.W.; Magnabosco, R.; Hincapie-Ladino, D.; Alonso-Falleiros, N. Effect of Niobium on Phase Transformations, Mechanical Properties and Corrosion of Supermartensitic Stainless Steel. *J. Mater. Eng. Perform.* **2017**, *26*, 1664–1672. [CrossRef]

42. Finkler, H.; Schirra, M. Transformation behaviour of the high temperature martensitic steels with 8–14% chromium. *Steel Res.* **1996**, *67*, 328–342. [CrossRef]

43. Zhang, Y.; Zhang, C.; Yuan, X.; Li, D.; Yin, Y.; Li, S. Microstructure Evolution and Orientation Relationship of Reverted Austenite in 13Cr Supermartensitic Stainless Steel During the Tempering Process. *Materials* **2019**, *12*, 589. [CrossRef] [PubMed]

44. Zou, D.; Liu, R.; Li, J.; Zhang, W.; Wang, D.; Han, Y. Corrosion Resistance and Semiconducting Properties of Passive Films Formed on 00Cr13Ni5Mo2 Supermartensitic Stainless Steel in Cl− Environment. *J. Iron Steel Res. Int.* **2014**, *21*, 630–636. [CrossRef]

45. Fu, X.; Bai, P.; Yang, J. Hot Deformation Characteristics of 18Cr-5Ni-4Cu-N Stainless Steel Using Constitutive Equation and Processing Map. *Metals* **2020**, *10*, 82. [CrossRef]

46. Roberts, W.N. Development of needles and sutures for microsurgery. *J. Biomed. Mater. Res.* **1975**, *9*, 399–405. [CrossRef] [PubMed]

47. Kevanlo, E.; Ebrahimi, G.R.; Sani, S.A.A.; Momeni, A. Dynamic recrystallization kinetics of AISI 403 stainless steel using hot compression test. *Iran. J. Mater. Form.* **2014**, *1*, 32–43. [CrossRef]

48. Mirzaee, M.; Keshmiri, H.; Ebrahimi, G.R.; Momeni, A. Dynamic recrystallization and precipitation in low carbon low alloy steel 26NiCrMoV 14-5. *Mater. Sci. Eng. A* **2012**, *551*, 25–31. [CrossRef]

49. Yan, T.; Yu, E.; Zhao, Y. Constitutive modeling for flow stress of 55SiMnMo bainite steel at hot working conditions. *Mater. Des.* **2013**, *50*, 574–580. [CrossRef]

50. Dehghan, H.; Abbasi, S.M.; Momeni, A.; Karimi Taheri, A. On the constitutive modeling and microstructural evolution of hot compressed A286 iron-base superalloy. *J. Alloys Compd.* **2013**, *564*, 13–19. [CrossRef]

51. Momeni, A.; Dehghani, K.; Heidari, M.; Vaseghi, M. Modeling the Flow Curve of AISI 410 Martensitic Stainless Steel. *J. Mater. Eng. Perform.* **2012**, *21*, 2238–2243. [CrossRef]

52. Momeni, A.; Dehghani, K.; Ebrahimi, G.R. Modeling the initiation of dynamic recrystallization using a dynamic recovery model. *J. Alloys Compd.* **2011**, *509*, 9387–9393. [CrossRef]

53. Bang, W.; Lee, C.S.; Chang, Y.W. Finite element analysis of hot forging with flow softening by dynamic recrystallization. *J. Mater. Process. Technol.* **2003**, *134*, 153–158. [CrossRef]

54. Huang, J.; Xu, Z. Evolution mechanism of grain refinement based on dynamic recrystallization in multiaxially forged austenite. *Mater. Lett.* **2006**, *60*, 1854–1858. [CrossRef]

55. Yang, Y.; Yan, Q.; Ge, C. Hot Deformation Behavior of Modified CNS- II F/M Steel. *J. Iron Steel Res. Int.* **2012**, *19*, 60–65. [CrossRef]

56. Li, C.; Liu, Y.; Tan, Y.; Zhao, F. Hot Deformation Behavior and Constitutive Modeling of H13-Mod Steel. *Metals* **2018**, *8*, 846. [CrossRef]

57. Xu, Y.; Liu, J.S.; Jiao, Y.X. Hot Deformation Behavior and Dynamic Recrystallization Characteristics of 12Cr Ultra-Super-Critical Rotor Steel. *Met. Mater. Int.* **2019**, *25*, 823–837. [CrossRef]

58. Murillo-Marrodán, A.; Puchi-Cabrera, E.S.; García, E.; Dubar, M.; Cortés, F.; Dubar, L. An Incremental Physically-Based Model of P91 Steel Flow Behaviour for the Numerical Analysis of Hot-Working Processes. *Metals* **2018**, *8*, 269. [CrossRef]

59. Wang, R.; Wang, M.; Li, Z.; Lu, C. Physics-based Constitutive Model for the Hot Deformation of 2Cr11Mo1VNbN Martensitic Stainless Steel. *J. Mater. Eng. Perform.* **2018**, *27*, 4932–4940. [CrossRef]

60. Safara, N.; Engberg, G.; Ågren, J. Modeling Microstructure Evolution in a Martensitic Stainless Steel Subjected to Hot Working Using a Physically Based Model. *Metall. Mater. Trans. A* **2019**, *50*, 1480–1488. [CrossRef]

61. Ren, F.; Chen, F.; Chen, J.; Tang, X. Hot deformation behavior and processing maps of AISI 420 martensitic stainless steel. *J. Manuf. Process.* **2018**, *31*, 640–649. [CrossRef]

62. Kishor, B.; Chaudhari, G.P.; Nath, S.K. Hot Deformation Characteristics of 13Cr-4Ni Stainless Steel Using Constitutive Equation and Processing Map. *J. Mater. Eng. Perform.* **2016**, *25*, 2651–2660. [CrossRef]

63. Taylor, T.; Danks, S.; Fourlaris, G. Dynamic Tensile Testing of Ultrahigh Strength Hot Stamped Martensitic Steels. *Steel Res. Int.* **2017**, *88*, 1600144. [CrossRef]

64. Burke, J.E.; Turnbull, D. Recrystallization and grain growth. *Prog. Met. Phys.* **1952**, *3*, 220–292. [CrossRef]

65. Doherty, R.D.; Hughes, D.A.; Humphreys, F.J.; Jonas, J.J.; Jensen, D.J.; Kassner, M.E.; King, W.E.; McNelley, T.R.; McQueen, H.J.; Rollett, A.D. Current issues in recrystallization: A review. *Mater. Sci. Eng. A* **1997**, *238*, 219–274. [CrossRef]

66. Seifert, M.; Siebert, S.; Huth, S.; Theisen, W.; Berns, H. New Developments in Martensitic Stainless Steels Containing C + N. *Steel Res. Int.* **2015**, *86*, 1508–1516. [CrossRef]

67. Sakai, T.; Belyakov, A.; Kaibyshev, R.; Miura, H.; Jonas, J.J. Dynamic and post-dynamic recrystallization under hot, cold and severe plastic deformation conditions. *Prog. Mater. Sci.* **2014**, *60*, 130–207. [CrossRef]

68. Miura, H.; Sakai, T.; Belyakov, A.; Gottstein, G.; Crumbach, M.; Verhasselt, J. Static recrystallization of SiO2-particle containing {011}<100> copper single crystals. *Acta Mater.* **2003**, *51*, 1507–1515. [CrossRef]

69. Humphreys, F.J.; Hatherly, M.B.T.-R. (Eds.) *Appendix 1 Texture*, 2nd ed.; Elsevier: Oxford, UK, 2004; pp. 527–540. ISBN 978-0-08-044164-1.

70. Ahlborn, H.; Hornbogen, E.; Köster, U. Recrystallisation mechanism and annealing texture in aluminium-copper alloys. *J. Mater. Sci.* **1969**, *4*, 944–950. [CrossRef]

71. Zeng, Z.; Chen, L.; Zhu, F.; Liu, X. Dynamic Recrystallization Behavior of a Heat-resistant Martensitic Stainless Steel 403Nb during Hot Deformation. *J. Mater. Sci. Technol.* **2011**, *27*, 913–919. [CrossRef]

72. Fang, Y.; Chen, X.; Madigan, B.; Cao, H.; Konovalov, S. Effects of strain rate on the hot deformation behavior and dynamic recrystallization in China low activation martensitic steel. *Fusion Eng. Des.* **2016**, *103*, 21–30. [CrossRef]

73. Ebrahimi, G.R.; Momeni, A.; Jahazi, M.; Bocher, P. Dynamic Recrystallization and Precipitation in 13Cr Super-Martensitic Stainless Steels. *Metall. Mater. Trans. A* **2014**, *45*, 2219–2231. [CrossRef]

74. Shafiei, E.; Dehghani, K. Prediction of Single-Peak Flow Stress Curves at High Temperatures Using a New Logarithmic-Power Function. *J. Mater. Eng. Perform.* **2016**, *25*, 4024–4035. [CrossRef]

75. Ahmadabadi, R.M.; Naderi, M.; Mohandesi, J.A.; Cabrera, J.M. Dynamic Recrystallization Behavior of AISI 422 Stainless Steel During Hot Deformation Processes. *J. Mater. Eng. Perform.* **2018**, *27*, 560–571. [CrossRef]

76. Samantaray, D.; Phaniraj, C.; Mandal, S.; Bhaduri, A.K. Strain dependent rate equation to predict elevated temperature flow behavior of modified 9Cr-1Mo (P91) steel. *Mater. Sci. Eng. A* **2011**, *528*, 1071–1077. [CrossRef]

77. Jonas, J.J.; Sellars, C.M.; Tegart, W.J.M. Strength and structure under hot-working conditions. *Metall. Rev.* **1969**, *14*, 1–24. [CrossRef]

78. Al-Samman, T.; Gottstein, G. Dynamic recrystallization during high temperature deformation of magnesium. *Mater. Sci. Eng. A* **2008**, *490*, 411–420. [CrossRef]

79. Ion, S.E.; Humphreys, F.J.; White, S.H. Dynamic recrystallisation and the development of microstructure during the high temperature deformation of magnesium. *Acta Metall.* **1982**, *30*, 1909–1919. [CrossRef]

80. Nesterenko, V.F.; Meyers, M.A.; LaSalvia, J.C.; Bondar, M.P.; Chen, Y.J.; Lukyanov, Y.L. Shear localization and recrystallization in high-strain, high-strain-rate deformation of tantalum. *Mater. Sci. Eng. A* **1997**, *229*, 23–41. [CrossRef]

81. Solberg, J.K.; McQueen, H.J.; Ryum, N.; Nes, E. Influence of ultra-high strains at elevated temperatures on the microstructure of aluminium. Part I. *Philos. Mag. A* **1989**, *60*, 447–471. [CrossRef]

82. Hales, S.J.; McNelley, T.R.; McQueen, H.J. Recrystallization and superplasticity at 300 °C in an aluminum-magnesium alloy. *Metall. Trans. A* **1991**, *22*, 1037–1047. [CrossRef]

83. Ponge, D.; Gottstein, G. Necklace formation during dynamic recrystallization: Mechanisms and impact on flow behavior. *Acta Mater.* **1998**, *46*, 69–80. [CrossRef]

84. Sakai, T.; Jonas, J.J. Plastic Deformation: Role of Recovery and Recrystallization. In *Encyclopedia of Materials: Science and Technology*; Buschow, K.H.J., Cahn, R.W., Flemings, M.C., Ilschner, B., Kramer, E.J., Mahajan, S., Veyssière, P., Eds.; Elsevier: Oxford, UK, 2001; pp. 7079–7084. ISBN 978-0-08-043152-9.

85. Qing, L.; Xiaoxu, H.; Mei, Y.; Jinfeng, Y. On deformation-induced continuous recrystallization in a superplastic Al Li Cu Mg Zr alloy. *Acta Metall. Mater.* **1992**, *40*, 1753–1762. [CrossRef]

86. Lv, J.; Zheng, J.-H.; Yardley, V.A.; Shi, Z.; Lin, J. A Review of Microstructural Evolution and Modelling of Aluminium Alloys under Hot Forming Conditions. *Metals* **2020**, *10*, 1516. [CrossRef]

87. Tsuji, N.; Matsubara, Y.; Saito, Y. Dynamic recrystallization of ferrite in interstitial free steel. *Scr. Mater.* **1997**, *37*, 477–484. [CrossRef]

88. Gourdet, S.; Montheillet, F. An experimental study of the recrystallization mechanism during hot deformation of aluminium. *Mater. Sci. Eng. A* **2000**, *283*, 274–288. [CrossRef]

89. Sitdikov, O.; Sakai, T.; Miura, H.; Hama, C. Temperature effect on fine-grained structure formation in high-strength Al alloy 7475 during hot severe deformation. *Mater. Sci. Eng. A* **2009**, *516*, 180–188. [CrossRef]

90. Sitdikov, O.; Kaibyshev, R. Dynamic Recrystallization in Pure Magnesium. *Mater. Trans.* **2001**, *42*, 1928–1937. [CrossRef]

91. Ito, Y.; Horita, Z. Microstructural evolution in pure aluminum processed by high-pressure torsion. *Mater. Sci. Eng. A* **2009**, *503*, 32–36. [CrossRef]

92. Kaibyshev, R.; Sitdikov, O.; Goloborodko, A.; Sakai, T. Grain refinement in as-cast 7475 aluminum alloy under hot deformation. *Mater. Sci. Eng. A* **2003**, *344*, 348–356. [CrossRef]

93. Mazurina, I.; Sakai, T.; Miura, H.; Sitdikov, O.; Kaibyshev, R. Effect of deformation temperature on microstructure evolution in aluminum alloy 2219 during hot ECAP. *Mater. Sci. Eng. A* **2008**, *486*, 662–671. [CrossRef]

94. Bolli, E.; Fava, A.; Ferro, P.; Kaciulis, S.; Mezzi, A.; Montanari, R.; Varone, A. Cr Segregation and Impact Fracture in a Martensitic Stainless Steel. *Coatings* **2020**, *10*, 843. [CrossRef]

95. Kassner, M.E.; Barrabes, S.R. New developments in geometric dynamic recrystallization. *Mater. Sci. Eng. A* **2005**, *410–411*, 152–155. [CrossRef]

96. Mironov, S.; Murzinova, M.; Zherebtsov, S.; Salishchev, G.A.; Semiatin, S.L. Microstructure evolution during warm working of Ti–6Al–4V with a colony-α microstructure. *Acta Mater.* **2009**, *57*, 2470–2481. [CrossRef]

97. Pettersen, T.; Holmedal, B.; Nes, E. Microstructure development during hot deformation of aluminum to large strains. *Metall. Mater. Trans. A* **2003**, *34*, 2737–2744. [CrossRef]

98. Blum, W.; Zhu, Q.; Merkel, R.; McQueen, H.J. Geometric dynamic recrystallization in hot torsion of Al-5Mg-0.6Mn (AA5083). *Mater. Sci. Eng. A* **1996**, *205*, 23–30. [CrossRef]

99. Miura, H.; Aoyama, H.; Sakai, T. Effect of Grain-Boundary Misorientation on Dynamic Recrystallization of Cu-Si Bicrystals. *J. Jpn. Inst. Met.* **1994**, *58*, 267–275. [CrossRef]

100. Wang, Z.; Wang, X.; Zhu, Z. Characterization of high-temperature deformation behavior and processing map of TB17 titanium alloy. *J. Alloys Compd.* **2017**, *692*, 149–154. [CrossRef]

101. Li, X.; Hou, L.; Wei, Y.; Wei, Z. Constitutive Equation and Hot Processing Map of a Nitrogen-Bearing Martensitic Stainless Steel. *Metals* **2020**, *10*, 1502. [CrossRef]

102. Großeiber, S.; Ilie, S.; Poletti, C.; Harrer, B.; Degischer, H.P. Influence of Strain Rate on Hot Ductility of a V-Microalloyed Steel Slab. *Steel Res. Int.* **2012**, *83*, 445–455. [CrossRef]

103. Lin, Y.C.; Liu, G. Effects of strain on the workability of a high strength low alloy steel in hot compression. *Mater. Sci. Eng. A* **2009**, *523*, 139–144. [CrossRef]

104. Prasad, Y.V.R.K.; Seshacharyulu, T. Modelling of hot deformation for microstructural control. *Int. Mater. Rev.* **1998**, *43*, 243–258. [CrossRef]

105. Huang, D.; Feng, W. Hot Deformation Characteristics and Processing Map of FV520B Martensitic Precipitation-Hardened Stainless Steel. *J. Mater. Eng. Perform.* **2019**, *28*, 2281–2291. [CrossRef]

106. Dehghan-Manshadi, A.; Barnett, M.R.; Hodgson, P.D. Recrystallization in AISI 304 austenitic stainless steel during and after hot deformation. *Mater. Sci. Eng. A* **2008**, *485*, 664–672. [CrossRef]

107. Bricknell, R.H.; Edington, J.W. Textures in a superplastic Al-6Cu-0.3Zr alloy. *Acta Metall.* **1979**, *27*, 1303–1311. [CrossRef]

108. Nes, E. Hot deformation behaviour of particle-stabilized structures in Zr-bearing Al alloys. *Met. Sci.* **1979**, *13*, 211–215. [CrossRef]

109. McQueen, H.J.; Solberg, J.K.; Ryum, N.; Nes, E. Evolution of flow stress in aluminium during ultra-high straining at elevated temperatures. Part II. *Philos. Mag. A* **1989**, *60*, 473–485. [CrossRef]

110. McQueen, H.J.; Ryan, N.D.; Konopleva, E.V.; Xia, X. Formation and application of grain boundary serrations. *Can. Metall. Q.* **1995**, *34*, 219–229. [CrossRef]

111. Zhu, Y.T.; Langdon, T.G. The fundamentals of nanostructured materials processed by severe plastic deformation. *JOM* **2004**, *56*, 58–63. [CrossRef]

112. Gourdet, S.; Montheillet, F. A model of continuous dynamic recrystallization. *Acta Mater.* **2003**, *51*, 2685–2699. [CrossRef]

113. McNelley, T.R.; Swisher, D.L.; Pérez-Prado, M.T. Deformation bands and the formation of grain boundaries in a superplastic aluminum alloy. *Metall. Mater. Trans. A* **2002**, *33*, 279–290. [CrossRef]

114. Mehtonen, S.V.; Karjalainen, L.P.; Porter, D.A. Hot deformation behavior and microstructure evolution of a stabilized high-Cr ferritic stainless steel. *Mater. Sci. Eng. A* **2013**, *571*, 1–12. [CrossRef]

115. Gao, F.; Liu, Z.; Misra, R.D.K.; Liu, H.; Yu, F. Constitutive modeling and dynamic softening mechanism during hot deformation of an ultra-pure 17%Cr ferritic stainless steel stabilized with Nb. *Met. Mater. Int.* **2014**, *20*, 939–951. [CrossRef]

116. Huang, K.; Logé, R.E. A review of dynamic recrystallization phenomena in metallic materials. *Mater. Des.* **2016**, *111*, 548–574. [CrossRef]

117. Lin, J.; Liu, Y.; Farrugia, D.C.J.; Zhou, M. Development of dislocation-based unified material model for simulating microstructure evolution in multipass hot rolling. *Philos. Mag.* **2005**, *85*, 1967–1987. [CrossRef]

118. Raabe, D. *Recovery and Recrystallization: Phenomena, Physics, Models, Simulation*, 5th ed.; Laughlin, D.E., Hono, K.B.T.-P.M., Eds.; Elsevier: Oxford, UK, 2014; pp. 2291–2397. ISBN 978-0-444-53770-6.

119. Yamagata, H. Dynamic recrystallization of single-crystalline aluminum during compression tests. *Scr. Metall. Mater.* **1992**, *27*, 727–732. [CrossRef]

120. Ebrahimi, G.R.; Keshmiri, H.; Maldad, A.R.; Momeni, A. Dynamic Recrystallization Behavior of 13%Cr Martensitic Stainless Steel under Hot Working Condition. *J. Mater. Sci. Technol.* **2012**, *28*, 467–473. [CrossRef]

121. Ebrahimi, G.R.; Keshmiri, H.; Momeni, A.; Mazinani, M. Dynamic recrystallization behavior of a superaustenitic stainless steel containing 16%Cr and 25%Ni. *Mater. Sci. Eng. A* **2011**, *528*, 7488–7493. [CrossRef]

122. Momeni, A.; Dehghani, K.; Keshmiri, H.; Ebrahimi, G.R. Hot deformation behavior and microstructural evolution of a super-austenitic stainless steel. *Mater. Sci. Eng. A* **2010**, *527*, 1605–1611. [CrossRef]

123. Ryan, N.D.; McQueen, H.J. Flow stress, dynamic restoration, strain hardening and ductility in hot working of 316 steel. *J. Mater. Process. Technol.* **1990**, *21*, 177–199. [CrossRef]

124. El Wahabi, M.; Gavard, L.; Montheillet, F.; Cabrera, J.M.; Prado, J.M. Effect of initial grain size on dynamic recrystallization in high purity austenitic stainless steels. *Acta Mater.* **2005**, *53*, 4605–4612. [CrossRef]

125. Huo, J.; Wu, H.; Yang, J.; Sun, W.; Li, G.; Sun, X. Multi-directional coupling dynamic characteristics analysis of TBM cutterhead system based on tunnelling field test. *J. Mech. Sci. Technol.* **2015**, *29*, 3043–3058. [CrossRef]

126. Momeni, A.; Dehghani, K. Prediction of dynamic recrystallization kinetics and grain size for 410 martensitic stainless steel during hot deformation. *Met. Mater. Int.* **2010**, *16*, 843–849. [CrossRef]

127. Cai, X.; Hu, X.-Q.; Zheng, L.-G.; Li, D.-Z. Hot Deformation Behavior and Processing Maps of 0.3C–15Cr–1Mo–0.5N High Nitrogen Martensitic Stainless Steel. *Acta Metall. Sin.* **2020**, *33*, 693–704. [CrossRef]

128. Mcqueen, H.J.; Jonas, J.J. Recent advances in hot working: Fundamental dynamic softening mechanisms. *J. Appl. Metalwork.* **1984**, *3*, 233–241. [CrossRef]

129. McQueen, H.J. The production and utility of recovered dislocation substructures. *Metall. Trans. A* **1977**, *8*, 807–824. [CrossRef]

130. Song, R.; Ponge, D.; Raabe, D.; Kaspar, R. Microstructure and crystallographic texture of an ultrafine grained C–Mn steel and their evolution during warm deformation and annealing. *Acta Mater.* **2005**, *53*, 845–858. [CrossRef]

131. Storojeva, L.; Ponge, D.; Kaspar, R.; Raabe, D. Development of microstructure and texture of medium carbon steel during heavy warm deformation. *Acta Mater.* **2004**, *52*, 2209–2220. [CrossRef]
132. Dehghan-Manshadi, A.; Hodgson, P.D. Effect of δ-ferrite co-existence on hot deformation and recrystallization of austenite. *J. Mater. Sci.* **2008**, *43*, 6272–6277. [CrossRef]
133. Tsuzaki, K.; Huang, X.; Maki, T. Mechanism of dynamic continuous recrystallization during superplastic deformation in a microduplex stainless steel. *Acta Mater.* **1996**, *44*, 4491–4499. [CrossRef]
134. Cizek, P. The microstructure evolution and softening processes during high-temperature deformation of a 21Cr–10Ni–3Mo duplex stainless steel. *Acta Mater.* **2016**, *106*, 129–143. [CrossRef]
135. Belyakov, A.; Miura, H.; Sakai, T. Dynamic recrystallization in ultra fine-grained 304 stainless steel. *Scr. Mater.* **2000**, *43*, 21–26. [CrossRef]
136. Mcqueen, H.J.; Jonas, J.J. Recovery and Recrystallization during High Temperature Deformation. In *Plastic Deformation of Materials*; Elsevier: Amsterdam, The Netherlands, 1975; Volume 6, pp. 393–493. ISBN 0161-9160.
137. Sakai, T. Dynamic recrystallization microstructures under hot working conditions. *J. Mater. Process. Technol.* **1995**, *53*, 349–361. [CrossRef]
138. Shaikh, A.; Churyumov, A.; Pozdniakov, A.; Churyumova, T. Simulation of the Hot Deformation and Fracture Behavior of Reduced Activation Ferritic/Martensitic 13CrMoNbV Steel. *Appl. Sci.* **2020**, *10*, 530. [CrossRef]
139. Opěla, P.; Schindler, I.; Kawulok, P.; Kawulok, R.; Rusz, S.; Navrátil, H.; Jurča, R. Correlation among the Power Dissipation Efficiency, Flow Stress Course, and Activation Energy Evolution in Cr-Mo Low-Alloyed Steel. *Materials* **2020**, *13*, 3480. [CrossRef] [PubMed]
140. Ma, X.P.; Wang, L.J.; Liu, C.M.; Subramanian, S.V. Microstructure and properties of 13Cr5Ni1Mo0.025Nb0.09V0.06N super martensitic stainless steel. *Mater. Sci. Eng. A* **2012**, *539*, 271–279. [CrossRef]
141. Chen, L.; Zeng, Z.; Zhao, Y.; Zhu, F.; Liu, X. Microstructures and High-Temperature Mechanical Properties of a Martensitic Heat-Resistant Stainless Steel 403Nb Processed by Thermo-Mechanical Treatment. *Metall. Mater. Trans. A* **2014**, *45*, 1498–1507. [CrossRef]
142. Hollner, S.; Piozin, E.; Mayr, P.; Caës, C.; Tournié, I.; Pineau, A.; Fournier, B. Characterization of a boron alloyed 9Cr3W3CoVNbBN steel and further improvement of its high-temperature mechanical properties by thermomechanical treatments. *J. Nucl. Mater.* **2013**, *441*, 15–23. [CrossRef]
143. Hollner, S.; Fournier, B.; Le Pendu, J.; Cozzika, T.; Tournié, I.; Brachet, J.-C.; Pineau, A. High-temperature mechanical properties improvement on modified 9Cr–1Mo martensitic steel through thermomechanical treatments. *J. Nucl. Mater.* **2010**, *405*, 101–108. [CrossRef]
144. Klueh, R.L.; Hashimoto, N.; Maziasz, P.J. New nano-particle-strengthened ferritic/martensitic steels by conventional thermo-mechanical treatment. *J. Nucl. Mater.* **2007**, *367–370*, 48–53. [CrossRef]
145. Klueh, R.L.; Hashimoto, N.; Maziasz, P.J. Development of new nano-particle-strengthened martensitic steels. *Scr. Mater.* **2005**, *53*, 275–280. [CrossRef]
146. Taneike, M.; Abe, F.; Sawada, K. Creep-strengthening of steel at high temperatures using nano-sized carbonitride dispersions. *Nature* **2003**, *424*, 294–296. [CrossRef] [PubMed]
147. Rojas, D.; Garcia, J.; Prat, O.; Agudo, L.; Carrasco, C.; Sauthoff, G.; Kaysser-Pyzalla, A.R. Effect of processing parameters on the evolution of dislocation density and sub-grain size of a 12%Cr heat resistant steel during creep at 650 °C. *Mater. Sci. Eng. A* **2011**, *528*, 1372–1381. [CrossRef]
148. Hald, J. Microstructure and long-term creep properties of 9–12% Cr steels. *Int. J. Press. Vessel. Pip.* **2008**, *85*, 30–37. [CrossRef]
149. Hu, Z.; He, D.; Mo, F. Carbides Evolution in 12Cr Martensitic Heat-resistant Steel with Life Depletion for Long-term Service. *J. Iron Steel Res. Int.* **2015**, *22*, 250–255. [CrossRef]
150. Hänninen, H.; Romu, J.; Ilola, R.; Tervo, J.; Laitinen, A. Effects of processing and manufacturing of high nitrogen-containing stainless steels on their mechanical, corrosion and wear properties. *J. Mater. Process. Technol.* **2001**, *117*, 424–430. [CrossRef]
151. Vedani, M.; Dellasega, D.; Mannuccii, A. Characterization of Grain-boundary Precipitates after Hot-ductility Tests of Microalloyed Steels. *ISIJ Int.* **2009**, *49*, 446–452. [CrossRef]
152. Ohadi, D.; Parsa, M.H.; Mirzadeh, H. Development of dynamic recrystallization maps based on the initial grain size. *Mater. Sci. Eng. A* **2013**, *565*, 90–95. [CrossRef]
153. Zhang, S.; Wang, P.; Li, D.; Li, Y. In situ investigation on the deformation-induced phase transformation of metastable austenite in Fe–13% Cr–4% Ni martensitic stainless steel. *Mater. Sci. Eng. A* **2015**, *635*, 129–132. [CrossRef]
154. Zou, D.; Liu, X.; Han, Y.; Zhang, W.; Li, J.; Wu, K. Influence of Heat Treatment Temperature on Microstructure and Property of 00Crl3Ni5Mo2 Supermartensitic Stainless Steel. *J. Iron Steel Res. Int.* **2014**, *21*, 364–368. [CrossRef]
155. Jia, N.; Cong, Z.H.; Sun, X.; Cheng, S.; Nie, Z.H.; Ren, Y.; Liaw, P.K.; Wang, Y.D. An in situ high-energy X-ray diffraction study of micromechanical behavior of multiple phases in advanced high-strength steels. *Acta Mater.* **2009**, *57*, 3965–3977. [CrossRef]
156. Fu, B.; Yang, W.Y.; Wang, Y.D.; Li, L.F.; Sun, Z.Q.; Ren, Y. Micromechanical behavior of TRIP-assisted multiphase steels studied with in situ high-energy X-ray diffraction. *Acta Mater.* **2014**, *76*, 342–354. [CrossRef]
157. Momeni, A.; Arabi, H.; Rezaei, A.; Badri, H.; Abbasi, S.M. Hot deformation behavior of austenite in HSLA-100 microalloyed steel. *Mater. Sci. Eng. A* **2011**, *528*, 2158–2163. [CrossRef]

158. Shabashov, V.A.; Korshunov, L.G.; Zamatovskii, A.E.; Litvinov, A.V.; Sagaradze, V.V.; Kositsyna, I.I. Deformation-induced dissolution of carbides of the Me(V, Mo)-C Type in high-manganese steels upon the friction effect. *Phys. Met. Metallogr.* **2012**, *113*, 914–921. [CrossRef]
159. Srivatsa, K.; Srinivas, P.; Balachandran, G.; Balasubramanian, V. Improvement of impact toughness by modified hot working and heat treatment in 13%Cr martensitic stainless steel. *Mater. Sci. Eng. A* **2016**, *677*, 240–251. [CrossRef]
160. Liu, Y.; Ye, S.L.; An, B.; Wang, Y.G.; Li, Y.J.; Zhang, L.C.; Wang, W.M. Effects of mechanical compression and autoclave treatment on the backbone clusters in the Al86Ni9La5 amorphous alloy. *J. Alloys Compd.* **2014**, *587*, 59–65. [CrossRef]
161. Momeni, A.; Dehghani, K. Characterization of hot deformation behavior of 410 martensitic stainless steel using constitutive equations and processing maps. *Mater. Sci. Eng. A* **2010**, *527*, 5467–5473. [CrossRef]
162. Zhou, P.; Ma, Q.-X. Dynamic Recrystallization Behavior and Processing Map Development of 25CrMo4 Mirror Plate Steel During Hot Deformation. *Acta Metall. Sin.* **2017**, *30*, 907–920. [CrossRef]
163. Zhao, C.; Zhang, J.; Yang, B.; Li, Y.F.; Huang, J.F.; Lian, Y. Hot Deformation Characteristics and Processing Map of 1Cr12Ni2Mo2WVNb Martensitic Stainless Steel. *Steel Res. Int.* **2020**, *91*, 2000020. [CrossRef]

Article

Hot Deformation Behavior of a Ni-Based Superalloy with Suppressed Precipitation

Franco Lizzi [1,2,3], Kashyap Pradeep [3,4], Aleksandar Stanojevic [5], Silvana Sommadossi [1,2] and Maria Cecilia Poletti [3,4,*]

1   Materials Characterization, Comahue National University, Neuquén 8300, Argentina; francolizzielectronica@gmail.com (F.L.); ssommadossi@conicet.gov.ar (S.S.)
2   IITCI CONICET-UNCo, Neuquén 8300, Argentina
3   Christian Doppler Laboratory for Design of High-Performance Alloys by Thermomechanical Processing, 8010 Graz, Austria; kasyap.venkata@tugraz.at
4   Institute of Materials Science, Joining and Forming, Graz University of Technology, 8010 Graz, Austria
5   Voestalpine Böhler Aerospace GmbH & Co KG, 8605 Kapfenberg, Austria; Aleksandar.Stanojevic@voestalpine.com
*   Correspondence: cecilia.poletti@tugraz.at; Tel.: +43-(316)-873-1676

Abstract: Inconel®718 is a well-known nickel-based super-alloy used for high-temperature applications after thermomechanical processes followed by heat treatments. This work describes the evolution of the microstructure and the stresses during hot deformation of a prototype alloy named IN718WP produced by powder metallurgy with similar chemical composition to the matrix of Inconel®718. Compression tests were performed by the thermomechanical simulator Gleeble®3800 in a temperature range from 900 to 1025 °C, and strain rates scaled from 0.001 to 10 s^{-1}. Flow curves of IN718WP showed similar features to those of Inconel®718. The relative stress softening of the IN718WP was comparable to standard alloy Inconel®718 for the highest strain rates. Large stress softening at low strain rates may be related to two phenomena: the fast recrystallization rate, and the coarsening of micropores driven by diffusion. Dynamic recrystallization grade and grain size were quantified using metallography. The recrystallization grade increased as the strain rate decreased, although showed less dependency on the temperature. Dynamic recrystallization occurred after the formation of deformation bands at strain rates above 0.1 s^{-1} and after the formation of subgrains when deforming at low strain rates. Recrystallized grains had a large number of sigma 3 boundaries, and their percentage increased with strain rate and temperature. The calculated apparent activation energy and strain rate exponent value were similar to those found for Inconel®718 when deforming above the solvus temperature.

Keywords: Inconel®718; hot deformation; Gleeble; recrystallization; flow modelling approach

Citation: Lizzi, F.; Pradeep, K.; Stanojevic, A.; Sommadossi, S.; Poletti, M.C. Hot Deformation Behavior of a Ni-Based Superalloy with Suppressed Precipitation. *Metals* **2021**, *11*, 605. https://doi.org/10.3390/met11040605

Academic Editors: Marcello Cabibbo and Maciej Motyka

Received: 7 March 2021
Accepted: 23 March 2021
Published: 8 April 2021

1. Introduction

Inconel®718 nickel-based superalloy is well known for its high mechanical and creep resistances. It presents high fatigue resistance at high temperatures and is suitable for use in corrosive environments [1]. These characteristics of Inconel®718 allow its use in a wide application range, such as in the aeronautical and energy production industries [2,3]. The alloy is used in aerospace engines, where creep resistance at the service temperature is required [4]. The most well-known feature that makes Inconel®718 very attractive for its best mechanical properties is the presence of different types of precipitates [5]. The alloy presents four distinct phases: the FCC matrix γ, and the precipitates γ', γ'' and δ. The γ' phase is $Ni_3(Al,Ti)$ with a cubic (L12) crystal structure; the γ'' phase has a composition Ni_3Nb and bct (D022) crystal structure; and the δ phase is described by Ni_3Nb and has an orthorhombic (D0a) crystal structure [6,7]. The first two precipitates are responsible for the hardening effect, while the δ phase can be used to control the grain size, thus

139

improving hardness and ductility [8]. The material can be cast and forged or produced by powder metallurgy (PM). Nickel-based superalloys produced by this technique have shown high fatigue resistance, hot-corrosion resistance, high-temperature mechanical properties, and high oxidation resistance up to 900 °C [9]

Inconel®718 is also characterized by being a material with low stacking fault energy [10]. During hot deformation, dynamic recovery (DRV) due to the annihilation and arrangement of dislocations into subgrain boundaries is fast overcome by the formation of new grains due to discontinuous dynamic recrystallization (dDRX) [11]. In the initial stages of deformation, dislocation multiplication results in strain hardening. In low stacking fault energy alloys, DRV hardly proceeds due to the restricted movement of dislocations and their decomposition into stable partial dislocations [12]. The material undergoes dDRX at strains over the critical strain by the formation of nuclei and their growth by the movement of the high-angle grain boundary [13,14]. In the specific case of Inconel®718, it the formation of annealing twins during recrystallization is widely accepted. It is reported that twins with boundaries type $\Sigma 3$ are promoted at higher temperatures and lower strain rates [15]. The main restoration process during forging is dynamic recrystallization (DRX), which, in competition with the DRV, reduces the stored energy [16,17]. However, it is precisely after hot forming at an industrial scale where meta-dynamic (mDRX) and post-dynamic (pDRX) recrystallization play an important role in nickel-based superalloys, especially at high strain rates [18]. The final microstructures then depend on the deformation parameters, such as temperature, strain, and strain rate, and also on the material history after deformation.

The presence of solutes and precipitates reduces the mobility of dislocations and boundaries in general. The most well-known temperature–time transformation (TTT) diagram of Inconel®718 [19] shows that precipitates can be formed before deformation below 930 °C during the soaking time. This precipitation is accelerated if the deformation is applied before the annealing treatment as shown by Renhof, L. et al. [20]. The solvus of the δ phase occurs between 990 and 1020 °C [21]. Other works [22,23] concluded that the δ phase can partially dissolve and fracture during deformation. The remaining particles can further stimulate the nucleation of DRX grains by inducing the generation and rotation of subgrains. In the presence of δ phase particles, grain growth can be hindered by the Zener pinning effect [24]. Many studies have been done to characterize the behavior of standard Inconel®718 alloys under different conditions. The behavior of the low-temperature deformation is affected by the presence of precipitates, especially of the δ phase with a solvus temperature between 990 and 1020 °C, depending on the chemical composition [25]. Primary γ' precipitates can also promote the recrystallization process by a mechanism called heteroepitaxial recrystallization [26]. In this case, the recrystallized grains grow with the same crystallographic orientation as the precipitate.

The present work focuses on the study of the behavior of an alloy where the precipitation is strongly reduced. In this context, the main objective of the present work is to describe the hot deformation behavior of a nickel-based superalloy with modified chemical composition to simulate the behavior of the matrix Inconel®718 without the massive interaction with δ, γ' and γ''. The outcome is compared with standard Inconel®718.

2. Materials and Methods

2.1. Materials

The innovative investigated material identified as "IN718WP" is a nickel-based superalloy based on the chemical composition of the γ phase of a standard Inconel®718 after precipitation of the other phases [27]. The precipitation of γ', γ'' and δ phases is therefore strongly reduced. To achieve this, two main requirements for fundamental deviations from the chemical composition of standard Inconel®718 were changed. Firstly, Cr and Fe contents were slightly increased as solid solution strengtheners. Secondly, Al, Ti, and Nb contents were significantly decreased to avoid the formation of γ'', γ' and δ phases. C was kept low to reduce the formation of carbides. With the above-mentioned changes in the chemical composition, an IN718WP disc of diameter 195 mm was produced by powder metallurgy at Voestalpine BÖHLER Edelstahl GmbH (Kapfenberg, Austria). The chemical composition of the IN718WP (in weight percentage) is shown in Table 1. The nominal chemical composition of Inconel®718 alloy is included for comparison [28]. Powder metallurgy was used to produce a material with a homogeneous small grain size, avoiding any grain coarsening expected to occur during upsetting. The hot compression trials studied here were meant to close the micro-porosity of the as-received condition.

Table 1. Chemical composition of both standard Inconel®718 and without precipitates IN718WP alloys.

Material	Elemental Composition (wt.%)										
	Ni	Cr	Fe	Mo	Al	Co	Si	Nb	C	Mn	Ti
IN718WP	50.18	23	22.07	3.71	max 0.28	0.22	0.1	max 0.28	max 0.03	0.1	max 0.03
Inconel®718	50–55	17–21	bal.	2.8–3.3	0.2–0.8	1	0.35	4.75–5.5	0.08	0.35	0.7–1.3

2.2. Compression Tests

Samples of 8 mm diameter and 15 mm length were compressed in the Gleeble®3800 simulator at 5 different strain rates (0.001, 0.01, 0.1, 1 and 10 s^{-1}) and 6 different temperatures (900, 925, 950, 975, 1000 and 1025 °C), up to a maximum strain of 0.8. A K-type thermocouple was spot-welded at the surface of the specimen to measure and control the temperature during the process. The samples were (i) heated by direct resistance at 5 °C/s to the selected temperature; (ii) homogenized for 2 min at the same temperature; (iii) compressed; and (iv) in situ water-quenched with water jets to freeze the microstructure. Graphite and tantalum foils were used between the edges of the specimens and the anvils to reduce the effects of friction during hot deformation and to decrease the temperature gradient along the sample. An Ar atmosphere was used to mitigate oxidation during the tests. The flow curves obtained experimentally were temperature-corrected to represent isothermal deformation conditions based on the literature [29,30] as described by P. Wang et al. [31]. The flow curves were not corrected by friction. Contrary to the tendency obtained in the work of G. Tan et al. [32] and in many other publications, the largest barrelling observed in our work occurred at the smallest strain rate. This must be related to a different mechanism of barrelling, such as wedge cracking promoted by diffusion, already observed by S.C. Medeiros et al. [33].

The interdependence of peak flow stress σ, strain rate $\dot{\varepsilon}$ and deformation temperature T was described using semi-empirical equations, as done for Mg alloys by C. Poletti et al. [34] and for steels by M. El Mehtedi et al. [35]. The analysis of the flow stress at the peak value was carried out using the universal constitutive equation proposed by Sellars and Tegart [36]:

$$\mathrm{Asinh}(\alpha\sigma)^n = \dot{\varepsilon}\,\exp(Q/RT) = Z \tag{1}$$

where A and α are material constants, R is the gas constant (8.314 J/kmol), T is the absolute temperature, Q is the apparent energy of activation, and n is the stress exponent. The last term is identified as the Zener–Hollomon parameter Z. The α constant was obtained from Equation (2):

$$\alpha = \frac{\beta}{n_1} \tag{2}$$

The coefficients β and n_1 are obtained from Equations (3) and (4)

$$\dot{\varepsilon}\,\exp(Q/RT) = A_1\sigma^{n_1} \tag{3}$$

$$\dot{\varepsilon}\,\exp(Q/RT) = A_2\exp(\beta\sigma) \tag{4}$$

The slope of $\ln(\dot{\varepsilon})$ vs. $\ln(\sigma)$ gives n_1, while the slope of $\ln(\dot{\varepsilon})$ vs. σ gives β.

The power law equation was additionally used for comparison:

$$A_P \sigma^{n_P} = \dot{\varepsilon} \, \exp(Q_P/RT) = Z_P \tag{5}$$

The subscript p refers to the power law. The power law is useful at low strain rates and high temperatures (low Z), but it deviates from this relationship at higher Z values.

The relative softening is calculated using the difference of the stress values at the peak and the stress value at a given strain and divided by the stress at the peak. This relative softening was then plotted as a function of the strain. Additionally, the relative softening at 0.6 of strain was compared with values obtained from the literature for Inconel®718.

2.3. Microstructure Characterization

The samples were cut in the compression direction at their center, embedded in resin, ground and polished, and chemically attacked with a mixture of ethanol, hydrochloric acid, and hydrogen peroxide (75 mL C_2H_5OH + 33 mL HCl + 3 mL H_2O_2) for optical microscopy. The selected samples were observed with an optical microscope provided by the company Zeiss, and the grain size was obtained following the interception method recommended by the American Society for Testing and Materials (ASTM) in the norm ASTM E112-12 [37].

Some samples without etching were analyzed with a field emission gun scanning electron microscope (FEG-SEM) model Mira3 of the TESCAN company (Brno, Czech Republic) using an accelerating voltage between 18 and 20 kV and step sizes of 0.1 μm (for high resolution) and 0.7 μm (for lower resolution). A backscattered electron (BSE) detector was used to identify carbides and pores within the matrix. A Hikari detector provided by EDAX (AMETEK GmbH, Weiterstadt, Germany) was used for electron backscattered diffraction (EBSD) to measure the local crystal orientation. OIM Analysis™ software (AMETEK GmbH, Weiterstadt, Germany) was used to determine the grain size and recrystallization percentages. Firstly, a cleaning process was applied to the EBSD measurements based on the CI (confidence index) standardization and neighborhood correlation. The following parameters were used: grain tolerance angle of 15°, a minimal grain size of 2 pixels, and a CI higher than 0.9. Recrystallized grains were identified by two features: a grain boundary angle larger than 15° and a grain orientation spread (GOS) smaller than 2.5 [38]. The average recrystallized grain size was obtained out of these criteria and the recrystallization fraction was obtained by dividing the area covered by the recrystallized grains with the inspected area (120 μm × 200 μm). Due to the formation of annealing twins during dynamic recrystallization (DRX), the reconstruction of the grains was performed once, considering all high-angle grain boundaries (with twins), and the second time by considering the annealing twins Σ3 and Σ9 to belong to the grain (without twins). Finally, the recrystallized grain size $\varnothing_{RX}$ was correlated with the Zener–Hollomon parameter, rearranging the equation presented by G. Shen et al. [39] for a Waspaloy as follows:

$$\ln Z = B + C \ln \varnothing_{RX} \tag{6}$$

where B and C are materials constants.

3. Results

3.1. Initial Microstructure

The initial microstructure of the as-received material was characterized by a mean grain diameter (without twins) of 20.4 μm. The microstructure showed annealing twins, as identified with the optical microscope in Figure 1a,b. Light optical images also showed the presence of micropores at some grain boundaries and carbides with sizes smaller than 1 μm decorating the grain boundaries. Micropores are typical features in powder metallurgy products [40]. Apart from the carbides, no other precipitates were detected, even at high resolution (Figure 1b,c). Additionally, Fe-rich regions were observed regularly in the sample with sizes between 60 and 100 μm (see Figure 1d,e), due to a lack of Fe mixing in the matrix during the sintering process. These Fe-rich regions occupied 1.25% of the volume.

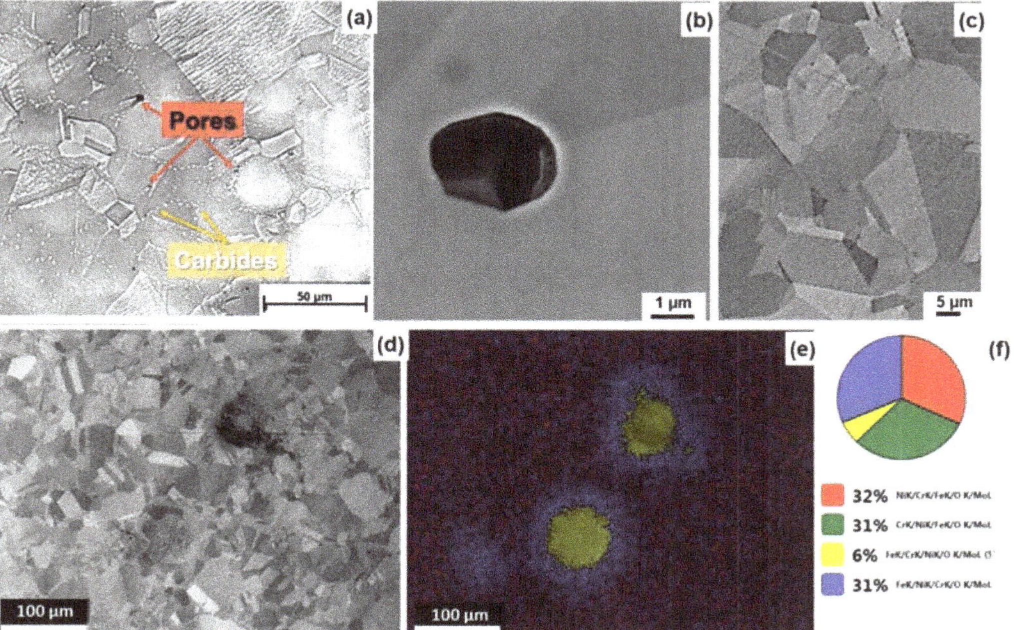

Figure 1. Microstructure of the as-received material: (**a**) light optical micrographs showing grains with annealing twins, carbides decorating some boundaries (light) and pores (black); (**b**) SEM picture in backscattered electron (BSE) mode showing a pore and no precipitate within the matrix; (**c**) SEM picture in BSE mode showing annealing twin carbides and pores (in black); (**d**) SEM micrographs in BSE mode; and (**e**) the corresponding EDS map, showing the grains of based material and the Fe-rich regions, respectively. The percentages are shown in (**f**).

The measured hardness of the as-received material was 199 HV, a value lower than the standard Inconel®718 alloy after solid solution treatment (245 HV), and much lower than the values of the standard Inconel®718 alloy after heat treatments (up to 500 HV) [41].

3.2. Flow Curves

The experimental flow curves corrected by temperature are shown in Figure 2. The curves present the typical characteristic of DRX: a single peak up to the peak stress, preceded by a rapid rise [18]. After the peak had been reached, the stress drops reached steady-state flow stress and suggested that recrystallization was not completed for all deformations. The strains at which the peak stress values are reached are a function of the temperature and the strain rate. The higher the temperature and the slower the deformation, the smaller the strains to start (just before peak stress) and finish (flow steady-state) the recrystallization. The increase with a large slope of the stress is a characteristic of alloys with low stacking fault energy (SFE) [11], such as the nickel alloy studied here.

In the whole temperature range, the peak stress decreased with increasing temperature and decreasing strain rate. Softening was more noticeable in the flow curves of the lowest strain rates (0.001–0.1 s^{-1}). The curves corresponding to the highest strain rates (1 and 10 s^{-1}) showed lower yield stress than the values reported for Inconel®718 at any temperature [42,43]. At high strain rates and low temperatures, a very small yield strength was calculated, and two hardening slopes were identified in Figure 2a. These phenomena could be related to the activity of deformation bands.

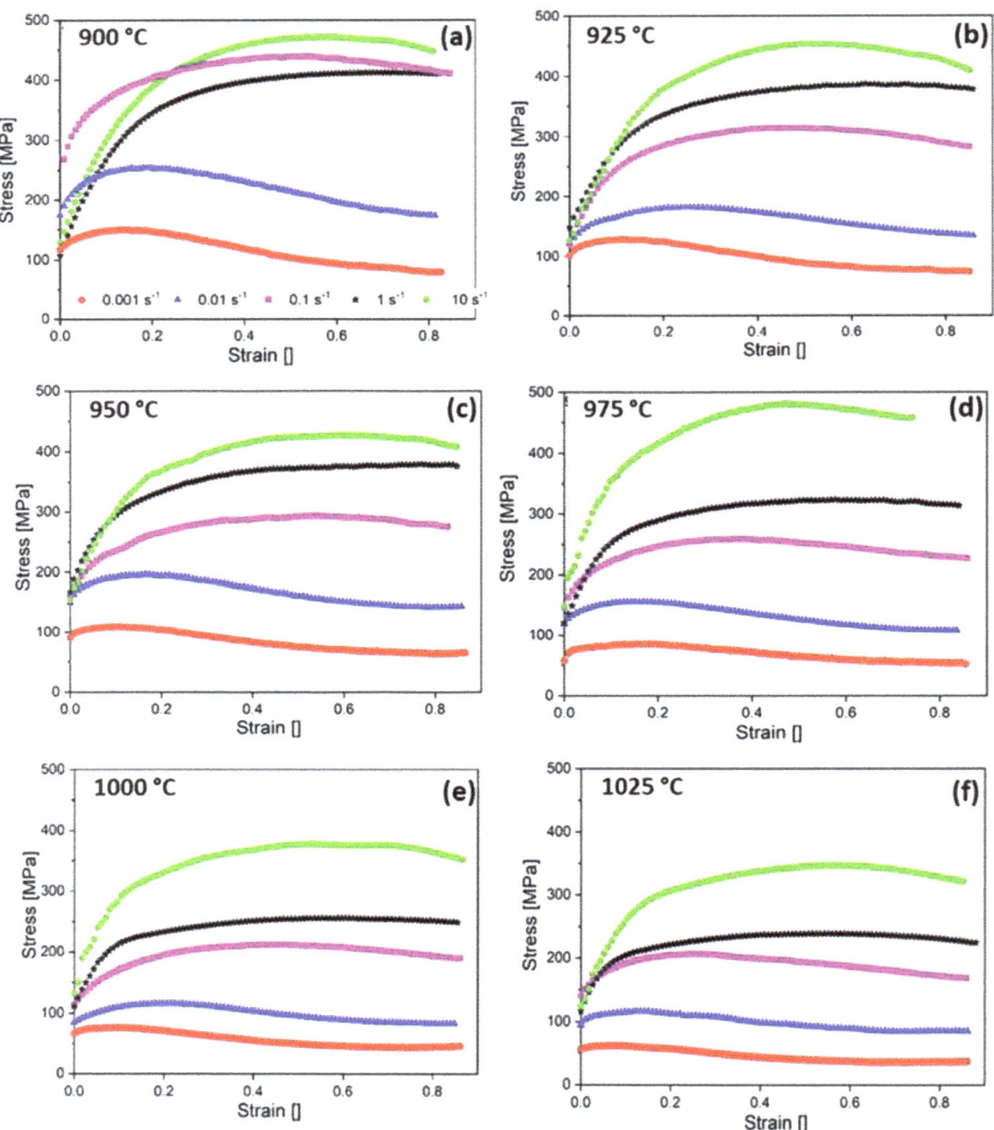

Figure 2. Flow curves of the IN718WP alloy for all strain rates (0.1–10 s^{-1}) at (**a**) 900 °C, (**b**) 925 °C, (**c**) 950 °C, (**d**) 1000 °C, (**e**) 1025 °C, and (**f**) 1050 °C.

The relative softening was plotted as a function of the strain in Figure 3. This softening was highly dependent on the strain rate, while the temperature had a negligible influence on the softening values and the shape of the curves. In general, it can be observed that at the strain rates between 0.1 and 10 s^{-1}, the hardening part consisted of a curve with two slopes. Then, the curves reached a minimum, corresponding to the peak stress. This minimum occurred at higher strains. The temperature plays a secondary role, as observed for comparison in Figure 4. It can be seen that the temperature has an influence on the hardening slope and on the strain at the peak stress, where the softening reaches zero. The higher the temperature, the lower the strain required to reach this condition. Furthermore, the softening rate, given by the slope after the zero value, is not influenced by the temperature at the measured conditions.

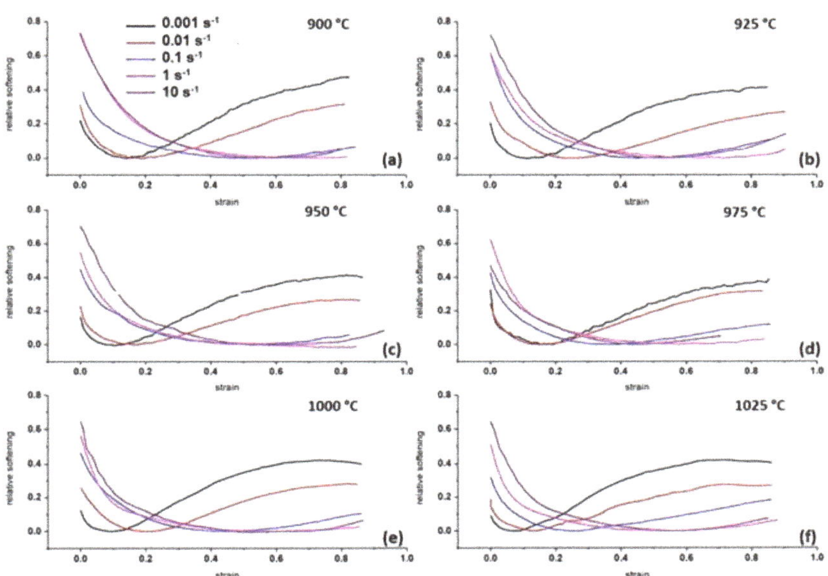

Figure 3. Relative softening as a function of the strain during deformation at (**a**) 900 °C, (**b**) 925 °C, (**c**) 950 °C, (**d**) 975 °C, (**e**) 1000 °C and (**f**) 1025 °C at different strain rates from 0.001 to 10 s^{-1}.

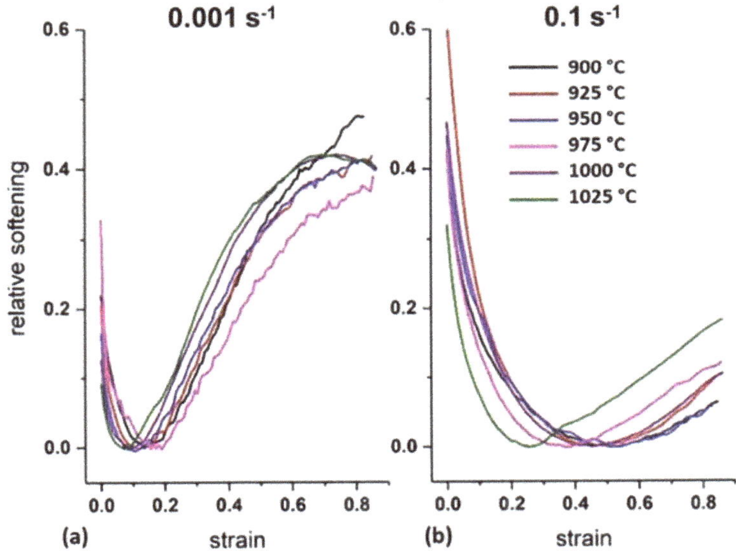

Figure 4. Relative softening showing the influence of the temperature at two different strain rates, at (**a**) 0.001 s^{-1} and (**b**) 0.1 s^{-1}.

While the softening at high strain rates is comparable with values in the literature for Inconel®718 [18,42,43], much larger values of softening are observed at strain rates below $0.01s^{-1}$. The softening of the model alloy IN718WP can reach 40% at the lowest strain rates, and only 5% at the highest strain rate (Figure 5).

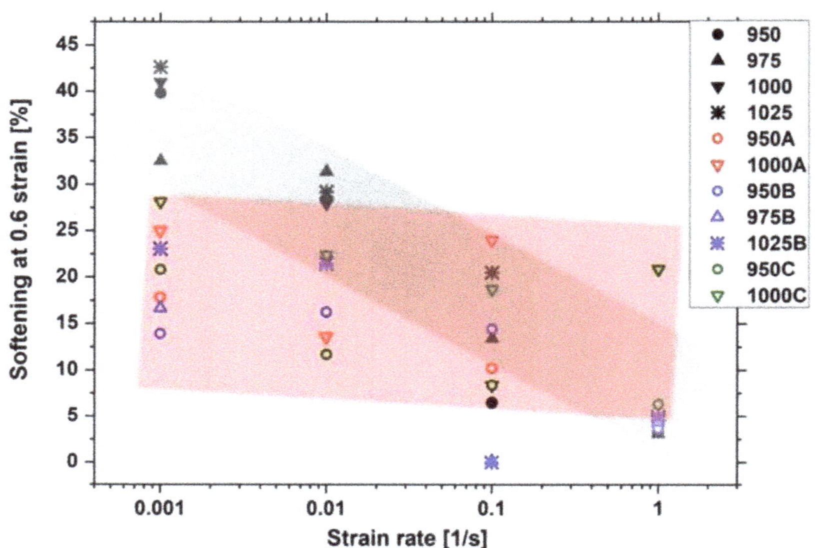

Figure 5. Comparison of softening at 0.6 of strain for different deformation conditions compared with standard Inconel® 718 obtained from the literature (A = [17] red, B = [42] blue, C = [43] green). The softening values are grouped in regions: grey, the values of the present work for IN718, and red the values of the literature for Inconel® 718.

3.3. Phenomenological Relationships

The peak stresses were correlated with the Zener–Hollomon parameter using Equations (1) and (4). The calculated average values of β and n_1 calculated from Equations (2) and (3) were 0.0275 and 60,282, respectively. The activation energy Q = 450 kJ/mol, the stress exponent n = 4.76, and α = 0.0046 were obtained using the sinh relationship (Equation (1)). The mean values of n and Q obtained for the creep law (Equation (4)) are n = 6 and Q = 400 kJ/mol. Using the creep relationship, Q has shown to be highly dependent on the strain rate $\dot{\varepsilon}$. Figure 6 shows the correlation of the peak stress with the Zener–Hollomon parameter calculated using both approaches. It can be seen that the correlation provided with Equation (1) deviates less at higher Z values than the power law (Equation (4)).

The values of the apparent activation energies Q obtained using Equations (1) and (4) are in agreement with the literature for Inconel®718 [8,23,43–45] (see Table 2) using similar α values for the sinh equation type. On the other hand, the stress exponents n found in the literature for Inconel®718 are similar to the value found here (4.76) when deforming above the solvus temperature. This must be related to a similar plastic deformation mechanism for both materials. A dependency of n with temperature was reported in the literature as well ([8,23]). The high values at low temperatures were attributed to the interaction of dislocations with particles [8]. This observation was in contradiction with the dependency of n with the temperature [23].

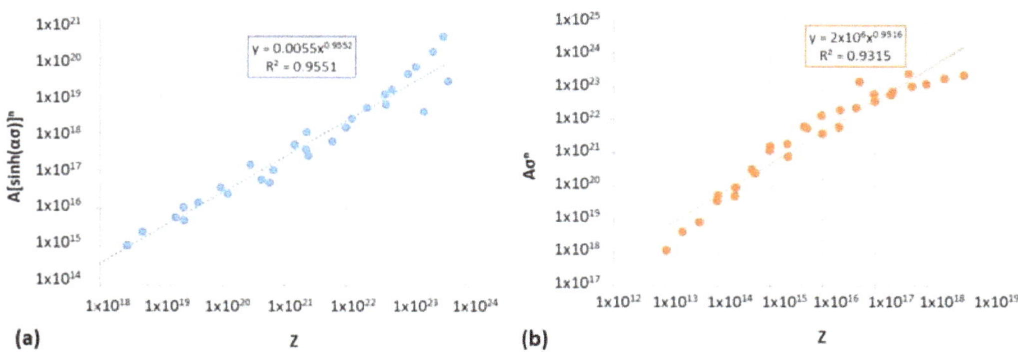

Figure 6. Correlation of the peak stress values with the Zener–Hollomon parameter using (**a**) Equation (1) and (**b**) Equation (4).

Table 2. Material constant values calculated from Equations (1) and (4) for IN718WP compared the literature for Inconel®718.

Alloy	α [MPa^{-1}]	Q [kJ/mol]	n	Source
IN718WP	0.0046	450	4.76	Equation (1)
	-	400	6	Equation (2)
Inconel®718	0.0042	364	4.57	[43]
	0.006104	461	3.52 (960 °C) 4.23 (1020 °C)	[23]
	NA	467	7.3, 6.3, 5.4 and 5.2	[8] (n varies with temperature)
	0.0054	468	5.2	[44] (at the peak)
	0.0054	427.6	4.12	[45] (above solvus)

3.4. Microstructure after Deformation and Damage

The light optical micrographs shown in Figure 7 depict the microstructure after deformation. Two main features can be distinguished: the presence of DRX grains, and the presence of damage (pore coarsening). Recrystallization twins are present within new recrystallized grains, and deformed annealed twins are observed inside deformed grains. As the strain rate increases and the deformation temperature decreases, the recrystallization grade is generally lower.

The evolution of the porosity was also studied. Figure 7 shows a larger number of pores for the lower strain rates. Coalescence and growth of micropores resulting in wedge cracking were produced favorably at high temperatures and long times of deformation [46]. This effect was also observed in other powder metallurgy-produced nickel-based alloys deformed at low strain rates [47]. In Figure 7, it can be seen that the damage is particularly large in the barrelled region due to multiaxial stress condition. On the other hand, the center of the sample shows much less porosity.

Figure 8 shows the grain orientation spread (GOS) maps of microstructures after deformation at different conditions of strain rate and temperature. Recrystallized grains were distinguished from the deformed ones by their level of internal misorientation, expressed in this case by the level of the GOS value. In general, the higher the strain rate and lower the temperature, the finer the recrystallized grains, in agreement with the literature for Inconel®718 [15]. Annealing twins were observed in all the recrystallized grains, as reported for Inconel®718 [48].

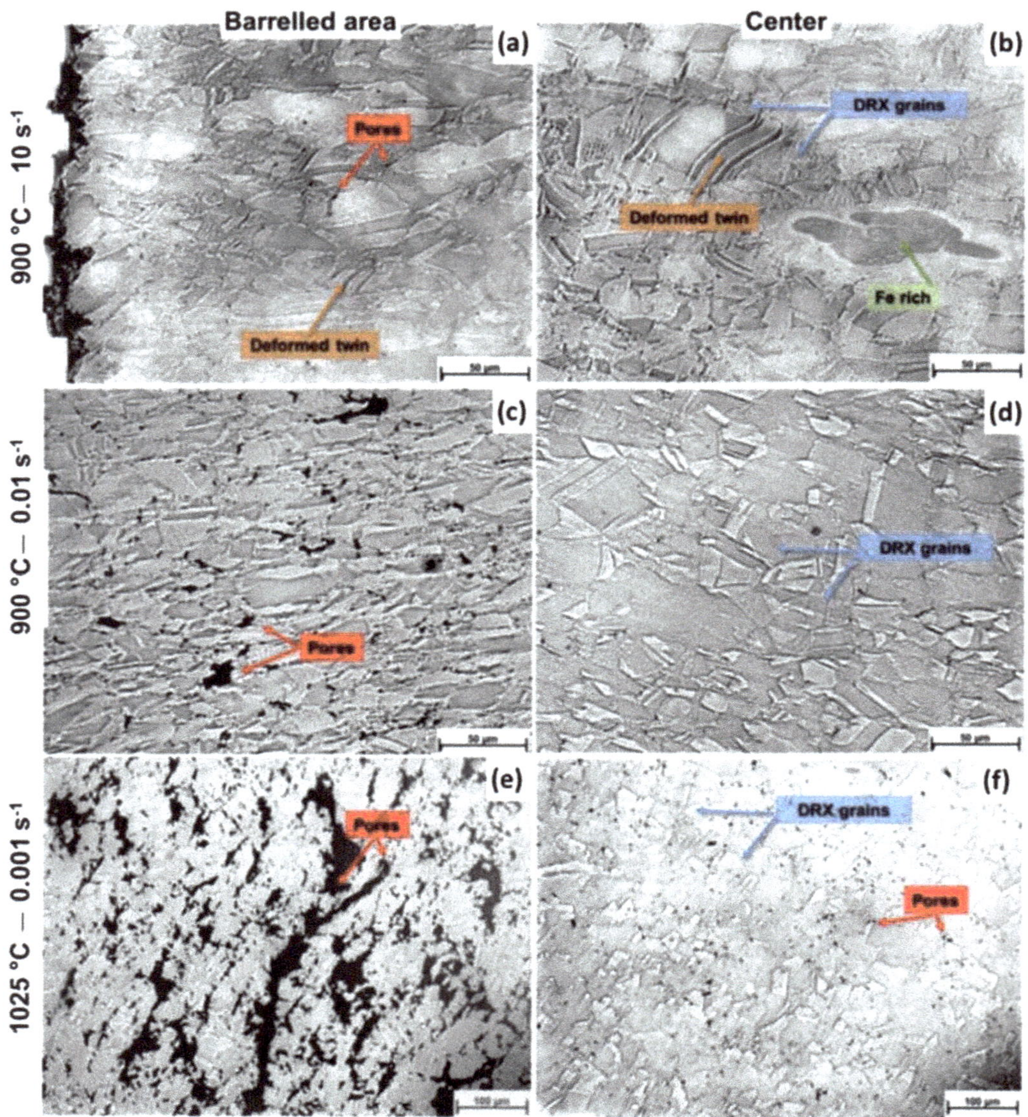

Figure 7. Light optical micrographs of IN718WP deformed after several conditions: (**a**,**b**) 900 °C and 10 s^{-1}, (**c**,**d**) 900 °C and 0.01 s^{-1} and (**e**,**f**) 1025 °C and 0.001 s^{-1}. The pictures correspond to the center of the sample (**b**,**d**,**f**) and the barrelled area (**a**,**c**,**e**). The micrographs show recrystallized and deformed grains, as well as pores at the center and wedge cracking at the barreled region.

The results of the recrystallized grain sizes are shown in Figure 9. The tendency shows that the larger the strain rate and the lower the temperature, the smaller the recrystallized grain [17]. The exceptions at high strain rates and low temperatures may be due to mDRX effects. An overestimation of the size occurs when using image analysis of light optical microscopy samples. Further analysis showed that measured recrystallized grain size at the lowest strain rates had a large standard deviation, meaning that there were multiple stages of the recrystallization happening simultaneously.

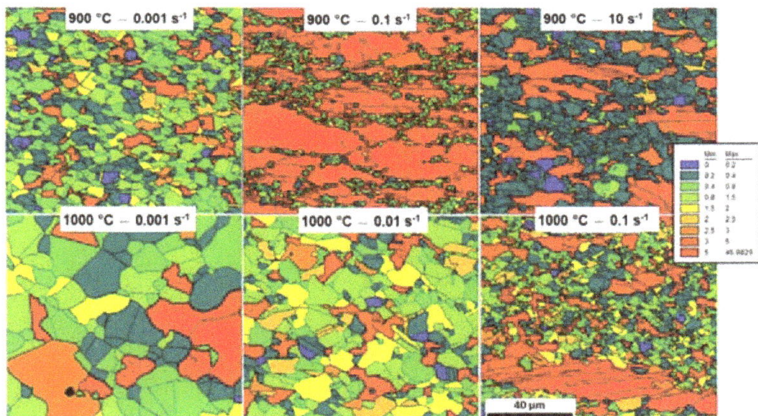

Figure 8. Grain orientation spread (GOS) obtained from SEM-EBDS measurements of samples deformed at several conditions. Grains can be considered as deformed (red and orange) and recrystallized (green, blue, and yellow).

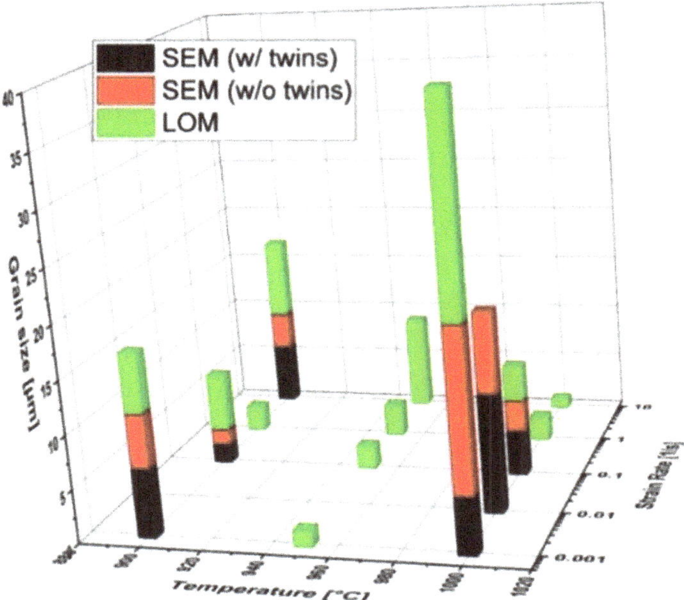

Figure 9. Grain size estimated using different methods: light optical microscopy (LOM), and scanning electron microscopy–electron backscattered diffraction (SEM-EBSD) without extracting the twin boundaries (w/twins) and extracting the twins (w/o twins) from the high-angle grain boundaries (HAGB).

4. Discussion on the Deformation Mechanisms

This section discusses the evolution of the microstructure and the correlation with the flow curves for the IN718WP. Furthermore, a comparison with Inconel®718 is conducted, where the deformation of the matrix takes place without interacting with precipitates.

The details of the microstructures of two samples partially recrystallized are shown in Figure 10. The grain reference orientation deviation (GROD) maps show the rotation of the grains, especially markedly close to the grain boundaries, and the formation of bands. The highly misoriented regions result in a fragmentation of the grains, especially at high strain rates and low temperatures. This effect corroborates the measurement of the misorientation cumulating along the deformed grain (see Figure 10c).

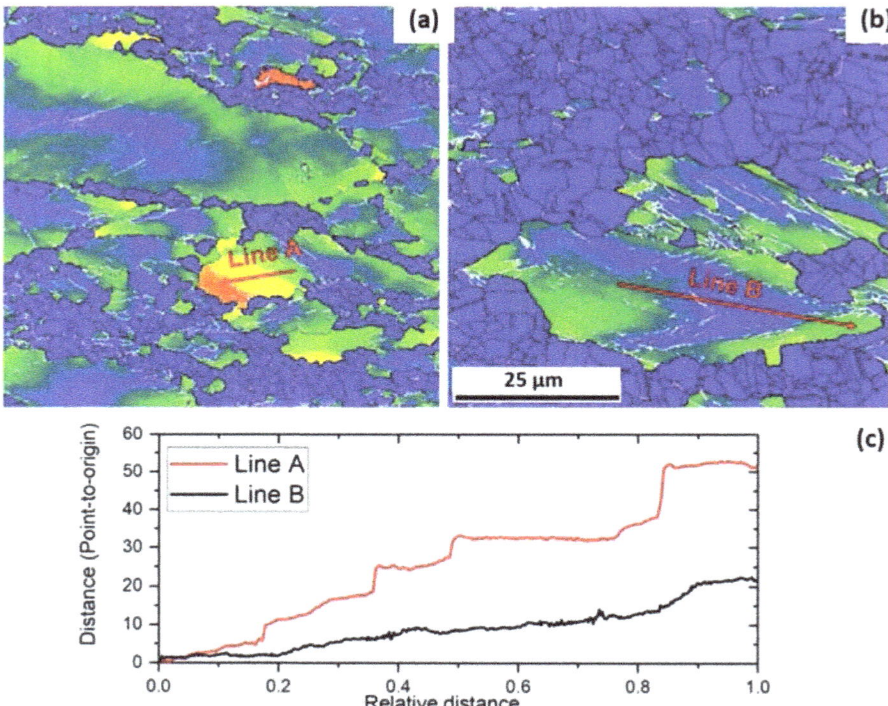

Figure 10. Detail of the microstructure of two samples after hot deformation at (**a**) 900 °C and 0.1 s^{-1} and (**b**) 900 °C and 10 s^{-1} showing partial recrystallization. Grain reference orientation deviation (min = 0° blue; max = 60° red). Black lines represent high-angle grain boundaries (above 15°), and white lines the low angle grain boundaries. (between 2° and 15°). The results of the misorientation scan through Line A and Line B can be seen in (**c**).

The intense crystal rotation shown in the deformed grains for strain rates higher than 0.1 s^{-1} is further illustrated in Figure 11a,b. The rotation of the grain and the change in the crystal orientation can be observed in the pole figure (PF) and the inverse pole figure (IPF). This microstructural aspect is related to the change in the slope during strain hardening shown in the flow curves Figure 2. The formations of bands at grains and existing annealing twins have the same amplitude as the serrations formed at all high-angle grain boundaries. This intense rotation and fragmentation of grains is responsible for the change in the hardening rate, as observed from the change in the slope of the softening curves up to the minimum value (Figure 3). A comparison of the microstructure can be observed when deformation occurs at low strain rate (Figure 11c). In this example, the non-recrystallized grains undergo dynamic recovery before recrystallization, thus reducing the stored energy.

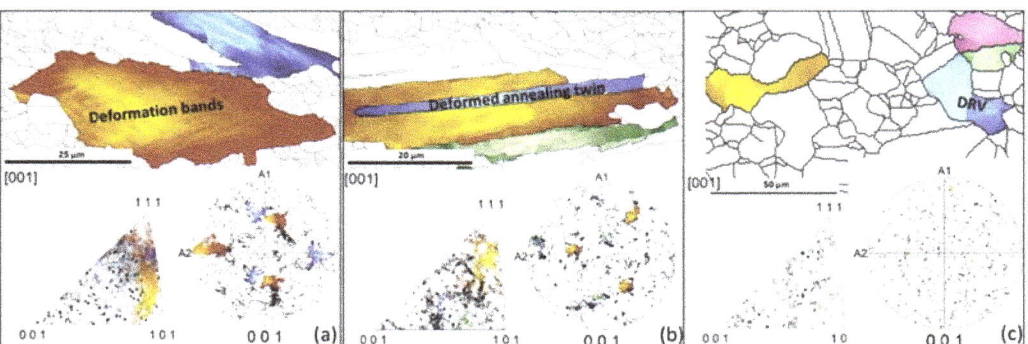

Figure 11. High-angle grain boundary maps (black lines) of samples after deformation at (**a**) 900 °C and 10 s^{-1}, (**b**) 1000 °C and 0.1 s^{-1} (**c**) 1000 °C and 0.001 s^{-1} (dynamic recovered grains—DRV are observed). Rotation of the grains can be observed in the inverse pole figures and pole figures. A1 and A2 refer to the reference axis.

The strain at the peak stress was plotted in Figure 12a. In general, the higher the strain rate, the larger the strain at the peak stress. The recrystallization percentage shows the opposite dependency on the strain rate than the strain at peak stress. The temperature has a negligible effect on the strain at the peak stress at the investigated conditions. Results in Figure 12 show that the percentage of recrystallization increases as the strain rate decreases in the 900–1000 °C range. Both diagrams have the same outcome: longer tests (slower strain rate) promote the formation of new grains, because the process is diffusion-controlled. Regardless of the deformation conditions, a minimum of 50% of recrystallization is always reached. It was shown that post-dynamic recrystallization pDRX can occur isothermally within some seconds [49]. Quenching seems to be fast enough to avoid post-dynamic recrystallization at high strain rates. The delay of 2–3 s as reported by A. Nicolaÿ et al. [50] was drastically reduced to 0.5 s in this work due to the in-situ water jet system. Therefore, the higher the strain rate, the lower the recrystallization grade. Concerning the temperature, lower recrystallization grades were observed at the middle-temperature range. This must be related with the competition of dynamic recovery and dynamic recrystallization: higher stored energy is achieved at lower temperatures, competing with the increment of the mobility of high-angle grain boundaries at high temperatures.

The types of boundaries at deformed and recrystallized grains were analyzed from the EBSD measurements. The fractions (f) of these boundaries are plotted in Figure 13 and are compared with the results obtained from the material in the as-received conditions. The deformed grains present a large number of low-angle grain boundaries (LAGB) due to lattice rotation and formation of deformation bands at high strain rates, and due to the formation of subgrains at smaller strain rates. The sigma 3 boundaries are nonexistent in the deformed grains, due to the progressive transformation of these boundaries into mobile boundaries [4]. Recrystallized grains show a negligible fraction of LAGB (fLAGB), and the microstructure mainly consists of a large fraction of HAGB (fHAGB) and sigma 3 grain boundaries (fsigma3) Figure 13.

Finally, the softening rate, identified as the slope of the relative softening from the peak stress, is plotted in Figure 14. The softening rate is highly dependent on the strain rate, while a temperature tendency could not be identified. The high values of the softening rate at low strain rates are a result of the pores coarsening as well as the fast growth of recrystallized grains without the interaction with any precipitate.

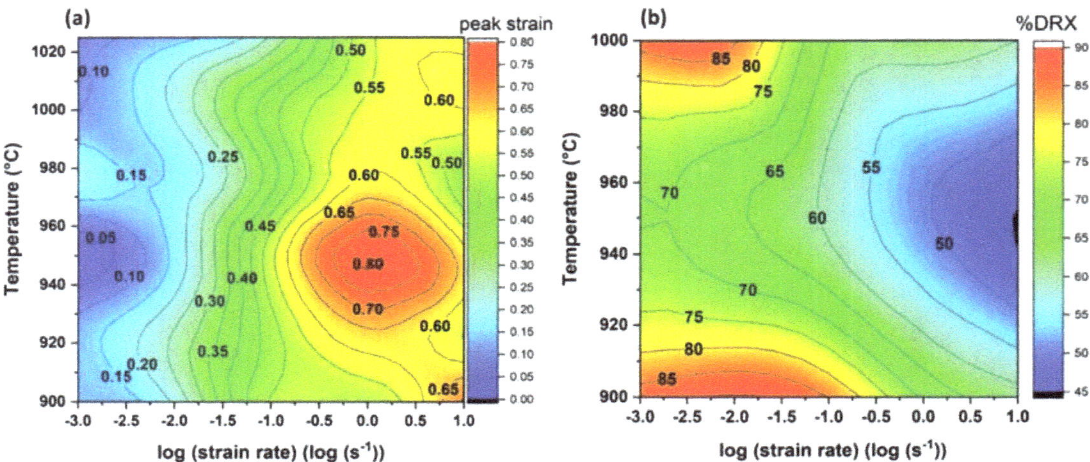

Figure 12. (**a**) Strain at the peak stress as a function of the temperature and the strain rate, and (**b**) recrystallization percentages of the IN718WP alloy for temperatures (900–1000 °C) and strain rates (0.1–10 s^{-1}) based on LOM and SEM analysis average values.

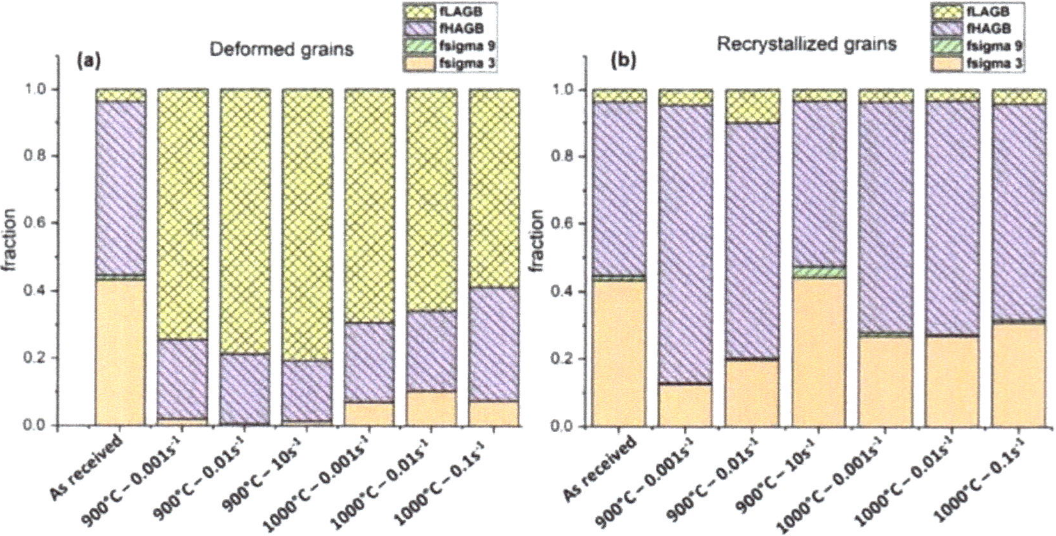

Figure 13. Fractions of the different type of boundaries: high-angle grain boundary (fHAGB), low-angle grain boundary (fLAGB); Σ3 (fsigma3) and Σ9 (fsigma9) measured at (**a**) deformed grains (**b**) recrystallized grains after different deformation conditions.

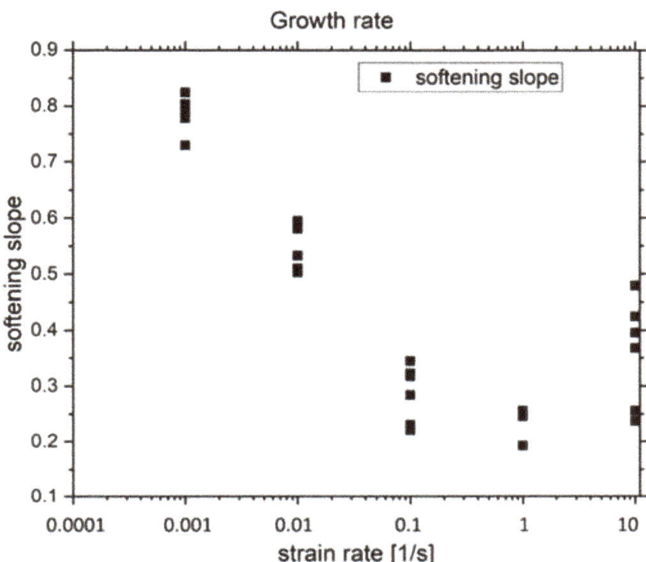

Figure 14. Slope of the softening region calculated from the softening curves shown in Figure 4.

The correlation of the recrystallized grain size with the Zener–Hollomon parameter was determined using both the sinh law (Equation (1)) and the power law (Equation (2)), by Equation (3). The measured grain sizes from EBSD data were analyzed using two approaches: the grain size as measured, and the grain size after extracting the twin boundaries. A linear correlation was obtained in a double logarithmic scale in all cases (Figure 15). When using the grain size, a slope of 3 was obtained using both approaches (Figure 15a), while a slope of 3.5 was obtained when extracting the twin boundaries from the grain size calculation (Figure 15b).

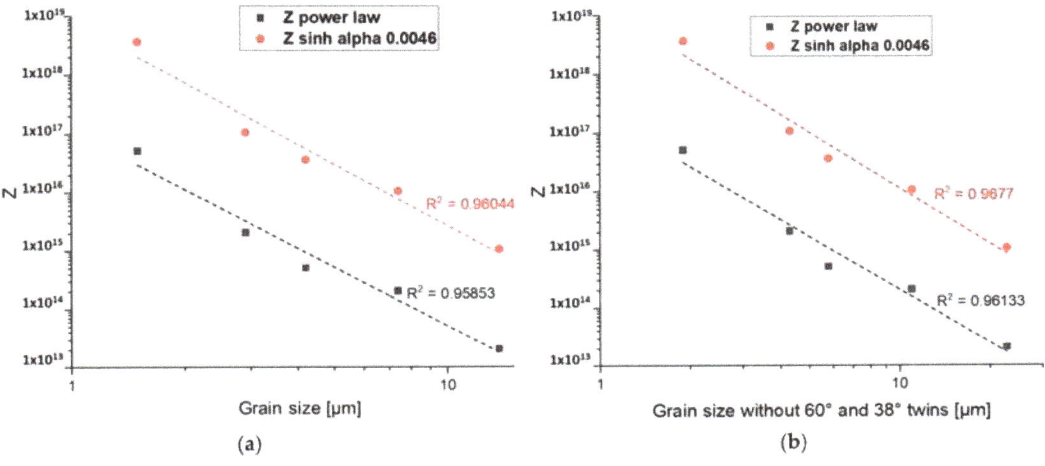

Figure 15. Correlation of the Zener–Hollomon parameter with the recrystallized grain size obtained by EBSD measurements (**a**) without taking into account the twins (**b**) extracting the twin boundaries.

5. Summary and Conclusions

The material free of precipitates, IN718WP produced by powder metallurgy, was tested under hot compression to understand the deformation behavior of this nickel-based superalloy. The following can be concluded:

- The material IN718WP deforms similarly to Inconel®718: after reaching peak stress, the stresses soften due to DRX;
- Lower stress values and larger softening than Inconel®718 may be related to the lack of precipitates;
- The relative stress softening is independent of the temperature, meaning that the growth is only dependent on the strain rate. The large stress softening at a low strain rate may be related to two phenomena: the fast recrystallization rate, and the coarsening of pores driven by diffusion;
- The growth rate of recrystallization is only dependent on the strain rate at the deformed conditions. This results in larger amounts of Σ3 boundaries within recrystallized grains at high strain rate for a given strain;
- The recrystallization grade %DRX increases from 45% at low strain rates up to 85% at the slowest deformation;
- The recrystallization grade reaches a minimum of 70% at 950 °C due to the competition of dynamic recovery and dynamic recrystallization occurring at low strain rates;
- Deformation takes place by first increasing the amount of low angle grain boundaries, while the Σ3 boundaries become mobile and their fraction decreases with deformation. Recrystallized grains have a large number of Σ3 boundaries, and their fraction increases with increasing strain rate and temperature;
- The apparent activation energy of 450 kJ/mol found for IN718WP and the strain rate exponent (n = 4.76) value are in agreement with the values found for Inconel®718 when deforming above the solvus temperature, suggesting similar deformation mechanisms.

Author Contributions: Conceptualization, F.L., A.S. and M.C.P.; methodology, F.L., K.P. and M.C.P.; formal analysis, F.L., K.P., A.S., S.S. and M.C.P.; investigation, F.L., K.P. and M.C.P.; resources, A.S. and M.C.P.; data curation, F.L., K.P. and M.C.P.; writing—original draft preparation, F.L.; writing—review and editing, F.L., K.P., A.S., S.S. and M.C.P.; visualization, F.L. and M.C.P.; supervision, S.S. and M.C.P.; project administration, A.S. and M.C.P.; funding acquisition, M.C.P. All authors have read and agreed to the published version of the manuscript.

Funding: This research was funded by Christian Doppler Forschungsgesellschaft, in the framework of CD-Laboratory for Design of High-Performance Alloys by Thermomechanical Processing. The authors accept Open Access Funding by the Graz University of Technology.

Institutional Review Board Statement: Not applicable.

Informed Consent Statement: Not applicable.

Data Availability Statement: The raw/processed data that support the findings of this study are available from the corresponding author, M.C.P. upon request.

Conflicts of Interest: The authors declare no conflict of interest.

References

1. Akca, E.; Gursel, A. A Review on Superalloys and IN718 Nickel-Based INCONEL Superalloy. *Period. Eng. Nat. Sci.* **2015**, *1*. [CrossRef]
2. Mouritz, A.M. (Ed.) Superalloys for Gas Turbine Engines. In *Introduction to Aerospace Materials*; Woodhead Publishing: Cambridge, UK, 2012; Chapter 12; pp. 251–267.
3. Reed, R.C. *The Superalloys: Fundamentals and Applications*; Cambridge University Press: Cambridge, UK, 2006.
4. Schafrik, R.E.; Ward, D.D.; Groh, J.R. Application of Alloy 718 in GE Aircraft Engines: Past, Present and Next Five Years. In *Superalloys 718, 625, 706 and Various Derivatives*; TMS: Pittsburgh, PA, USA, 2001; pp. 1–11.
5. Ahmadi, M.R.; Povoden-Karadeniz, E.; Whitmore, L.; Stockinger, M.; Falahati, A.; Kozeschnik, E. Yield strength prediction in Ni-base alloy 718Plus based on thermo-kinetic precipitation simulation. *Mater. Sci. Eng. A* **2014**, *608*, 114–122. [CrossRef]

6. Cozar, R.; Pineau, A. Morphology of γ′ and γ″ precipitates and thermal stability of inconel 718 type alloys. *Metall. Trans.* **1973**, *4*, 47–59. [CrossRef]

7. Sundararaman, M.; Mukhopadhyay, P.; Banerjee, S. Some aspects of the precipitation of metastable intermetallic phases in INCONEL 718. *Metall. Trans. A* **1992**, *23*, 2015–2028. [CrossRef]

8. Wang, Y.; Shao, W.Z.Z.; Zhen, L.; Zhang, B.Y.Y. Hot deformation behavior of delta-processed superalloy 718. *Mater. Sci. Eng. A* **2011**, *528*, 3218–3227. [CrossRef]

9. Gasson, P.C. The Superalloys: Fundamentals and Applications. *Aeronaut. J.* **2008**, *112*, 291. [CrossRef]

10. Zhou, L.; Baker, T. Effects of strain rate and temperature on deformation behaviour of IN 718 during high temperature deformation. *Mater. Sci. Eng. A* **1994**, *177*, 1–9. [CrossRef]

11. Zhang, H.Y.; Zhang, S.H.; Li, Z.X.; Cheng, M. Hot die forging process optimization of superalloy IN718 turbine disc using processing map and finite element method. *Proc. Inst. Mech. Eng. Part B J. Eng. Manuf.* **2009**, *224*, 103–110. [CrossRef]

12. Huang, K.; Logé, R.E. A review of dynamic recrystallization phenomena in metallic materials. *Mater. Des.* **2016**, *111*, 548–574. [CrossRef]

13. Humphreys, F.J.; Hatherly, M. *Recrystallization and Related Annealing Phenomena*; Elsevier: Amsterdam, The Netherlands, 2004.

14. Doherty, R.D.; Hughes, D.A.; Humphreys, F.J.; Jonas, J.J.; Jensen, D.J.; Kassner, M.E.; King, W.E.; McNelley, T.R.; McQueen, H.J.; Rollett, A.D. Current issues in recrystallization: A review. *Mater. Sci. Eng. A* **1997**, *238*, 219–274. [CrossRef]

15. Azarbarmas, M.; Aghaie-Khafri, M.; Cabrera, J.M.M.; Calvo, J. Dynamic recrystallization mechanisms and twining evolution during hot deformation of Inconel 718. *Mater. Sci. Eng. A* **2016**, *678*, 137–152. [CrossRef]

16. *Madeleine Durand-Charre, The Microstructure of Superalloys*; CRC Press: Boca Raton, FL, USA, 1998.

17. Azarbarmas, M.; Aghaie-Khafri, M.; Cabrera, J.M.; Calvo, J. Microstructural evolution and constitutive equations of Inconel 718 alloy under quasi-static and quasi-dynamic conditions. *Mater. Des.* **2016**, *94*, 28–38. [CrossRef]

18. Zhou, L.X.; Baker, T.N. Effects on dynamic and metadynamic recrystallization on microstructures of wrought IN-718 due to hot deformation. *Mater. Sci. Eng. A* **1995**, *196*, 89–95. [CrossRef]

19. Oradei-Basile, A.; Radavich, J.F. A current T-T-T diagram for wrought alloy 718. In *Superalloys 718, 625 and Various Derivates*; Loria, E.A., Ed.; The Minerals, Metals & Materials Society: Pittsburgh, PA, USA, 1991; pp. 325–335.

20. Renhof, L.; Krempaszky, C.; Werner, E.; Stockinger, M. Analysis of Microstructural Properties of IN 718 after High Speed Forging. *Proc. Int. Symp. Superalloys Var. Deriv.* **2005**. [CrossRef]

21. Cai, D.Y.; Zhang, W.H.; Nie, P.L.; Liu, W.C.; Yao, M. Dissolution kinetics and behavior of δ phase in Inconel 718. *Trans. Nonferrous Met. Soc. China* **2003**, *13*, 1338–1341.

22. Lin, Y.C.; He, D.G.; Chen, M.S.; Chen, X.M.; Zhao, C.Y.; Ma, X.; Long, Z.L. EBSD analysis of evolution of dynamic recrystallization grains and δ phase in a nickel-based superalloy during hot compressive deformation. *Mater. Des.* **2016**, *97*, 13–24. [CrossRef]

23. Páramo-Kañetas, P.; Özturk, U.; Calvo, J.; Cabrera, J.M.; Guerrero-Mata, M. High-temperature deformation of delta-processed Inconel 718. *J. Mater. Process. Technol.* **2018**, *255*, 204–211. [CrossRef]

24. Song, K.; Aindow, M. Grain growth and particle pinning in a model Ni-based superalloy. *Mater. Sci. Eng. A* **2008**, *479*, 365–372. [CrossRef]

25. John, S.R.; Tien, K. (Eds.) *Refractory Alloying Elements in Superalloys*; American Society for Metals and Associacao Brasileira de Metais: Araxá, Brazil, 1984.

26. Charpagne, M.-A.; Billot, T.; Franchet, J.-M.; Bozzolo, N. Heteroepitaxial recrystallization: A new mechanism discovered in a polycrystalline γ-γ′ nickel based superalloy. *J. Alloy. Compd.* **2016**, *688*, 685–694. [CrossRef]

27. Drexler, A.; Oberwinkler, B.; Primig, S.; Turk, C.; Povoden-Karadeniz, E.; Heinemann, A.; Ecker, W.; Stockinger, M. Experimental and numerical investigations of the γ″ and γ′ precipitation kinetics in Alloy 718. *Mater. Sci. Eng. A* **2018**, *723*, 314–323. [CrossRef]

28. Special Metals INCONEL® Alloy 718. 2007. Available online: https://www.specialmetals.com/ (accessed on 18 March 2021).

29. Goetz, R.L.; Semiatin, S.L. The adiabatic correction factor for deformation heating during the uniaxial compression test. *J. Mater. Eng. Perform.* **2001**, *10*, 710–717. [CrossRef]

30. Yan, L.; Shen, J.; Li, Z.; Li, J.; Yan, X. Microstructure evolution of Al-Zn-Mg-Cu-Zr alloy during hot deformation. *Rare Met.* **2010**, *29*, 426–432. [CrossRef]

31. Wang, P.; Hogrefe, K.; Piot, D.; Montheillet, F.; Poletti, M.C. A flow instability criterion for alloys during hot deformation. *Proced. Manuf.* **2019**, *37*, 319–326. [CrossRef]

32. Tan, G.; Li, H.Z.; Wang, Y.; Yang, L.; Huang, Z.Q.; Qiao, S.C.; Liu, M.X. Physical-Based Constitutive Modeling of Hot Deformation in a Hot-Extruded Powder Metallurgy Nickel-Based Superalloy. *J. Mater. Eng. Perform.* **2021**, *30*, 794–804. [CrossRef]

33. Medeiros, S.C.; Frazier, W.G.; Prasad, Y.V.R.K. Hot deformation mechanisms in a powder metallurgy nickel-base superalloy IN 625. *Metall. Mater. Trans. A* **2000**, *31*, 2317–2325. [CrossRef]

34. Poletti, C.; Dieringa, H.; Warchomicka, F. Local deformation and processing maps of as-cast AZ31 alloy. *Mater. Sci. Eng. A* **2009**, *516*. [CrossRef]

35. El Mehtedi, M.; Gabrielli, F.; Spigarelli, S. Hot workability in process modeling of a bearing steel by using combined constitutive equations and dynamic material model. *Mater. Des.* **2014**, *53*, 398–404. [CrossRef]

36. Sellars, C.M.; McTegart, W.J. On the mechanism of hot deformation. *Acta Metall.* **1966**, *14*, 1136–1138. [CrossRef]

37. *ASTM E112-12 Standard Test Methods for Determining Average Grain Size*; ASTM International: Conshohocken, PA, USA, 2012.

38. Mitsche, S.; Poelt, P.; Sommitsch, C. Recrystallization behaviour of the nickel-based alloy 80 a during hot forming. *J. Microsc.* **2007**, *227*, 267–274. [CrossRef]
39. Shen, G.; Semiatin, S.L.; Shivpuri, R. Modeling microstructural development during the forging of Waspaloy. *Metall. Mater. Trans. A* **1995**, *26*, 1795–1803. [CrossRef]
40. Dobrzanski, L.A. (Ed.) *Powder Metallurgy—Fundamentals and Case Studies*; InTech Open: London, UK, 2017.
41. Slama, C.; Servant, C.; Cizeron, G. Aging of the Inconel 718 alloy between 500 and 750 °C. *J. Mater. Res.* **1997**, *12*, 2298–2316. [CrossRef]
42. Chen, F.H.; Liu, J.; Ou, H.; Lu, B.; Cui, Z. *Long Flow Characteristics and Intrinsic Workability of IN718 Superalloy*; Elsevier BV: Amsterdam, The Netherlands, 2015; Volume 642, pp. 279–287.
43. Gujrati, R.; Gupta, C.; Jha, J.S.; Mishra, S.; Alankar, A. Understanding activation energy of dynamic recrystallization in Inconel 718. *Mater. Sci. Eng. A* **2019**, *744*, 638–651. [CrossRef]
44. Si, J.; Liao, X.; Xie, L.; Lin, K. Flow Behavior and Constitutive Modeling of Delta-Processed Inconel 718 Alloy. *J. Iron Steel Res. Int.* **2015**, *22*, 837–845. [CrossRef]
45. Zhang, H.; Zhang, K.; Lu, Z.; Zhao, C.; Yang, X. Hot deformation behavior and processing map of a γ′-hardened nickel-based superalloy. *Mater. Sci. Eng. A* **2014**, *604*, 1–8. [CrossRef]
46. Ahmadi, M.R.R.; Sonderegger, B.; Yadav, S.D.D.; Poletti, M.C.C. Modelling and Simulation of Diffusion Driven Pore Formation in Martensitic Steels during Creep. *Mater. Sci. Eng. A* **2018**, *712*, 466–477. [CrossRef]
47. Higashi, M.; Kanno, N. Effect of initial powder particle size on the hot workability of powder metallurgy Ni-based superalloys. *Mater. Des.* **2020**, 108926. [CrossRef]
48. Wang, Y.; Shao, W.Z.; Zhen, L.; Lin, L.; Cui, Y.X. Investigation on Dynamic Recrystallization Behavior in Hot Deformed Superalloy Inconel 718. *Mater. Sci. Forum* **2007**, *546–549*, 1297–1300. [CrossRef]
49. Zouari, M.; Logé, R.; Bozzolo, N. In Situ Characterization of Inconel 718 Post-Dynamic Recrystallization within a Scanning Electron Microscope. *Metals* **2017**, *7*, 476. [CrossRef]
50. Nicolaÿ, A.; Fiorucci, G.; Franchet, J.M.; Cormier, J.; Bozzolo, N. Influence of strain rate on subsolvus dynamic and post-dynamic recrystallization kinetics of Inconel 718. *Acta Mater.* **2019**, *174*, 406–417. [CrossRef]

Article

Influence of Grain Orientation Distribution on the High Temperature Fatigue Behaviour of Notched Specimen Made of Polycrystalline Nickel-Base Superalloy

Benedikt Engel [1,*], Sebastian Ohneseit [2], Lucas Mäde [3] and Tilmann Beck [4]

1. Gasturbine and Transmission Research Center (G2TRC), University of Nottingham, Nottingham NG7 2RD, UK
2. Institute of Applied Materials—Applied Materials Physics, Karlsruhe Institute of Technology, 76344 Eggenstein-Leopoldshafen, Germany; sebastian.ohneseit@kit.edu
3. Gas and Power Division, Department for Technology & Innovation, Siemens AG, 10553 Berlin, Germany; lucas.maede@siemens.com
4. Institute of Materials Science and Engineering, University of Kaiserslautern, 67655 Kaiserslautern, Germany; beck@mv.uni-kl.de
* Correspondence: benedikt.engel@nottingham.ac.uk

Abstract: Two different material batches made of random and textured orientated polycrystalline nickel-base superalloy René80 were investigated under isothermal low cycle fatigue tests at 850 °C for a notched specimen geometry. In contrast to a smooth specimen geometry, no significant improvement in fatigue behaviour of the notched specimen could be observed for the textured material. Finite element simulations reveal an area along the notch where high stiffness evolves for the textured material, which lead to nearly similar shear stresses in the slip systems compared to a random orientation distribution and therefore to no distinct differences in the lifetime.

Keywords: nickel-base superalloy; LCF; grain orientation distribution; texture; resulting shear stress; polycrystalline finite element simulation; high temperature

Citation: Engel, B.; Ohneseit, S.; Mäde, L.; Beck, T. Influence of Grain Orientation Distribution on the High Temperature Fatigue Behaviour of Notched Specimen Made of Polycrystalline Nickel-Base Superalloy. *Metals* **2021**, *11*, 731. https://doi.org/10.3390/met11050731

Academic Editor: Stefano Spigarelli

Received: 12 April 2021
Accepted: 26 April 2021
Published: 29 April 2021

Publisher's Note: MDPI stays neutral with regard to jurisdictional claims in published maps and institutional affiliations.

1. Introduction

The worldwide demand for the reduction of CO_2 emissions during power generation and the associated increased usage of renewable energy sources has lead to fluctuation in the power supply and can cause problems with regard to power grid stability. In particular, the generation of wind and solar power is strongly dependent on the current weather conditions, which result in volatile energy input into the power grid. In order to fill in these gaps, gas power plants are used due to their ability to generate power in short times. Modern gas turbines, as a main part of gas power plants, can be started from cold to maximum power output in less than 30 min [1]. Due to frequent starts, stops and load changes from part to full load, the materials in the hot gas section and especially the turbine blades have to withstand significantly fluctuating mechanical and thermal loads. Hence, thermo–mechanical fatigue, high-temperature fatigue behavior and creep effects of the turbine material must be investigated in order to reach high efficiencies with simultaneous economic design aspects. Since component testing is very complex and expensive, generally, a standardized specimen of the used material is examined under high-temperature fatigue conditions. To represent uniaxial stress states, specimens with cylindrical gauge sections are used; moreover, notched specimens are used to determine the fatigue behavior of the material under multiaxial stress states as well as stress gradients to represent complex component structures. Besides experimental material testing, numerical models are indispensable during the design process in order to predict the deformation and fatigue behavior under various load and temperature conditions. Based on experimental observed and measured mechanical material responses, models such as the commonly used Ramberg–Osgood or the Chaboche model are used to describe the nonlinear stress strain relationship for materials under high-temperature conditions [2–4].

Polycrystalline nickel-base superalloys, such as René80, are used as blades or vanes in the rear hot gas sections of gas turbines with maximum operating temperatures up to 982 °C [5,6]. Due to the conventional vacuum casting process of the turbine blades, different grains sizes and orientation distributions arise depending on the component geometry and cooling conditions. Coarse grains and low grain numbers located in highly stressed component areas in combination with high elastic anisotropies of nickel-base superalloys lead to large scattering in fatigue lifetimes and mechanical properties [7–9], which lead to high safety factors for deterministic design approaches. In order to achieve higher efficiencies for the components, these conservative deterministic lifetime models are successfully replaced by probabilistic approaches [10–13]. For this purpose, the influencing factors and their statistical distributions on high-temperature fatigue behavior must be identified and investigated for the material. Seibel et al. [11,14] investigated the fatigue behavior of conventionally casted polycrystalline nickel-base superalloy René80 with an assumed random grain orientation distribution under low cycle fatigue (LCF) conditions at 850 °C. For the same total strains, notched specimen shows a distinct increased fatigue lifetime compared to the smooth specimen. This can be mainly attributed to the statistical size effect, where the notched specimen has a significantly reduced highly loaded volume and therefore the probability of grains which tend to crack initiation is decreased. In addition, the formation of stress gradients, caused by the notch geometry, lead to decreased stresses into the bulk material, which could result in lower crack growth rates and therefore to longer lifetimes compared to cylindrical specimens. For different notch geometries, an increase in lifetime with an increase of the notch factor could be shown. A similar behavior could also be proven for different material such as steels [15].

Numerical evaluations in [16], considering the influence of grain orientation on the mechanical behavior of nickel-base superalloys, reveal a relationship between the local grain orientation and the resulting shear stress within the slip system, responsible for plastic deformation. Due to high elastic anisotropies of nickel-base superalloys, the highest shear stresses for the <111>{110} slip systems were achieved for grain orientations with corresponding Young's moduli between 180 GPa and 230 GPa (at 850 °C), and Schmid factors between 0.38 and 0.46 for uniaxial loading cases. Orientations with a maximum in Schmid factors of 0.5 show distinct lower shear stresses, due to low correlated elastic properties at around 120 GPa.

Engel, Mäde et al. showed in [17] the influence of the local grain orientation on the probability of fatigue failure for smooth cylindrical specimen made of polycrystalline René80 with a random and textured grain orientation distribution. By determining that the grain orientation depended on the resulting shear stress within the slip systems of the material by polycrystalline finite element simulation, reliable probabilistic lifetime predictions could be made for randomly orientated material as well as textured material. As a fatigue life describing parameter, a modified Schmid factor approach was introduced. For the nodes of a polycrystalline finite element simulation, the modified Schmid factor is defined as the quotient of von Mises stress and the resulting shear stress in the activated slip systems. It could be shown that the determined texture of the investigated material lead to a lower Young's moduli for the cylindrical specimen and therefore to lower stresses during uniaxial strain controlled tests compared to the randomly orientated material. Furthermore, the texture influences the orientation of the slip system and lead to lower modified Schmid factors. Thus follows the requirement of higher local stresses in the slip system which results in a delay in crack initiation whereby higher fatigue lifetimes were achieved.

In addition to the results of different smooth-specimen grain orientation distributions in [17,18], the following paper investigates the high-temperature LCF behavior of the same material batch of nickel-base superalloy René80 for a notched specimen geometry. Isothermal LCF tests at 850 °C were carried out for a random and a textured grain orientation distribution, and the lifetimes compared to the smooth specimen from previous investigations. To explain the observations and differences in the lifetimes as well as the variation in mechanical behavior, polycrystalline finite element simulations for notched geometries were carried out to analyze the influence of local grain orientation on the local mechanical behavior during cyclic high-temperature fatigue tests.

2. Materials and Methods

The composition of the polycrystalline nickel-base superalloy René80 used in this work is shown in the following Table 1 and was measured by the manufacturer.

Table 1. Composition of René80 in wt.%.

Element	Ni	Cr	Co	Ti	Mo	W	Al	C	B	Zr
René80	Bal.	14.04	9.48	5.08	4.03	4.02	2.93	0.17	0.015	0.011

The material was vacuum casted from the same melt and heat treated after [19] at Doncaster in Bochum/Germany as bars of 150 mm in length with 20 mm and 12 mm in diameter. The different diameters of the bars lead to different cooling conditions and temperature transients during solidification in the radial direction (x−direction, see Figure 1). In addition, the casting process as a standing mould leads to a further temperature gradient in the axial direction (y−direction). Due to the faster cooling conditions in the 12 mm bar, grain growth is inhibited, which leads to smaller elongated grains compared to the 20 mm bar, as Figure 1 shows. The white dashed line represents the gauge section of the smooth specimen, and the green dashed line the gauge section of the notched specimen.

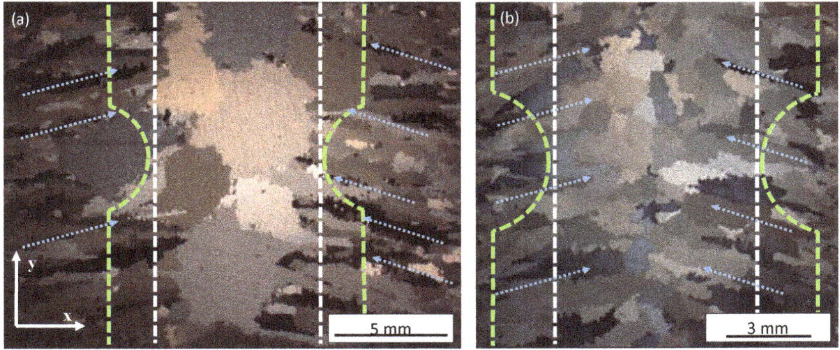

Figure 1. Longitudinal cut of the 20 mm bar (**a**) and 12 mm bar (**b**).

The preferred growth of the [100]−orientation towards the temperature gradients known for nickel-based superalloys leads to dendritic solidification along the edge areas of the bars [20] represented as blue dotted lines. During manufacturing of the sample geometries, the dendritic edge area is almost completely removed from the gauge section of the 20 mm rod. However, the samples, which were manufactured from the 12 mm rod, still show dendritic structures within the gauge section, as seen in Figure 1. As described in [17], the dendritic solidification and the resulting preferential direction of the grains, with an average alignment of $\vartheta = 25°$, result in a different material behavior compared to the specimen from the 20 mm bar material (with random orientation). To determine the grain size, the grains in the gauge section were evaluated by the equivalent diameter using a light microscope and an image processing software. The ratio of the grain sizes within the gauge section of the specimen between the 12 mm bar and 20 mm bar is about 1:3. In the following, the specimen taken from the middle of the 20 mm bar will be called coarse-grained and consist of a random orientation distribution of the grains, since no obvious crystallographic texture could be determined using electron backscatter diffraction.

The specimens made from the 12 mm bar are called fine-grain textured material, due to the preferred solidification in direction of the resulting temperature gradient. The René80 material from the isothermal LCF tests results taken from [14] for comparison was also manufactured and heat-treated by Doncaster but casted as solid plate with a thickness of 20 mm. The grain sizes are in a very good accordance to the coarse-grain material as well as the grain orientation distribution was also described as random. In order to compare the results to [14], the same notch geometry as well as cylindrical gauge section were chosen and are shown in Figure 2. The notch factor was calculated to a value of 1.62.

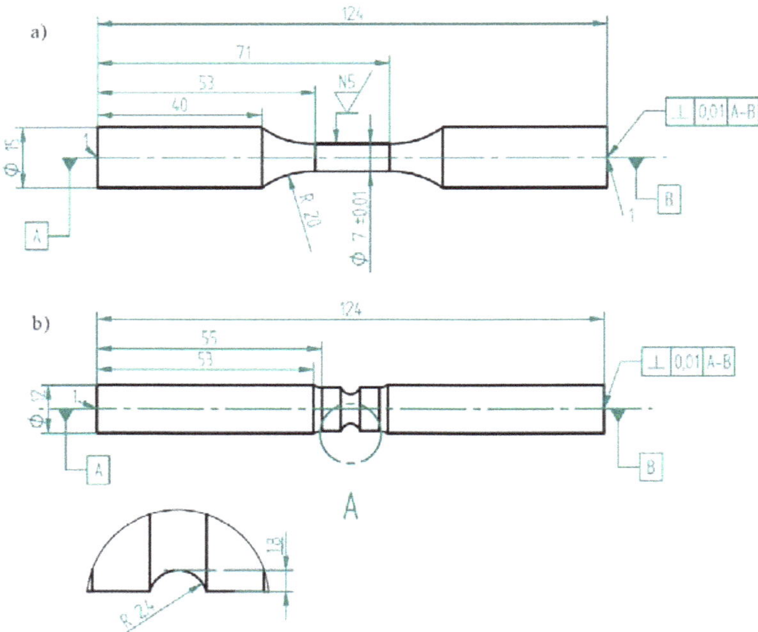

Figure 2. Geometry of the smooth (**a**) and notched (**b**) specimen in accordance to [14,21].

Isothermal LCF tests were carried out at 850 °C at a servo hydraulic MTS test rig with a maximum load capacity of 100 kN combined with an Hüttinger TrueHeat (TRUMPF Hüttinger GmbH, Germany, Freiburg) MF 5000 induction heating system. Strain-controlled tests were performed at R = −1, using an MTS high-temperature extensometer type 632.53 with ceramic rods, a measurement length of 12 mm and a constant air flow cooling to prevent temperature influences on the measurement. All tests were carried out with a test frequency of 1 Hz. Details to the tests conducted with the smooth samples can be found in [17,21].

Since the used type K ribbon thermocouple could not be applied in the notch due to geometrical conditions, it was attached above and calibrated for each specimen to the temperature of 850 °C by means of a wire thermocouple within the notch. The thermal gradient between these two thermocouples was below 8 °C for 850 °C, whereas a controlling by only one wire thermocouple attached at the notch was not practicable due to contact problems during testing. A drop of the measured and stabilized stress amplitude of 2.35% indicates the failure of the notched specimen, which is equal to a drop of 2.5% for the smooth specimen if a penny-shaped crack surface of 0.96 mm^2 in both specimen types is assumed [22].

It should be noted here that due to confidential agreements with the project partners, diagrams which contain lifetime and material data of the investigated material René80 are normalized. Wöhler diagrams are labelled with "low" and "high", which refers to lifetimes and total strain, and stress amplitudes for the low cycle fatigue and high cycle fatigue regime. Diagrams which contains material properties are normalized by the highest occurring value.

3. Modelling Approaches

3.1. Modelling the Notch Strain with Finite Element Analysis for an Isotropic Material Assumption

Due to the usage of the MTS 12-mm high-temperature extensometer for both specimen types, the axial strain within the notch must be determined numerically, in order to compare the fatigue tests to smooth specimen fatigue tests in strain amplitude Wöhler diagrams. Due to the notch geometry, stress concentrations occur within the notch root and lead to local plastic deformation. Therefore, an isotropic elastic plastic material model, based on a Ramberg–Osgood relationship (material parameters provided by Siemens), for the coarse-grained and fine-grain textured René80 at 850 °C, was implemented into the finite element solver ABAQUS (Version 2018, Dessault systemes, Vélizy-Villacoublay, France). A virtual extensometer with a measurement length of 12 mm was attached symmetrically around the notch of a full specimen model. As a global loading, the end face of the model was displaced step by step, where the front face was locked in the axial direction. For each step, the measured axial strain on the extensometer was determined as well as the axial strain within the notch root. The following Figure 3 shows the relationship between the axial strain applied to the extensometer ε_{global} (respectively to the specimen) and the corresponding strain in the notch root ε_{notch} for the coarse- and fine-grain textured material (normalized with the maximum occurring global strain ε_{global}).

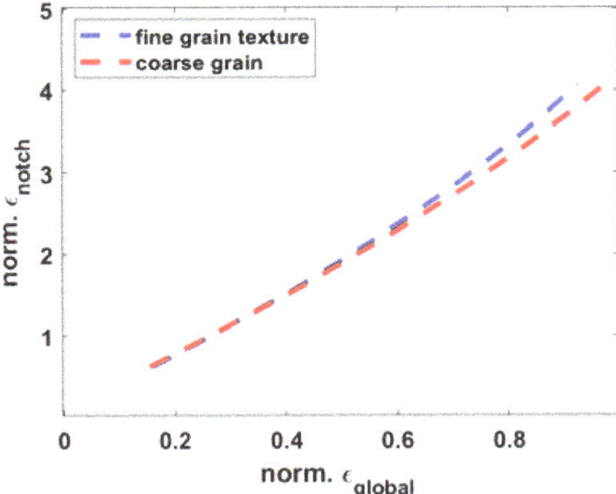

Figure 3. Normalized relationship between the global strain at the extensometer ε_{global} and the local strain at the notch root ε_{notch} for the coarse- and fine-grain textured material.

To estimate the required strain amplitude within the notch root, fitted quadratic functions (from Figure 3) were used to adjust the total strain control value at the fatigue test controlling system.

3.2. Modelling the Local Material Behaviour with Finite Element Analysis for Polycrystalline Nickel-Base Superalloys

In order to simulate the material behaviour in dependence of different grain orientation distributions, the open-source software NEPER (Version 3.3, by Romain Quey, MINES Saint-Étienne, France) was used to generate random grain morphologies using the 3D Voronoi tessellation method [23–25]. Since the notched section is of great importance for the mechanical behaviour of the specimen, a cuboid with the dimension 10 mm × 11 mm × 5.5 mm and two notches, according to Figure 4, was modelled. The thickness of the model was chosen to ensure that the surface of the two notches is equal to the circumferential surface of the round notch on the test specimen in order to avoid statistical size effects. The cuboid type of model was selected because it is not practicable to perform rotationally symmetrical cuts on cylindrical models within the used version of NEPER. Due to the considerations of a flat notched specimen in the finite element models, differences in the stress state of the notch root occur when compared to a circumferential notched specimen, as used in the experiment. For example, the tangential stress component of the circumferential notched specimen is not existent in the flat notched specimen. However, because of the low notch factor of 1.62, the axial stress component is significantly higher than the remaining stress components and therefore these differences in stress state can be neglected.

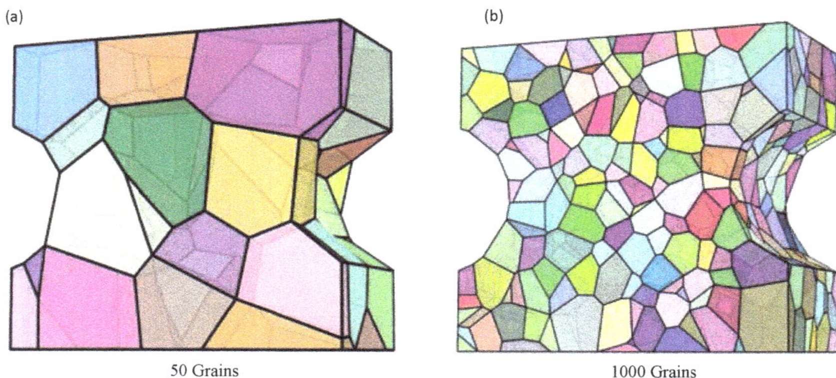

Figure 4. Polycrystalline models of the notched specimen, coarse-grain with 100 grains (**a**), fine grain with 1000 grains (**b**).

According to the grain size distribution mentioned in Section 2, the coarse-grain model with random orientation distribution contains of 100 grains, which result in an equivalent grain diameter of 2.76 ± 0.55 mm, evaluated using the NEPER grain statistics tool. The fine-grain textured model contains of 1000 grains with an equivalent grain diameter of 1.03 ± 0.16 mm. Due to the very high computing time for generation, meshing and calculation, three models with three textured grain orientation distributions each were created for the fine-grain textured sample. The coarse-grained models cover four different models with three different grain orientation distributions each. For both types of models, the value for the characteristic length of the elements for the average cell size was set to rcl = 0.3 (relative element length) within the NEPER meshing tool. As a result, the mesh consists of a uniform distribution of approximately 250,000 quadratic tetrahedral elements (C3D4) for the coarse-grain model and approximately 700,000 quadratic tetrahedral elements (C3D4) for the fine-grain textured model. Both polycrystalline specimen models are shown in Figure 4. Note, the edges within the notch appear in the visualization and are removed by the meshing procedure.

A detailed description of the model properties of the smooth specimen as well as generation is described in [17]. In order to expand the simulation database, six coarse-grain smooth models with random orientation distributions and six fine-grain smooth models with a textured orientation distribution were added and analysed. Table 2 summarises all conducted numerical simulations with the corresponding orientation distributions.

Table 2. Overview of the different numerical simulations.

Specimen Type	Number of Grains	Grain Orientation Distribution	Number of Models with Different Grain Morphologies	Number of Simulations per Model
Notched, flat	100	coarse random	4	3
Notched, flat	1000	fine textured	3	3
Smooth, round	49	coarse random	6	1
Smooth, round	500	fine textured	6	1

In order to create local material properties, each grain was rotated using rotational matrices U (U is a function of the Euler angles φ_1, φ_2 and ϑ) according to an orientation distribution in a pre-processing step and the respective data written to the input file. The grains in the coarse-grain René80 batch show no preferential direction in orientation, which is why the rotational matrices U used in this pre-processing step are distributed according to an isotropic measure, mathematically given by the Haar measure at the SO3 group of rotations [10]. The rotational matrices U which were assigned to the fine-grain textured models were generated by random distribution of the Euler angles φ_1, φ_2 where ϑ kept constant with 25° to represent the examined texture. To calculate the elastic properties of the grains in dependence of their local coordinate system, a global, anisotropic, linear-elastic material law was defined in ABAQUS. As there is no data concerning the elastic constants of René80 in dependence of the temperature in the literature, elastic values for an IN738LC were taken from [26]. Since both the composition and content of the γ' phase are very similar in IN738LC and René80, it can be assumed that the elastic behaviour of both alloys are qualitatively comparable. A linear interpolation from the data at 800 °C and 898 °C results in the elastic constants C_{11} = 225.83 MPa, C_{12} = 161.45 MPa and C_{44} = 98.79 MPa for a temperature of 850 °C. During the simulation, the local material properties in dependence of the local grain orientation were transferred to the globally defined material law and the grains interact according to their local stiffness. All simulations were carried out at T = 850 °C in displacement control, in order to achieve consistency to the LCF experiments. The nodes of one face were locked only in the axial direction, in order to enable transverse contraction, whereas the nodes of the opposite face were displaced by 0.01 mm. Due to the cyclic stability of the material behaviour during high-temperature LCF testing, only one single loading cycle was simulated.

A python-based tool was developed to calculate the maximum resulting shear stress within the slip systems (i,j) on each node of the finite element models in dependence of the local stress tensor σ and the assigned local grain orientation U. Since the crystallographic structure of nickel-base superalloys is face-centred cubic (fcc), dislocation movement occurs usually in slip systems of the type {111}<110> if a critical value τ_{crit} is exceeded. The resulting shear stress on each node is determined by calculating the shear stresses for all 12 slip systems, where the slip system with the highest value is the activated one—see Equation (1).

$$\tau_{res,ij} = U \cdot n_i \, \underline{\sigma} \cdot U \cdot s_{i,j} \tag{1}$$

Details for the calculation procedure can be found in [10,17,21]. Since all simulations are purely elastic, only linear elastic shear stresses are calculated within the slip systems. Still, their distribution at the model surface, i.e., the notch root, gives an indication about the expected onset of plasticity and therefore the possibility of a fatigue crack initiation.

4. Results

4.1. Experimental Results of Isothermal LCF Tests at 850 °C for Smooth and Notched Specimen

In the following section, the results of the total strain-controlled isothermal LCF experiments at 850 °C of the notched specimen are presented in comparison to the results of the smooth specimen. Figure 5 shows the influence of grain orientation distribution and specimen geometry on the high-temperature LCF behaviour of René80 as the total strain amplitude Wöhler curve. For the notched specimen, the shown total strain amplitude $\varepsilon_{a,t,notch}$ represents the total strain amplitude within the notch root (see Section 3.1), determined by finite element simulation using a Ramberg–Osgood relationship. The investigated coarse-grained specimens are displayed in green, whereas, in addition, the results for coarse-grain smooth and notched specimens (taken from [14]) are shown in grey. Smooth and notched specimens made from the fine-grain textured material are marked in blue. Smooth specimens are shown as circular data points, whereas notched specimens are represented by a rhombus. Run out tests are marked with an arrow.

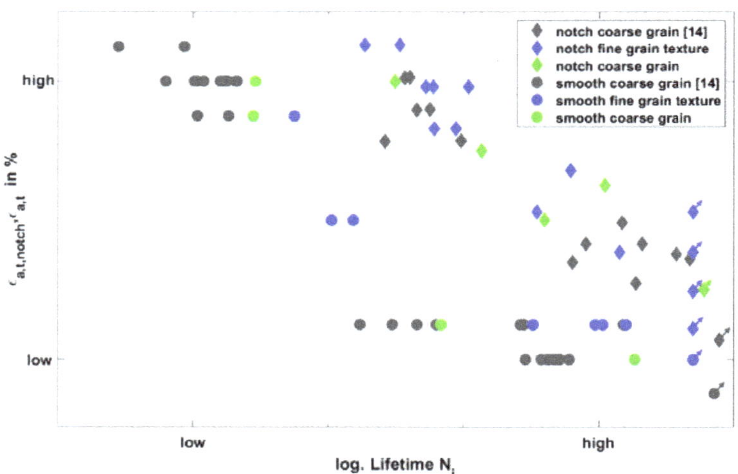

Figure 5. Total strain amplitude Wöhler curve for smooth and notched specimen with different grain orientation distributions for isothermal LCF tests at 850 °C.

Figure 5 clearly shows, for all total strain amplitudes and grain orientation distributions, that fatigue lifetimes of the notched specimen are higher compared to the smooth specimen. As already explained in [17,21], smooth specimens with a fine-grain textured material show higher lifetimes compared to the coarse-grain material, especially for low total strains. However, for the notched specimens, no unambiguous improvement regarding fatigue life can be found for the fine-grain textured material. It is noticeable that over a relatively large range of low total strains, both failure and run out occur for the notched geometry, which leads to distinct lifetime scattering. At high total strains, the lifetime scatter appears slightly decreased.

Figure 6 shows the measured stress amplitude in order to represent the influence of stiffness of the individual specimens. For the notched specimen, the stress amplitude corresponds to the maximum stress in the notch root ($\sigma_{a,notch}$).

For the notched specimen, the measured stress amplitudes are in general lower compared to the smooth specimen for the same total strains. While the smooth specimen shows a clearly lower stress amplitude for the textured fine-grain material, no clear influence of grain orientation on the stress amplitude for the notched specimen can be determined. Additionally, the scatter in stress amplitude seems, for the smooth specimen, much higher compared to the notched specimen, where the highest scatter could be found for the coarse-grain smooth specimen.

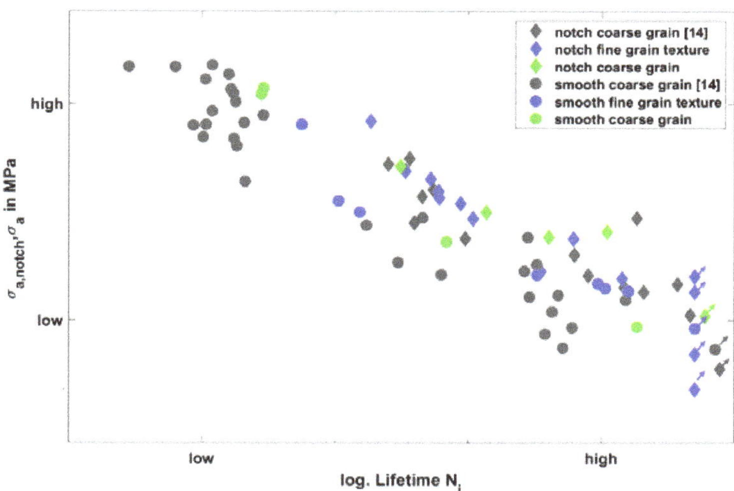

Figure 6. Stress amplitude Wöhler curve for smooth and notched specimen with different grain orientation distributions.

4.2. Microscopic Analysis of the Fracture Aurfaces

The fracture surfaces of the notched specimen will be evaluated in the following, where a detailed analysis of the fracture surface for the smooth specimen can be found in [18]. In general, the typical high-temperature fatigue cracking behaviour is observed, consisting of a smooth and oxidized part which represents the fatigue crack surface, which was developed during cycling and a rough non-oxidized part which represents the residual fracture surface (caused by the shutdown of the heating system after specimen failure). Moreover, scanning electron microscope (SEM) images show the typical honeycomb structure and prominent cracks in the residual fracture area. Conversely, the detailed examination of the fatigue crack surface clearly shows the crack initiation sites at the surface with the typical lens around the initiation spot. Subsequently, river patterns are visible in the fatigue rupture surface by SEM investigations, as visible in Figure 7.

Independent of specimen geometry, the fractured surfaces show an influence of the total strain amplitude. At low total strain amplitudes, single crack initiation with large fatigue surfaces were observed, as seen in Figure 8a. Almost half of the fracture surface is represented by the fatigue crack originating from the single crack initiation site, while the other half is dominated by final rupture surface. In contrast, at high total strain amplitudes, multiple crack initiation sites distributed all around the notch root were observed, as shown in Figure 8b. High strains and therefore higher stresses lead to multiple crack initiation sites and faster crack growth. As a result, the fatigue surfaces are significantly smaller and irregularly shaped. Partially, the merging of cracks can be observed. The red arrows in Figure 8 illustrate the crack initiation site.

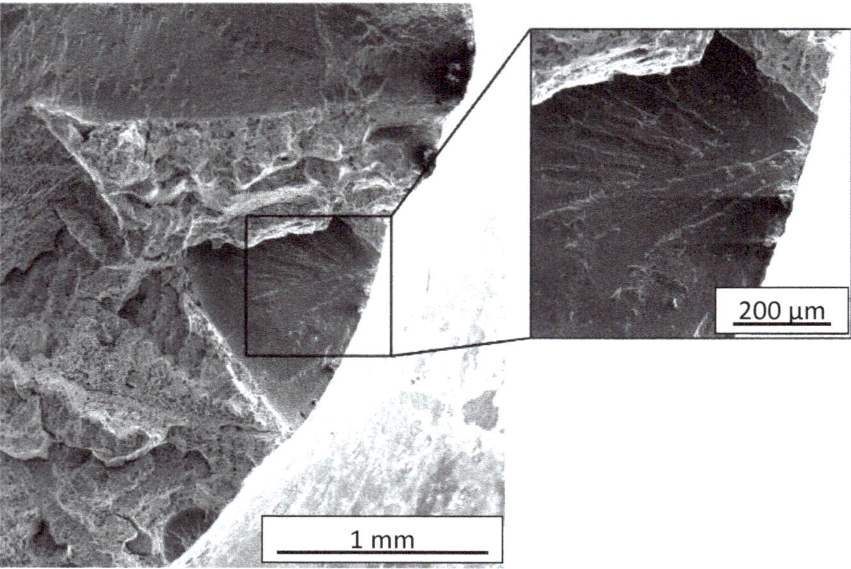

Figure 7. Fatigue and residual fracture surface of a notched specimen.

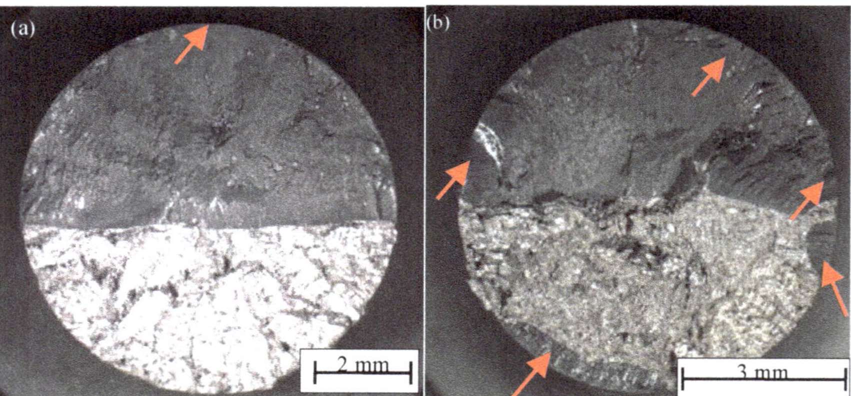

Figure 8. Crack initiation sites of a notched specimen fatigued with (**a**) low total strain amplitude, (**b**) high total strain amplitude at 850 °C.

4.3. Cyclic Deformation Behaviour of the Notched Specimen under Total Strain Control

The development of stress amplitude $\sigma_{a,notch}$ for the notched specimen over the cycle number is presented in Figure 9 (total strain amplitudes are normalized).

As expected, the stress amplitude $\sigma_{a,notch}$ decreases with a decrease of total strain amplitude, which results in longer lifetimes. All curves show a stable cyclic behaviour during their full lifetime until a crack initiation occurs; therefore, their cyclic behaviour is comparable to the smooth specimen [18]. After crack initiation, some samples show a steep drop in stress amplitude $\sigma_{a,notch}$. However, stress amplitude increases are also possible. After [27], this effect can be explained by the position of the crack relative to the extensometer. Since the duration of the crack growth phase is distinctly smaller compared to the total lifetime, the effect of the orientation of the crack to the extensometer can be neglected.

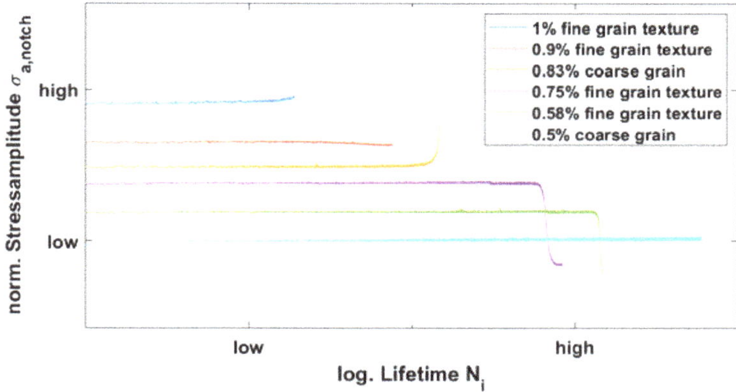

Figure 9. Development of stress amplitude over lifetime during strain-controlled LCF testing at 850 °C for coarse-grain and fine-grain textured material.

4.4. Results of the Finite Element Simulation

For the determination of the elastic axial strains within the notch, results of the isotropic simulations based on the Ramberg–Osgood relationship (mentioned in Section 3.1) are shown in Figure 10. The axial displacement of 0.1% in the y–direction in the following illustrations was chosen to compare the results with the anisotropic approaches and to avoid plastic deformation (the polycrystalline simulations are calculated fully elastic). In order to differentiate the two material batches, the respective Young's moduli (taken from the Ramberg–Osgood relationship) were assigned to the models.

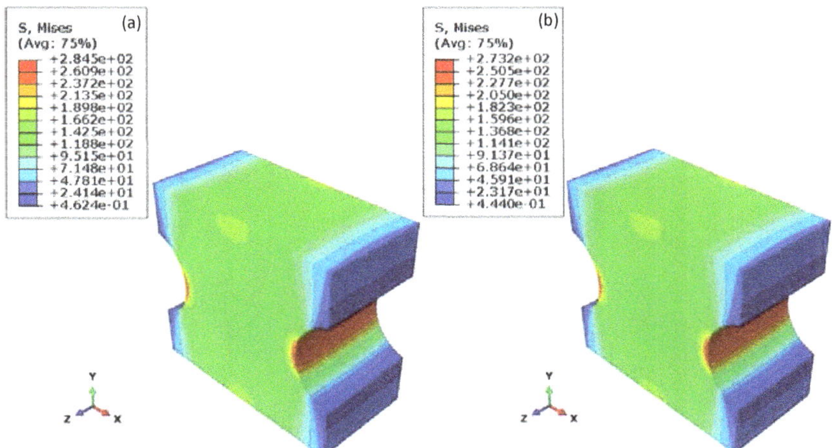

Figure 10. Isotropic elastic simulation coarse-grain (**a**) material and for the fine-grain textured material (**b**), using a Ramberg–Osgood approach.

The von Mises stress distribution reveals the anticipated inhomogeneous distribution within the notch where the maximum von Mises stress is located within the notch root, with 284.5 MPa for the coarse-grain material and 273.2 MPa for the fine-grain textured material. Due to the slightly lower Young's moduli of the fine-grain textured material, the local stresses are slightly decreased compared to the coarse-grained material. In addition, the isotropic simulation shows a geometrically caused slight reduction of the stress at the edges of the notch.

The results of the finite element analysis for the polycrystalline notched specimen at 850 °C and an applied total strain of 0.1% will be presented in the following. Figure 11 shows the distribution of the von Mises stress for a model with coarse-grain material (100 grains, random orientation distribution) and a model with the fine-grain textured material (1000 grains).

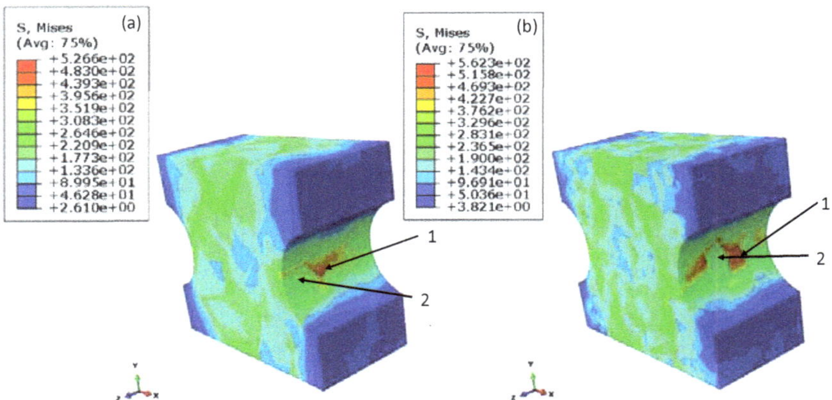

Figure 11. Distribution of von Mises stress for notched specimen for the coarse-grain material (100 grains, (**a**)) and fine-grained textured material (1000 grains, (**b**)).

It can be clearly seen that compared to an isotropic approach the stress distribution is highly inhomogeneous within the notch area, as well as in the bulk material caused by grains with different orientations and therefore varying elastic properties. Within the notch root local maxima of stress, located close to grain boundaries, areas that 1 indicate can be found. At the same time, areas with significantly reduced stresses can occur even in the notch root (areas 2). A statistical evaluation of all simulations revealed that, on average, the maximum von Mises stress for the coarse-grained and randomly orientated specimen is 508 ± 58 MPa (evaluated from 12 simulations). The average maximum von Mises stress for the fine-grain and textured material is with 634 ± 45 MPa significantly increased (evaluated from nine simulations). It should be noted here that both models were calculated with the same applied total strain of 0.1% and the same material model; only the grain orientation distribution and grain numbers were different.

Comparing the von Mises stress results to the isotropic approach presented in Figure 10 reveals distinct higher von Mises stresses for the polycrystalline model. However, they appear on a much smaller scale within the notch root, whereas, for the isotropic approach, the whole notch root shows a maximum in von Mises stress.

Figure 12 shows the distribution of the strain component E22 in the loading direction (y–direction) for the coarse-grain and fine-grain textured material.

Similar to the von Mises stress distribution, the strain distribution is also inhomogeneously distributed along the notch area, which results in a distinct strain scatter between 0.1% and 0.3%, where local maxima (area 1) as well as minima (area 2) can be found in the notch root. Moreover, some models show, in contrast to isotropic calculated models, that strain maxima can also occur outside the notch root, as area 3 shows. For all nine calculated models, the maximum strain E22 in loading direction for the fine-grain textured material is slightly higher with $0.279 \pm 0.015\%$ than for the coarse-grain randomly orientated material with $0.271 \pm 0.033\%$ (12 models).

The calculation of the resulting shear stress distribution is carried out with the python tool mentioned in Section 3.2. Figure 13 shows the distribution of the maxima in the resulting shear stress within the slip system of the grains and gives an indication of where plastic deformation, and therefore fatigue crack initiation, occurs favourably.

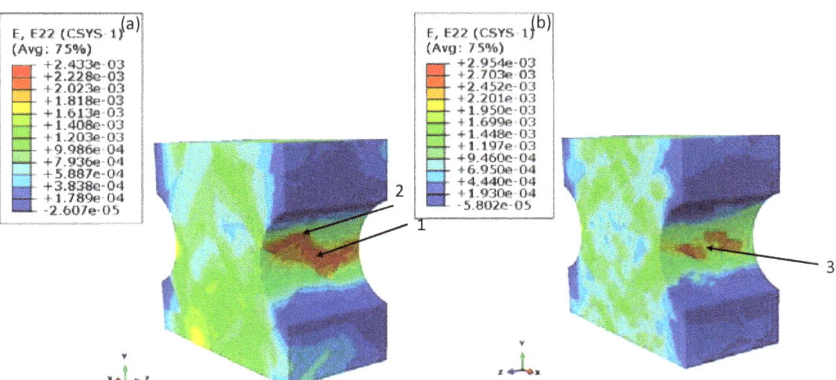

Figure 12. Distribution of the strain component E22 in loading direction for notched specimen for the coarse-grain material (100 grains, (**a**)) and fine-grained textured material (1000 grains, (**b**)).

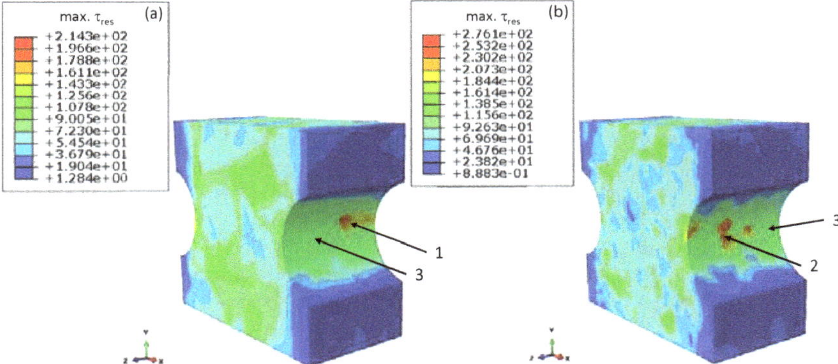

Figure 13. Distribution of the maximum shear stress within the slip system of the grains for the coarse-grain material (100 grains, (**a**)) and fine-grained textured material (1000 grains, (**b**)).

Due to the inhomogeneous distribution of stresses and strains within the notch, caused by the polycrystalline modelling approach, the maximum resulting shear stresses are also inhomogeneously distributed. Local maxima can be found within the notch root (area 2) and also outside the notch root (see area 1). Since the shear stress distribution also depends on the stress distribution, areas with low shear stresses can also be found in the notch root, as shown for the areas 3. In these areas, a predominantly elastic deformation occurs, whereas plastic deformation can be expected first in the areas with high resulting shear stresses. Figure 13 already reveals local higher-shear stresses for the fine-grain textured material which can be confirmed for all simulations. The average value for the maximum resulting shear stress is 287 ± 9.8 MPa for the fine-grained textured material, while the average value for the coarse-grain randomly orientated material is 236 ± 26 MPa.

In conclusion, the notched models with a fine-grain textured material reveal, on average, higher mechanical properties, i.e., stress, strains and shear stresses, compared to notched models with a random orientation distribution (coarse-grain) for the same loading condition.

4.5. Shear Stress Distribution of the Smooth Specimen

In addition to [17], further smooth-specimen simulations were carried out in order to increase the database. Figure 14 shows the distribution of shear stress for coarse-grain and fine-grain textured models at 850 °C and an applied total strain of 0.25%.

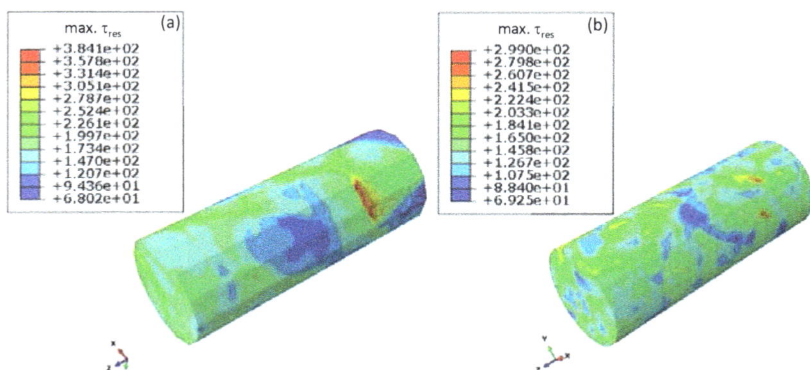

Figure 14. Distribution of the maximum shear stress within the slip system for smooth specimen with coarse-grain material (49 grains, (**a**)) and fine-grained textured material (500 grains, (**b**)).

For the simulation of the smooth specimen, areas with low resulting shear stress as well as areas with locally significant increased shear stresses occurring on the surface can be clearly seen. An evaluation of the average maximum resulting shear stress of all calculated models (including the results from [17]) shows, for the coarse-grain, random orientated specimen, an average resulting shear stress of 397 ± 25 MPa. The average value for fine-grain textured material is clearly lower with about 315 ± 21 MPa.

A summary of the finite element simulations reveals for the fine-grain textured material higher local mechanical properties (stress, strain and shear stresses) when compared to the coarse-grain material for the case of a notched specimen. Interestingly, in the case of a smooth sample, the fine-grain textured material led to distinct lower mechanical properties. The following section will explain where these differences in mechanical behaviour result from and how they affect the fatigue behaviour.

5. Discussion

A general difference of lifetime behaviour between the smooth and notched specimen can be statistically attributed to the size effect. For the smooth specimen, the possible area for crack initiation, i.e., the gauge section surface, is about 6.5 (the whole surface of the notch) times higher for the notched specimen. However, just around a third of the notch surface lies within the notch root; therefore, the surface relation between the notched and smooth specimen where fatigue crack probably occurs is much higher. Despite more grains on the surface of the fine-grain textured smooth specimen which from a statistical aspect should increase the possibility of a crack initiation, a longer lifetime can be observed for the same total strains in comparison to the coarse-grain smooth specimen. The reason lies in the preferred orientation of the grains in combination with the global uniaxial stress state. This results in lower Young's moduli of the specimen compared to a randomly orientated material. It follows for strain-controlled test conditions, lower stress amplitudes and lower shear stresses within the slip systems and therefore longer lifetimes [17]. A significantly improved lifetime of the fine-grain textured material, however, cannot be determined for the notched samples. Due to the multiaxial character of the notch and the resulting multiaxial stress states, an evaluation of Young's moduli as a ratio of stress and strain in loading direction is not appropriate. To determine the influence of stiffness by using finite element, the reaction forces in the loading direction of all nodes were calculated and summarized to calculate the resulting force on the specimen caused by the displacement of 0.1%. Related to the cross section within the notch, the averaged nominal stress which occurs for a displacement of 0.1% for the fine-grain textured material is 240 ± 1.5 MPa. For the randomly orientated coarse-grain material, the average stress in the notch can be calculated to 190 ± 7 MPa. It could be assumed here that, for the coarse-grain material, an insufficient number of grains with perhaps stiff orientations could influence the results. As mentioned in [28], a minimum of six to eight grains are required within the cross section to minimize the influence of different grain orientations. For the coarse-grain notched specimen, more than 10 grains are located in the cross section of the notch for every model. In addition, the number of simulations (12 in total) with random grain orientation distributions, generated with the Haar measure, ensures a sufficient amount of data to provide secure statistical evaluations.

Thus, the averaged values reveal a general higher stress required for the textured material to achieve a total strain of 0.1% than for the randomly oriented material. Therefore, the notched specimens show a contrary behaviour to the smooth specimens with regard to stiffness. Since all models, notched and smooth, were calculated with the same material models as well as orientation distribution, the main the differences in mechanical behaviour can be attributed to the influence of the notch.

5.1. Development of Stiffness along the Notch

The following section explains where the stiffness differences arise from by investigating the elements of the finite elements models and assuming that elements can have different elastic properties due to their local orientation in relation to the loading axis. For all the explanations given below, an anisotropic material behaviour is assumed due to the texture and differences in the alignment of the grains (i.e., elements) between the local principal stress axis and loading axis, but the microstructure of the material is assumed to be homogenous (non-consideration of grains). Therefore, a grain interaction caused by differently shaped and orientated grains is not considered here. For the smooth specimen, the loading axis aligns with the axis of maximum principal stress σ_I for all elements (σ_{II}, $\sigma_{III} << \sigma_I$). Assuming the determined crystallographic texture for all elements, with φ_1, φ_2 random and ϑ constant with 25°, the Young's moduli, i.e., stiffness, are nearly uniformly distributed with an average value of 128 ± 0.9 GPa, calculated for 850 °C. This outcome results from Monte Carlo simulations in which a single element was rotated 10,000 times while the stiffness in loading direction (direction of maximum principal stress) was evaluated. For the notched specimen, the direction of maximum principal stress is not constant and changes for each element along the notch due to its geometry, as Figure 15 illustrates.

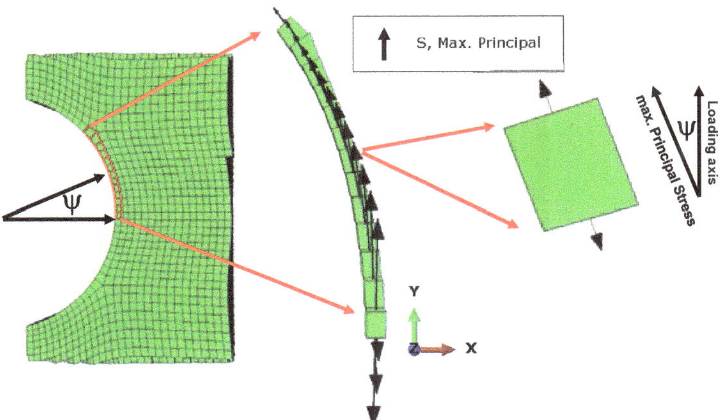

Figure 15. Directions of maximum principle stress along the notch compared to loading direction y.

It can be clearly seen that along the notch, an angle ψ between the maximum principal stress and global loading axis forms. If now a constant texture, i.e., constant orientation of each element is assumed, the principal stress direction of the elements aligns tangentially along the notch with regard to the global loading axis y. This assumption is valid for notches with a circular or elliptical shape (no sharp notches). For elements in the notch root, the difference angle is $\psi = 0°$ and therefore the results of the Monte Carlo simulation to determine the possible stiffness at this point are similar to the results for a uniaxial case in the smooth specimen (the direction of principal stress align with the loading direction). For calculating the stiffness for elements with angles $\psi \neq 0°$, the following method was applied. Instead of calculating the stiffness for the constant texture with changing directions of the maximum principal stress (to represent the notch), the principal stress direction was kept constant, and the orientation of the elements were changed. For an angle of 5° between the loading direction and principal stress axis, the element was rotated about $\psi = 5°$, in such a way that the principle stress axis and global loading direction align. For the textured material with $\vartheta = 25°$ follows as the new angle $\vartheta + \psi = 30°$. Monte Carlo simulations were carried out for angles $\psi = 5°, 10°, \ldots, 60°$ (3000 samples each). Figure 16 shows the development of a stiffness increase in dependence of the angle ψ for a textured orientation distribution with $\vartheta = 25°$ for all elements along the notch (φ_1, φ_2 = random).

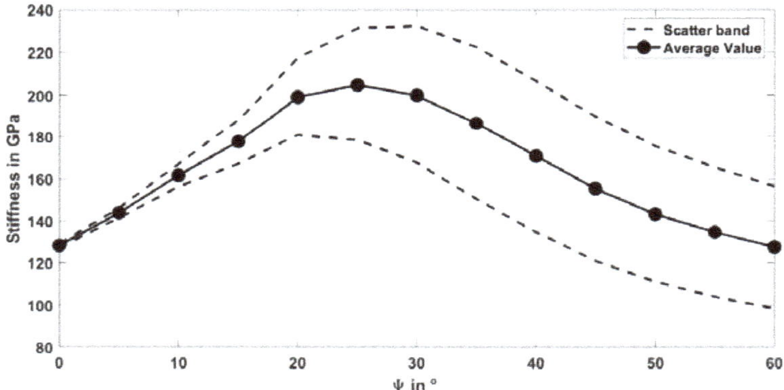

Figure 16. Development of stiffness along the notch for the textured material.

The analysis of the Monte Carlo simulation reveals an increase in stiffness until a maximum value with 200 ± 18 GPa is reached for $\psi = 25°$ at the notch. With a further increase in ψ, the stiffness decreases with simultaneously increasing scatter, which can be explained by plotting the inverse pole figure for different angles ψ over the Young's moduli at 850 °C, as presented in Figure 17.

Possible orientations (black dotted line) with a fixed angle of $\psi = 0°$ (equal to the determined texture with $\vartheta = 25°$) show low corresponding Young's moduli and are located in a region where the Young's moduli itself is nearly homogeneously distributed, which results in a negligible scatter. With an increase in ψ, the possible orientations cross through areas where the width of possible values for Young's moduli increases, which also increases the scatter. As shown in Figure 16, the highest average values occur for ψ between 20° and 30°. This is due to the case that some orientations align close to the [111] axis, which result in the highest Young's moduli values of about 250 GPa. For $\psi > 30°$, the scatter remains nearly constant with decreasing average value. As can be seen at the pole figures, with further increasing ψ, the possible Young's moduli values move closer to areas close to the [100] and [110] orientations, which results in a decrease of the average Young's moduli. For these orientations, the scatter remains quite high, since the possible orientations cross areas with a broad range of possible Young's moduli.

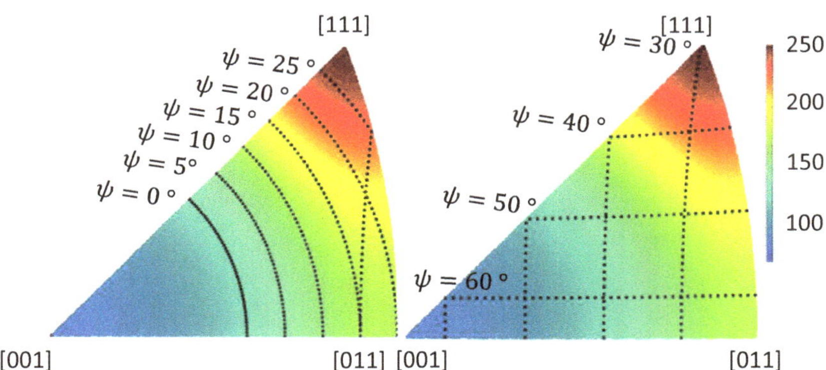

Figure 17. Inverse pole figure plot for different orientation distributions with constant value of ψ over the Young's moduli at 850 °C.

From this behaviour follows a huge stiffness discrepancy with an average 70 GPa higher stiffness at $\psi = 25°$ compared to 128 GPa at $\psi = 0°$. It should be noted that this behaviour is only valid for elements on the surface of the notch root. For elements in the direction of the bulk material (x–direction), the orientation changes of principal stress are shown on a cross section through the notch in Figure 18.

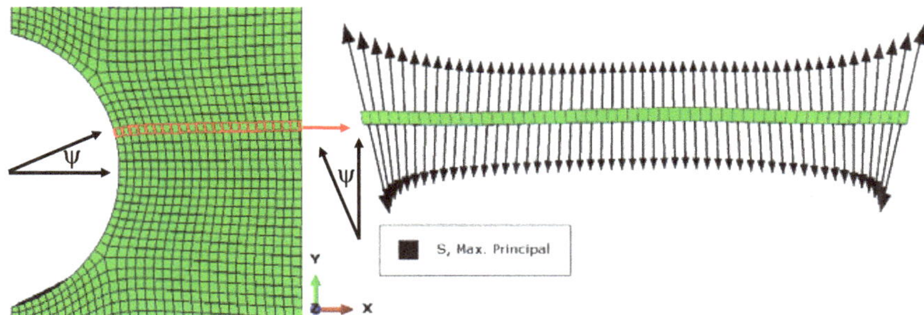

Figure 18. Cross section of elements through the notch with directions of principle stress.

For the first couple of elements close to the notch, a distinct difference angle ψ between the global loading axis y and principal stress direction can be identified. With increasing distance from the notch (in x–direction), the maximum principal stress aligns more and more with the global loading direction. Transferred to the results for the elements in the notch, a stiffness gradient follows with high stiffness at the notch root and decreasing values into the bulk material. Figure 19 shows the development of stiffness into the bulk material through a cross section calculated by Monte Carlo simulations. The cross section is taken horizontally at $\psi = 25°$ since it is the found maximum—see Figure 16.

As can be seen on the direction of the principal stress for each chosen element in Figures 18 and 19, a uniaxial stress state is already reached after a couple of elements, which is less than a millimetre into the bulk material. This can lead to high stiffness differences between the notch area and the bulk material which reach up to 104 GPa in the worst case. The majority of bulk material, on the other hand, has only low stiffness due to the predominant uniaxial stress state in combination with the examined texture (comparable to the stress state in the smooth specimen).

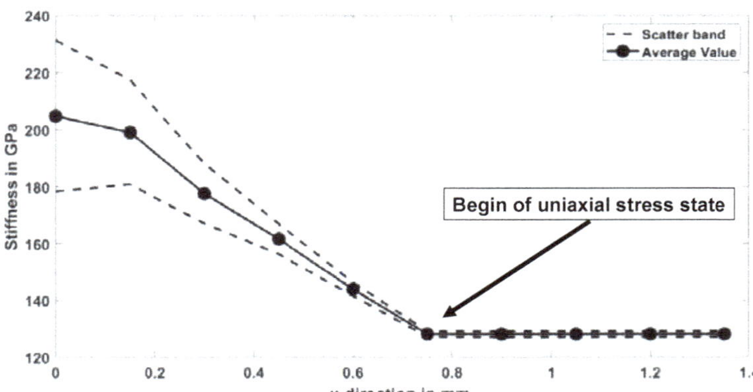

Figure 19. Development of stiffness into the bulk material for the fine-grain textured material.

Interestingly, the highest resulting stiffness (associated with an angle $\psi = 25°$) lies still in the range of the highest stresses, calculated for the isotropic approach caused by the notch and therefore in the critical area for crack initiation. In summary, it can be stated for the notched specimen with the examined textured orientation distribution that high stiffness appears in the highly loaded and therefore critical areas of the notch due to a stiffness increase along the notch. This results in high local stresses as well as shear stresses, as seen by the finite element simulations in Section 4.3. The bulk material, where the loading direction and maximum principal stress direction align, show significantly lower stiffness due to the texture. Thus follows for the examined textured in combination with a notched specimen geometry, that the total stiffness can be seen as a composite of the low stiffness of the bulk and potentially high stiffness of the notched area.

For a real considered polycrystalline and anisotropic material, the local mechanical behaviour is much more complex. Besides the previous described differences in alignment between maximum principal stress and global loading axis caused by the geometric character of the notch, further effects occur. Especially, the interaction between grains of different sizes and orientations where resulting constraints lead to inhomogeneous stress states along the notch as well as in the bulk material, as seen in the finite element simulations. Due to the complexity, grain interactions are not considered for the evaluation of stiffness properties.

5.2. Influence of Grain Orientation Distribution on the Shear Stress in Slip Systems

Besides the influence on stiffness, the texture of the fine-grain textured material also leads to a preferred orientation of the slip systems. To determine the onset of plasticity, the maximum resulting shear stress within the slip system was calculated according to Equation (1) and the modified Schmid factor obtained by dividing the resulting shear stress by the von Mises stress of the node. Due to the node-based formulation of the python code, the polycrystalline finite element models can be considered, whereby the local grain interaction is taken into account for the resulting shear stress evaluation. While for the smooth sample all nodes on the surface were evaluated, for the notch sample only nodes in the area of the notch root (location of possible crack initiation) were considered. Figure 20 shows the distribution of the modified Schmid factor for the smooth and notched specimen for the two different investigated grain orientation distributions.

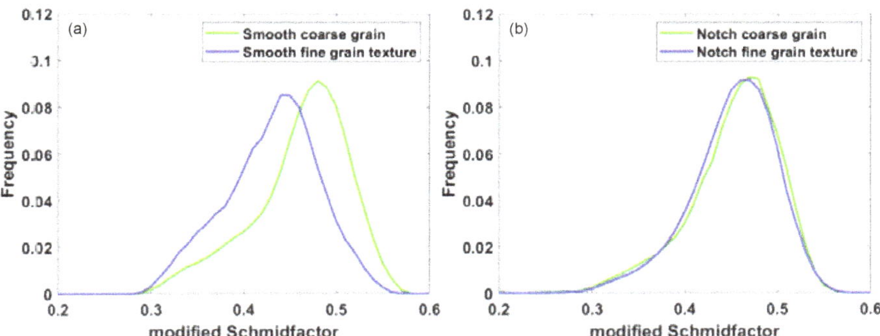

Figure 20. Distribution of modified Schmid factor for random-orientated coarse-grain and fine-grain textured smooth (**a**) (according to [17]) and notched specimen (**b**).

The averaged modified Schmid factor for the smooth specimen is 0.431 ± 0.05 for the fine-grain textured material and therefore significantly lower than for the coarse-grain material with 0.468 ± 0.05. Figure 20b shows the distribution of the modified Schmid factor for all simulations carried out with a notched model. It can be clearly seen that the distributions of the modified Schmid factor show similar profiles for the notched specimen for both orientation distributions. The average modified Schmid factor for the fine-grain textured material is 0.448 ± 0.05 and nearly similar to the value for the coarse-grained randomly orientated material with 0.454 ± 0.06. The nearly similar modified Schmid factor distribution can be explained by the change of the direction of the principal stress along the notch. For the smooth specimen, the loading direction is constant and lead by the constant texture to an improvement of the modified Schmid factor. For the notched specimen, the texture is also constant, but the direction of maximum principle stress changes along the notch, which leads to a variation in Schmid factors due to constant texture with varying principal stress direction (analogue to the stiffness). Since nearly similar modified Schmid factor distributions were found for the notched specimen, the increased values of the resulting shear stress of about 50 MPa (see Figure 13) for the fine-grained textured material can be attributed to the evaluated higher stiffness along the notch root. The increased stiffness in combination with slightly higher strains (see Figure 12) leads to higher stresses in the notch root and finally to higher local shear stresses.

A slightly shorter lifetime associated with the slightly higher shear stress of the fine-grain textured and notched specimen cannot be clearly evaluated regarding the stress and strain Wöhler curves shown in Figures 5 and 6. Due to the low amount of data, it is very difficult to make reliable statements about differences in lifetime due to the large scatter.

6. Conclusions and Outlook

In the present work, the influence of a random and textured grain orientation distribution on the mechanical as well as fatigue behaviour was investigated for the polycrystalline nickel-base superalloy René80. Isothermal LCF tests at 850 °C with notched specimen show a stable cyclic behaviour over the lifetime independent of the applied strain amplitude. Analogous to the fatigue behaviour of smooth specimen, low total strain leads to single crack initiation, whereas high total strain results in multi-crack initiation sites. Overall, for same total strain amplitudes, the notched specimen reveals distinct higher lifetimes compared to the smooth specimen, which can be mainly attributed to a significantly smaller highly loaded volume with distinct smaller grain numbers. In contrast to the smooth specimen, where the fine-grain textured improved the fatigue life due to lower Young's moduli and modified Schmid factor, no improvement was found for the notched specimen. This can mainly be attributed to the development of an area with high stiffness along the notch, caused by constant grain texture with varying principal stress directions generated by the notch geometry. Under loading, this results in slightly higher stresses and therefore to higher shear stresses in the notch root area compared to the random orientated material. Since the distribution of modified Schmid factors show no differences for the notched random and textured orientated specimen, the increased shear stress results mainly from the increased stiffness development along the notch. According to minor differences in shear stress for the random and textured material with about 50 GPa, no deviations in lifetime could be observed, which can be attributed to the small number of tests as well as to the occurring scatter in lifetime.

In summary, it can be stated that the examined material texture is beneficial for uniaxial fatigue loading in combination with smooth specimens due to its improvement in elastic properties as well as the onset of plasticity. However, for notched specimens, these advantages disappear and a nearly similar fatigue behavior evolves between the textured and random-orientated material.

Using the presented modeling approach, it is furthermore possible to design grain orientation distributions which generate minima in stress as well as resulting in shear stress for complex geometries, to improve the component performance under high-temperature LCF conditions.

Author Contributions: B.E. wrote the present paper and executed the material modelling. Experimental work and analysis was carried out by B.E. and supported S.O. L.M., and T.B. were mainly involved in discussing and interpreting the results as well as in supervising the work. All authors have read and agreed to the published version of the manuscript.

Funding: The investigations were conducted as part of the joint research program COOREFLEX TURBO in the frame of AG Turbo. The work was supported by the Federal Ministry for Economic Affairs and Energy as per resolution of the German Federal Parliament under grant number 03ET7041K.

Data Availability Statement: Data not available due to commercial restrictions.

Acknowledgments: The authors gratefully acknowledge AG Turbo and Siemens AG for their support and permission to publish this paper. The responsibility for the content lies solely with its authors.

Conflicts of Interest: The authors declare no conflict of interest.

References

1. Walsh, P.P.; Fletcher, P. *Gas Turbine Performance*; Blackwell Science Ltd.: Oxford, UK, 2004.
2. Ramberg, W.; Osgood, W.R. *Description of Stress-Strain Curves by Three Parameters*; National Advisory Committee for Aeronautics: Edwards, CA, USA, 1943.
3. Zhao, L.G.; Tong, J.; Vermeulen, B.; Byrne, J. On the uniaxial mechanical behaviour of an advanced nickel base superalloy at high temperature. *Mech. Mater.* **2001**, *33*, 593–600. [CrossRef]
4. Zhang, Z.; Yu, H.; Dong, C. LCF behavior and life prediction method of a single crystal nickel-based superalloy at high temperature. *Front. Mech. Eng.* **2015**, *10*, 418–423. [CrossRef]
5. Neidel, A.; Riesenbeck, S.; Ullrich, T.; Völker, J.; Yao, C. Hot cracking in the HAZ of laser-drilled turbine blades made from René 80. *Mater. Test.* **2005**, *47*, 553–559. [CrossRef]
6. Siavashani, R.S.; Novinrooz, A.J.; Taherkhani, A. The Effect of Different Continuous Cooling Rates on Lattice Constant and Morphology of the Precipitates in Nickel-Base Super Alloy Rene 80. *J. Basic. Appl. Sci.* **2013**, *3*, 14–19.
7. Donachie, M.J.; Donachie, S. *Superalloys: A Technical Guide*, 2nd ed.; ASM International: Russell Township, OH, USA, 2002.
8. Kulawinski, D. *Biaxial-Planare Isotherme und Thermo-Mechanische Ermüdung an polykristallinen Nickelbasis-Superlegierungen*; Logos Verlag: Berlin, Germany, 2015.
9. Holländer, D. Experimentelles Verfahren zur Charakterisierung des Einachsigen Ermüdungsverhaltens auf Basis Miniaturisierter Prüfkörper und Anwendung auf Hochtemperatur-Legierungen der Energietechnik. Ph.D. Thesis, Logos Verlag, Berlin, Germany, 2017.
10. Moch, N. From Microscopic Models of Damage Accumulation to the Probability of Failure of Gas Turbines. Ph.D. Thesis, Universitätsbibliothek Wuppertal, Wuppertal, Germany, 2019.
11. Schmitz, S.; Seibel, T.; Beck, T.; Rollmann, G.; Krause, R.; Gottschalk, H. A probabilistic model for LCF. *Comput. Mater. Sci.* **2013**, *79*, 584–590. [CrossRef]
12. Gottschalk, H.; Schmitz, S.; Seibel, T.; Rollmann, G.; Krause, R.; Beck, T. Probabilistic Schmid factors and scatter of low cycle fatigue (LCF) life. *Materialwissenschaft und Werkstofftechnik* **2015**, *46*, 156–164. [CrossRef]
13. Mäde, L.; Kumar, K.; Schmitz, S.; Gundavarapu, S.; Beck, T. Evaluation of component-similar rotor steel specimens with a local probabilistic approach for LCF. *Fatigue Fract. Eng. Mater. Struct.* **2019**, *91*, 319. [CrossRef]

14. Seibel, T. Einfluss der Probengröße und der Kornorientierung auf die Lebensdauer einer polykristallinen Ni-Basislegierung bei LCF-Beanspruchung. Ph.D. Thesis, Forschungszentrum Jülich, Jülich, Germany, 2014.
15. Mäde, L.; Schmitz, S.; Gottschalk, H.; Beck, T. Combined notch and size effect modeling in a local probabilistic approach for LCF. *Comput. Mater. Sci.* **2018**, *142*, 377–388. [CrossRef]
16. Engel, B.; Beck, T.; Moch, N.; Gottschalk, H.; Schmitz, S. Effect of local anisotropy on fatigue crack initiation in a coarse grained nickel-base superalloy. In Proceedings of the 12th International Fatigue Congress (FATIGUE 2018), Poitiers, France, 27 May–1 June 2018. [CrossRef]
17. Engel, B.; Mäde, L.; Lion, P.; Moch, N.; Gottschalk, H.; Beck, T. Probabilistic Modeling of Slip System-Based Shear Stresses and Fatigue Behavior of Coarse-Grained Ni-Base Superalloy Considering Local Grain Anisotropy and Grain Orientation. *Metals* **2019**, *9*, 813. [CrossRef]
18. Engel, B.; Beck, T.; Schmitz, S. High temperatue Low Cylce Fatigue of the Ni-base Superallor René80. In Proceedings of the Eighth International Conference on Low Cycle Fatigue (LCF8), Dresden, Germany, 27–29 June 2018.
19. Safari, J.; Nategh, S. On the heat treatment of Rene-80 nickel-base superalloy. *J. Mater. Process. Technol.* **2006**, *176*, 240–250. [CrossRef]
20. Takaki, T.; Sakane, S.; Ohno, M.; Shibuta, Y.; Aoki, T.; Gandin, C.-A. Competitive grain growth during directional solidification of a polycrystalline binary alloy: Three-dimensional large-scale phase-field study. *Materialia* **2018**, *1*, 104–113. [CrossRef]
21. Engel, B. Einfluss der Lokalen Kornorientierung und der Korngröße auf das Verformungs- und Ermüdungsverhalten von Nickelbasis Superlegierungen. Ph.D. Thesis, Technische Universität Kaiserslautern, Kaiserslautern, Germany, 2019.
22. ISO International Organization for Standardization. *Metallic Materials—Fatigue Testing—Axial-Strain-Controlled*; ISO International Organization for Standardization: Geneva, Switzerland, 2017.
23. Quey, R.; Dawson, P.R.; Barbe, F. Large-scale 3D random polycrystals for the finite element method: Generation, meshing and remeshing. *Comput. Meth. Appl. Mech. Eng.* **2011**, *200*, 1729–1745. [CrossRef]
24. Ghazvinian, E.; Diederichs, M.S.; Quey, R. 3D random Voronoi grain-based models for simulation of brittle rock damage and fabric-guided micro-fracturing. *J. Rock Mech. Geotech. Eng.* **2014**, *6*, 506–521. [CrossRef]
25. Quey, R.; Renversade, L. Optimal polyhedral description of 3D polycrystals: Method and application to statistical and synchrotron X-ray diffraction data. *Comput. Meth. Appl. Mech. Eng.* **2018**, *330*, 308–333. [CrossRef]
26. Hermann, W.; Sockel, H.G.; Han, J.; Bertram, A. Elastic Properties and Determination of Elastic Constants of Nickel-Base Superalloys by a Free-Free Beam Technique. In Proceedings of the Eighth International Symposium on Superalloys, Champion, PA, USA, 22–26 September 1996; pp. 229–238.
27. Gollmer, M. Schadensakkumulationsverhalten der Superlegierung René 80 unter Zweistufiger Low Cycle Fatigue Beanspruchung. Ph.D. Thesis, Technische Universität Kaiserslautern, Kaiserslautern, Germany, 2018.
28. Hyde, T.H.; Sun, W.; Williams, J.A. Requirements for and use of miniature test specimens to provide mechanical and creep properties of materials: A review. *Int. Mater. Rev.* **2007**, *52*, 213–255. [CrossRef]

Article

Temperature Resistance of Mo₃Si: Phase Stability, Microhardness, and Creep Properties

Olha Kauss [1,2,*], **Susanne Obert** [3], **Iurii Bogomol** [4], **Thomas Wablat** [1], **Nils Siemensmeyer** [1], **Konstantin Naumenko** [2] **and Manja Krüger** [1]

[1] Faculty of Mechanical Engineering (FMB), Institute of Material and Joining Technology (IWF), Otto von Guericke University (OVGU) Magdeburg, Universitätsplatz 2, 39106 Magdeburg, Germany; thomas.wablat@st.ovgu.de (T.W.); nils.siemensmeyer@st.ovgu.de (N.S.); manja.krueger@ovgu.de (M.K.)

[2] Faculty of Mechanical Engineering (FMB), Institute of Mechanics (IFME), Otto von Guericke University (OVGU) Magdeburg, Universitätsplatz 2, 39106 Magdeburg, Germany; konstantin.naumenko@ovgu.de

[3] Institute for Applied Materials (IAM-WK), Karlsruhe Institute of Technology (KIT), Engelbert-Arnold-Straße 4, 76131 Karlsruhe, Germany; susanne.obert@kit.edu

[4] Faculty of Physical Engineering (ІФФ), Department of High-Temperature Materials and Powder Metallurgy (ВТМТаПІМ), National Technical University of Ukraine "Igor Sikorsky Kyiv Polytechnic Institute" (NTUU "KPI"), Peremogy Ave. 37, 03056 Kyiv, Ukraine; ubohomol@iff.kpi.ua

* Correspondence: olha.kauss@ovgu.de

Citation: Kauss, O.; Obert, S.; Bogomol, I.; Wablat, T.; Siemensmeyer, N.; Naumenko, K.; Krüger, M. Temperature Resistance of Mo₃Si: Phase Stability, Microhardness, and Creep Properties. *Metals* **2021**, *11*, 564. https://doi.org/10.3390/met11040564

Academic Editor: Stefano Spigarelli

Received: 26 February 2021
Accepted: 24 March 2021
Published: 30 March 2021

Publisher's Note: MDPI stays neutral with regard to jurisdictional claims in published maps and institutional affiliations.

Abstract: Mo-Si-B alloys are one of the most promising candidates to substitute Ni based superalloys in gas turbines. The optimization of their composition can be achieved more effectively using multi-scale modeling of materials behavior and structural analysis of components for understanding, predicting, and screening properties of new alloys. Nevertheless, this approach is dependent on data on the properties of the single phases in these alloys. The focus of this investigation is Mo₃Si, one of the phases in typical Mo-Si-B alloys. The effect of 100 h annealing at 1600 °C on phase stability and microhardness of its three near-stoichiometric compositions—Mo-23Si, Mo-24Si and Mo-25Si (at %)—is reported. While Mo-23Si specimen consist only of Mo₃Si before and after annealing, Mo-24Si and Mo-25Si comprise Mo₅Si₃ and Mo₃Si before annealing. The latter is then increased by the annealing. No significant difference in microhardness was detected between the different compositions as well as after annealing. The creep properties of Mo₃Si are described at 1093 °C and 1300 °C at varying stress levels as well as at 300 MPa and temperatures between 1050 °C and 1350 °C. Three constitutive models were used for regression of experimental results—(i) power law with a constant creep exponent, (ii) stress range dependent law, and (iii) power law with a temperature-dependent creep exponent. It is confirmed that Mo₃Si has a higher creep resistance than Mo$_{ss}$ and multi-phase Mo-Si-B alloys, but a lower creep strength as compared to Mo₅SiB₂.

Keywords: molybdenum silicide; microstructure; annealing; phase stability; microhardness; creep; Mo-Si-B alloys

1. Introduction

The development of new materials, that can operate at increasing temperatures in combustion chambers of gas turbine engines, is getting more important. At the moment the single crystal Ni based superalloys are used for this application at ~1100 °C with hot spots of ~1200 °C, which is ~90% of their melting point [1]. Thereby, the heat resistance of Ni based superalloys could not have been increased significantly in recent years. This causes huge challenges in the development of gas turbine engines to meet the demanding requirements. Thus, alternative materials have to be developed in order to achieve a significant increase in operating temperatures in combustion chambers. The high solidus temperature of Mo-Si-B alloys makes them one of the most promising material classes for this approach. Detailed reviews on the properties of Mo-Si-B alloys and their phases are given in [1–3]. The best

179

combination of properties was found in three-phase alloys consisting of Mo_{ss}, Mo_3Si, and Mo_5SiB_2. Mo_{ss} provides the ductility and both intermetallic phases are responsible for adequate creep and oxidation resistance under certain circumstances [1–3].

Multi-phase materials for structural application beyond the capability of Ni base superalloys should provide an adequate balance of ambient and ultra-high temperature properties for critical components. The fracture toughness of polycrystalline Mo_3Si and Mo_5SiB_2 at room temperature is ~3 MPa·m$^{1/2}$ [4,5]. The low plasticity of these phases is caused by low symmetry of the crystal structures of these phases, which inhibits plastic deformation through dislocation formation and motion [3]. Another reason for low fracture toughness of Mo_5SiB_2 is anisotropy of the coefficient of thermal expansion along c- and a-axes, which encourages the development of internal residual stresses [3]. However, the multi-phase Mo-Si-B alloys with volume fractions of ductile $Mo_{ss} \geq 30\%$ provide improved fracture toughness, because the plastic deformation of the ductile phase consumes significantly more energy, which leads to toughening by crack arrest and bridging. The common value of fracture toughness for Mo-Si-B alloys is in the range 7–15 MPa·m$^{1/2}$ [2,6–13]. The alloys with coarse Mo_{ss} particles forming a continuous network possess even improved fracture toughness of 15–21 MPa·m$^{1/2}$ [9,11,14], which meets the required minimum fracture toughness for materials beyond Ni base superalloys of 20 MPa·m$^{1/2}$ stablished by Bewlay et al. [15].

High temperature application of new material is often limited by its creep properties. Different Mo-Si-B alloys have a creep strength of 50–190 MPa at a creep rate of 2×10^{-8} s^{-1} at 1200 °C [12,16–18]. Creep properties are mainly determined by on the fraction of the intermetallic phases and microstructure. Alloying can extremely increase the creep strength. Thus, alloying with Nb increased creep strength of Mo-12Si-8.5B (at %) from (extrapolated) 75 MPa to 260 MPa for Mo-19.5Nb-12Si-7.5B (at %) at a creep rate of 2×10^{-8} s^{-1} at 1200 °C. For more detailed comparison creep properties in wider range should be considered, especially at higher stresses. Such comparison for some well-known alloys and results of this study are given in Section 3.3.1. Bewlay et al. [19,20] suggested that the creep strength should be greater than 170 MPa at a creep rate of 2×10^{-8} s^{-1} at 1200 °C for Nb alloys beyond Ni based superalloys for turbine applications. Schneibel [21] adjusted this goal for alloys with different density and set the value of 250 MPa at a creep rate of 2×10^{-8} s^{-1} at 1200 °C for Mo-Si-B alloys for high temperature application beyond Ni base superalloys. Thus Mo-Si-B alloys can reach the required creep strength. Nevertheless, different microstructure strategies are optimal for balanced ductility vs. creep strength. Therefore, a profound material design is needed for Mo-Si-B alloys to achieve the optimal properties.

However, the development of new alloys is extremely time- and cost-consuming. To reduce the costs required for experimental analysis, multi-scale modeling of materials behavior and structural analysis of components of creep behavior can be performed [22–24]. In addition, appropriate creep models are required for the structural analysis of turbine blades. The constitutive behavior of new alloys can be predicted with material modeling using different kinds of homogenization of their per-phase properties. For this approach, the properties of these phases are often in demand. Additionally, there is still a lack in property data for the phases being present in Mo-Si-B alloys. Therefore, this work is dedicated to the investigation of the creep properties of the Mo_3Si phase in order to overcome the lack in available data.

A section of the binary Mo-Si phase diagram is shown in Figure 1. According to it the solid solution Mo_{ss} has a maximum solubility of approximately 4 at % Si at the peritectic temperature of 2025 °C. The melting point of Mo is 2623 °C. The cubic intermediate phase Mo_3Si with 25 at % Si decomposes peritectically in reaction $L + Mo_{ss} \leftrightarrow Mo_3Si$ at 2025 °C with isotherm between 4 and 25.72 at %. The eutectic reaction $L \leftrightarrow Mo_3Si + Mo_5Si_3$ occurs at 2020 °C with eutectic point of 26.4 at % Si and isotherm between 25.72 and 37 at %. The tetragonal intermediate phase Mo_5Si_3 with 37.5 at % Si is solidified congruently in reaction $L \leftrightarrow Mo_5Si_3$ at 2180 °C. The phase diagram range with higher Si content is not relevant for this study.

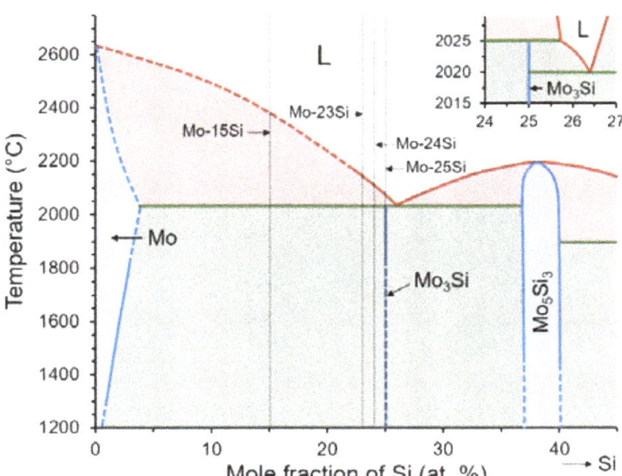

Figure 1. A section of the binary Mo-Si phase diagram reprinted according to data from [25], representing the alloy compositions investigated in the present study as well as Mo$_3$Si and Mo$_5$Si$_3$ phases. Inset—a section of the binary Mo-Si phase diagram representing peritectic reaction L + Mo$_{ss}$ ↔ Mo$_3$Si and eutectic reaction L ↔ Mo$_3$Si + Mo$_5$Si$_3$. Grey vertical lines—alloy compositions investigated in the present study; red curves—solidus; blue curves—liquidus; green lines—reaction isotherm. Solid curves—experimental results. Dashed curves—extrapolated phase boundaries into metastable regions of the phase diagram.

Therefore Rosales and Schneibel [4] considered, that Mo$_3$Si is not a real line compound and slightly off-stoichiometric as it is single-phase only in a small composition range near Mo-24Si. Thus, the dependence of the phase stability on slight variations in the Mo to Si ratio was assessed by analyzing the microstructure of different near-stoichiometric specimens. Additionally, the microhardness was determined in order to comment on the presence of internal stresses, which might influence the mechanical properties. Also, Gnesin, I. and Gnesin, B. [26] indicated Mo$_3$Si with X-ray microanalysis and reported, that it exists in a small compositional region of 22.33 ± 0.34 – 23.6 ± 0.56 at %.

2. Materials and Methods

First, three near-stoichiometric compositions Mo-23Si, Mo-24Si, and Mo-25Si (Figure 1) were produced. All elemental and phase proportions are indicated in atomic percent in this paper. The intention of selecting these compositions was to analyze the phase stability after annealing as well as potential internal stresses caused by arc melting by determining the microhardness before and after annealing. These results allowed the selection of the appropriate chemical composition and manufacturing route of the Mo$_3$Si specimens intended for the investigation of the creep properties.

Buttons with a mass of ~5 g with compositions Mo-23Si, Mo-24Si, and Mo-25Si were processed from high purity bulk materials of Mo and Si by arc melting and re-melting five times in an arc melter MAM-1 (Edmund Buehler GmbH, Bodelshausen, Germany). Therefore, a water-cooled copper crucible was used and a high purity Ar atmosphere established. Prior to melting, a Zr getter was melted to reduce residual oxygen from the process chamber. A specimen after arc-melting is shown in Figure 2b. Afterwards, these buttons were cut in four pieces using electrical discharge machining (EDM). One specimen of each composition was annealed at 1600 °C for 100 h in a flowing Ar atmosphere. Macroscopic surface of specimens before and after annealing has not changed.

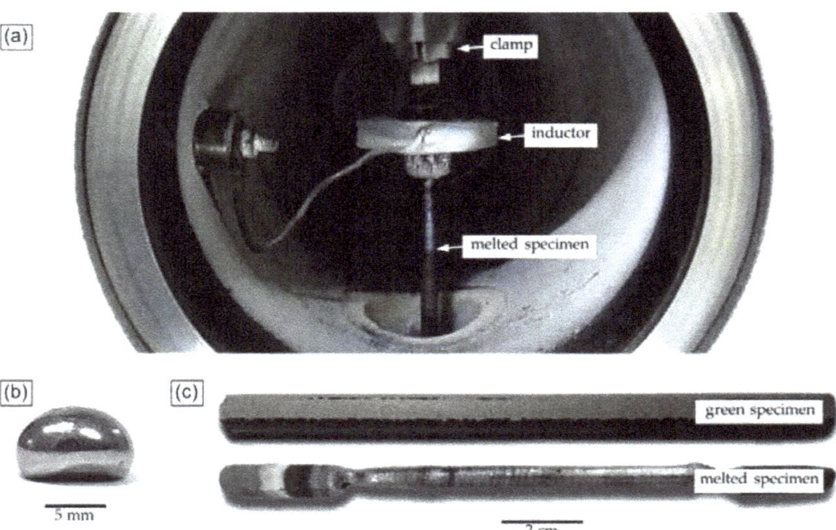

Figure 2. Processing: (**a**) crucible-free zone melting chamber, (**b**) specimen after arc-melting, (**c**) specimen before and after zone melting.

The specimen with composition Mo-15Si was produced from the mixture of elemental high purity powders of Mo and Si by crucible-free zone melting of compacted powders [27,28]. This composition leading to a multi-phase microstructure (Figure 1) was selected to investigate the microhardness of the Mo_3Si phase in an alloy and to compare it with the microhardness of the above-mentioned single-phase specimens. Thus, the floating zone melting method was selected to achieve a coarse microstructure providing the possibility for micro indentation in single grains. A green rod was cold pressed from the powder mixture in a hydraulic press in sectional powder compaction tool for extra-long specimens. After that it was processed in a heated zone melting device Crystal-206 (Yoshkar-Ola, ME, Russian Federation) equipped with an induction-type heater in a high purity He atmosphere using solidification rates of 60 mm/h (Figure 2a). The resulting specimen was ~80 mm in length, having a diameter of ~7 mm. A specimen before and after crucible-free zone melting is shown in Figure 2c. The specimens were cut by EDM parallel and perpendicular to the growth direction in order to investigate the microstructure and determine the microhardness.

For the characterization of the microstructure, the specimens were embedded in cold hardening epoxy Demotec 15 Plus (DEMOTEC, idderau, Germany). Then, they were ground with SiC abrasive paper of grit P600, P800, P1200, and P2400 and polished with diamond suspensions with a particle size of 3 μm and 1 μm. In order to improve phase contrast, chemical polishing was carried out with a solution consisting of 98 mL O.P.S. (oxide polishing suspension) + 1 mL H_2O_2 + 1 mL NH_3.

The microstructure was quantitatively evaluated using scanning electron microscopes (SEM) FEI XL30 (Hillsboro, OR, USA) and Zeiss EVO 15 (Oberkochen, Germany) both equipped with EDX (energy dispersive X-ray). The SEM images were typically obtained in the backscattered electron (BSE) mode, but the images of cracks were obtained in secondary electrons (SE) mode for morphological representation of the surface. Macrostructure analysis was carried out on a ZEISS Stemi 2000 C stereo microscope (Oberkochen, Germany) equipped with ZEISS AxioCam MRC camera (Oberkochen, Germany). X-ray diffraction (XRD) measurements were performed using an PANalytical X'pert Powder X-ray diffractometer (Almelo, The Netherlands) with a Cu-Kα beam. The phase identification was obtained using the analysis software *HighScore Plus*, version: 3.0e (3.0.5) dated 30/01/2012; produced by: PANalytical B.V. (Almelo, The Netherlands).

Vickers microhardness measurements were carried out with a Leica VM HT hardness tester (Vienna, Austria) at a load of 500 g and a holding time of 15 s. Standard deviations were calculated from 30 measurements each.

For compressive creep tests specimens of Mo-23Si were manufactured with the same technique as described first in this section, which is same as for the previous analyses of this composition. The specimens were cut to dimensions of $2.5 \times 2.5 \times 3.75$ mm^3 by EDM. All creep tests were carried out under constant true stresses in a Zwick universal testing machine Z100 (Ulm, Germany) equipped with a Maytec furnace (Singen, Germany). The creep strength was evaluated in a flowing Ar/H$_2$ atmosphere for the tests under 1200 °C and in vacuum ($<10^{-4}$ Pa) for the tests at 1200 °C and above. In this study, the secondary creep stage characterized with minimum creep rate is discussed. Although this approach does not allow an estimate of the time to creep fracture, it provides a comparable parameter to assess the creep performance and, thus, is frequently used in the early stages of alloy and component design. Consequently, the investigation of the creep properties using step loading is permissible. This means that if the samples were free of any sign of damage after reaching the constant steady-state creep rate for the current constant stress level, another stress level was applied for further investigation. In this way, the specimens were loaded with 1 to 3 loading steps depending on the stress level.

3. Results and Discussion

3.1. Microstructure

The microstructures of Mo-25Si, Mo-24Si, and Mo-23Si after arc-melting (before annealing) are shown in Figure 3a,c,d, respectively. The results of the EDX area analysis of the elemental composition of the specimens were in good agreement with the alloy composition weighted in.

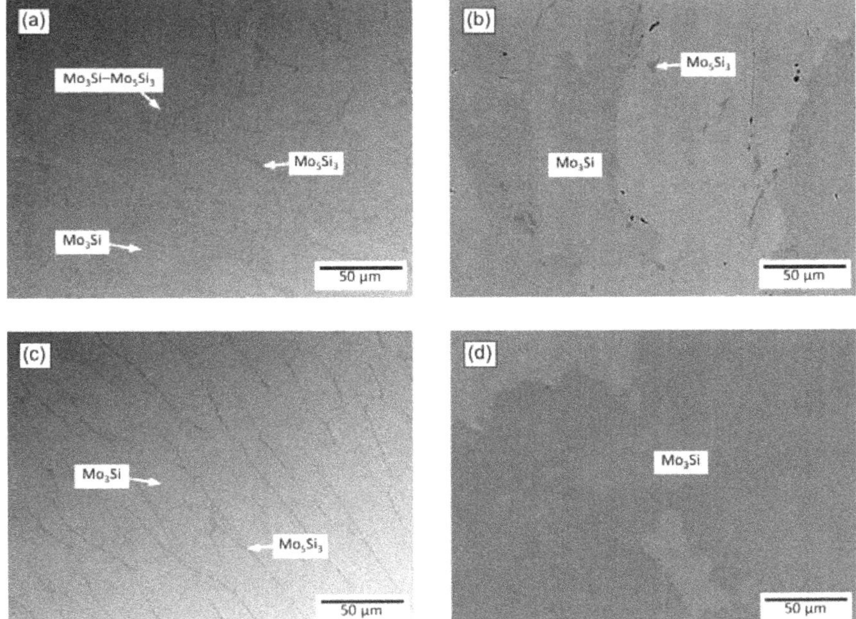

Figure 3. Microstructures of Mo-25Si after (**a**) arc-melting and (**b**) 100 h annealing at 1600 °C as well as (**c**) Mo-24Si and (**d**) Mo-23Si after arc-melting.

It is revealed by EDX analysis that the lighter phase consists of ~23% Si and the darker of ~35% Si in all specimens, which correlates with the phases Mo_3Si and Mo_5Si_3 but is not exactly stoichiometric. This is in accordance to [4] reporting that the Mo_3Si phase is not stoichiometric. Besides the fact that the Mo_5Si_3 phase may be non-stoichiometric as well, it should be considered that the inclusions of Mo_5Si_3 are exceedingly small. Thus, it is likely that ad- and subjacent regions are detected as well during the EDX point analyses influencing the results.

Note that the amount of Mo_5Si_3 phase along the grain boundaries is different for Mo-25Si, Figure 3a, and Mo-24Si, Figure 3c, namely increased in the latter alloy. This is caused by higher content of silicon in Mo-25Si specimen. Additionally, Mo-25Si has eutectic regions of Mo_5Si_3–Mo_3Si. The microstructure of Mo-23Si, Figure 3d reveals only Mo_3Si phase as also confirmed by XRD analysis, which shows only intensities typical for of Mo_3Si. Figure 4 shows measured XRD pattern of Mo-23Si. Figure 3b represents Mo-25Si after 100 h annealing at 1600 °C comprising significantly less Mo_5Si_3 phase than before annealing. Mo-24Si consists solely of Mo_3Si phase after annealing and have the similar microstructure as Mo-23Si before annealing (Figure 3d). The microstructure of Mo-23Si has not changed after annealing and, as before, consists solely of Mo_3Si phase. Thus, the composition Mo-23Si is used for creep testing since it corresponds to single-phase Mo_3Si and it seems to have reached a thermodynamically stable state after annealing.

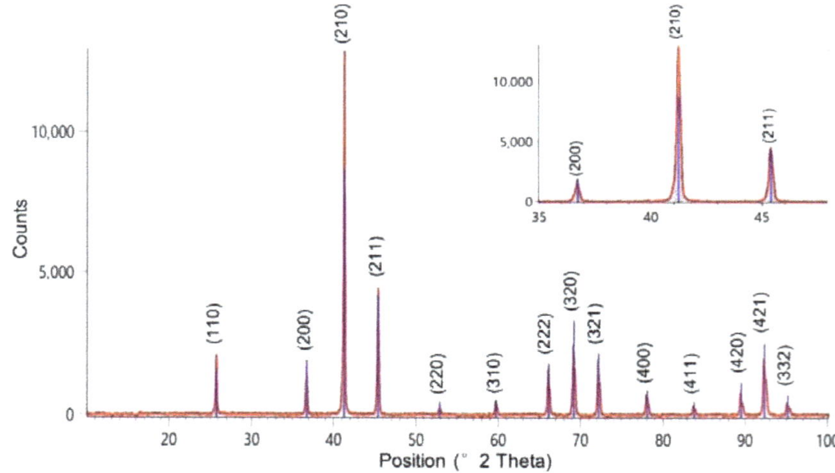

Figure 4. XRD pattern of Mo-23Si. Red curve—measured pattern of Mo-23Si. Inset—highest peaks. Blue lines—reference pattern for Mo_3Si in ICDD 2004 database (Ref. Code 00-051-00764, ICDD Grant in Aid, Baker, I.).

Mo-15Si specimen was manufactured by zone melting in order to achieve a coarse microstructure enabling the determination of the microhardness of the Mo_3Si phase in the alloy. The achieved microstructure meets the requirements (Figure 5) and consist of coarse grains of primarily solidified Moss and the matrix phase Mo_3Si. There is no significant difference observed between samples oriented parallel and perpendicular to the growth direction. The results of the EDX area analysis were in good agreement with the alloy composition weighted in. According to the EDX point analyses, the Si content in the darker phase is found to be ~22% and in the brighter ~7.4%, which might be related to Mo_3Si and supersaturated Mo_{ss} [29], the latter is expected to be reduced by homogenization heat treatment.

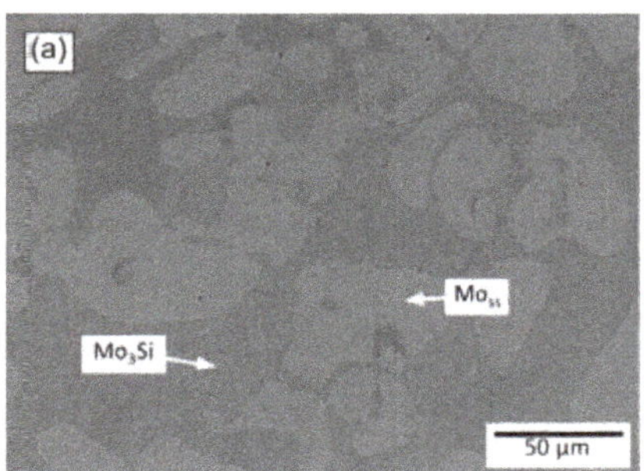

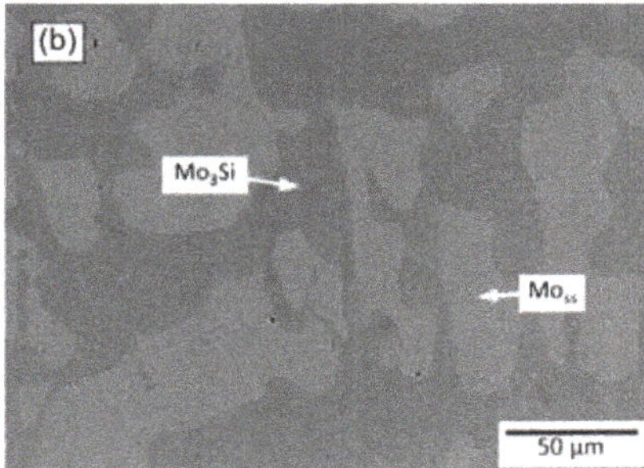

Figure 5. The microstructure of Mo-15Si after zone melting: (**a**) perpendicular and (**b**) parallel to the growth direction.

3.2. Microhardness

The results of the microhardness measurements are presented in Figure 6. The slight difference in microhardness observed for Mo-23Si and Mo-24Si does not allow for sound interpretations as significant scatter in data needs to be considered. Additionally, there is no significant change in microhardness before and after 100 h annealing at 1600 °C, which is in accordance with the non-significant change in microstructures due to the annealing. In general, the measured value of the Mo$_3$Si microhardness for all specimens is HV 0.5 = 1336 ± 87, which is in very good agreement to the data reported in [4] after 24 h annealing at 1600 °C as well as after 2 h annealing at 1200 °C. The independence of the microhardness on different annealing regimes indicates that no significant modifications of the local chemical compositions as well as no internal stress relief occurs during annealing, which leads to the assumption that no significant internal stresses are present after the arc melting, which might influence the mechanical properties.

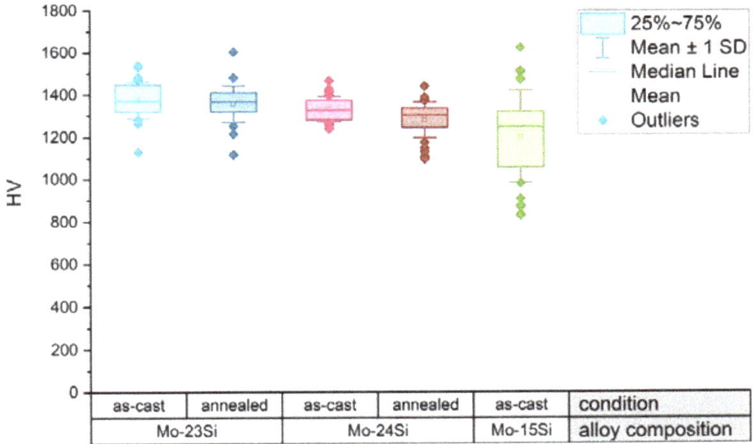

Figure 6. Microhardness of Mo$_3$Si in different alloy compositions in as-cast state and after annealing for 100 h at 1600 °C.

In the two-phase alloy Mo-15Si, the measured microhardness in the Mo_3Si grains is HV 0.5 = 1201 ± 217. Note the considerably increased standard deviation, which might be caused by subjacent, comparatively weak Mo_{ss} grains affecting the measurement (microhardness, measured in Mo_{ss} grains in the same specimen with alloy composition Mo-15Si, is HV 0.5 = 702 ± 45). Generally, the microhardness of Mo_3Si measured in this alloy is similar to the results for the single-phase specimens.

3.3. Constant Load Compressive Creep Tests

3.3.1. Experimental Results

The investigated variation in the minimum creep rates of the Mo_3Si specimens is represented in Figure 7 and compared to pure Mo [30,31], Mo_{ss} [18], and Mo_5SiB_2 [32] as well as to some Mo-Si-B alloys [33,34]. Note that in study [18] the creep properties were studied in tensile mode. Nevertheless, similar compressive and tensile creep behavior was reported [16,18]. Thus, it is assumed that the creep response of Mo_3Si, which is one of the phases in Mo-Si-B alloys with low plasticity, does not differ in tensile and compressive tests and can be compared to those results.

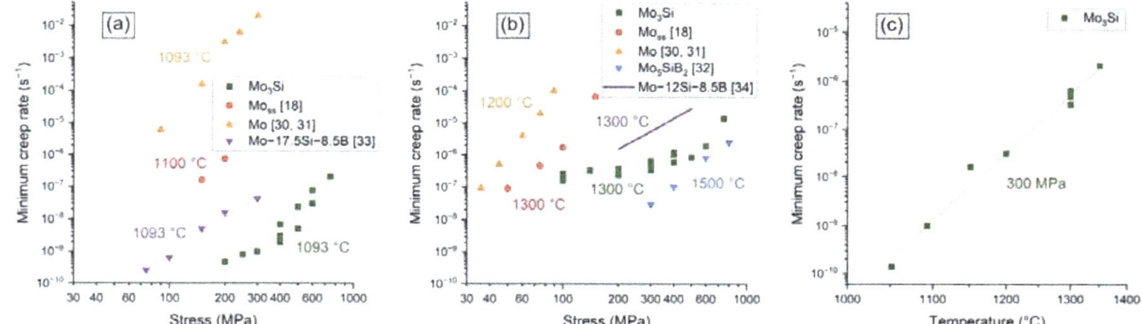

Figure 7. Creep properties of Mo_3Si at (**a**) intermediate and (**b**) high temperatures in comparison to pure Mo [30,31], Mo_{ss} [18], Mo_5SiB_2 [32], and Mo-Si-B alloys [33,34], including fitting with power creep law with constant creep exponent. (**c**) Arrhenius plot for Mo_3Si at 300 MPa.

The achieved results confirm the conception on the properties of the phases in Mo-Si-B alloys [1–3], which was described in the introduction. So, Mo_3Si exhibits significantly lower creep rates than Mo_{ss}, which in turn are much lower than the creep rates of the pure Mo. Additionally, Mo_3Si has lower creep rates than Mo-Si-B alloys, which typically include the weaker Mo_{ss} phase. The creep properties of polycrystalline Mo_5SiB_2 have only been studied at 1500 °C, but even at this temperature Mo_5SiB_2 has lower creep rates than Mo_3Si at 1300 °C. In all specimens, a brittle failure type was observed. It is clear that the damaged samples were not used for further stress-step-tests. Some examples of the cracks in the specimens and damaged samples after creep tests are displayed in the Figure 8. The brittle manner of fracture indicates high brittleness of Mo_3Si even at high temperatures. The problem is that brittle materials absorb less energy before breaking or fracturing. Thus, material fails suddenly instead of deforming or straining under load when a brittle failure occurs and this can happen with only a small load, impact force, or shock. Therefore Mo-Si-B alloys should have enough ductile Mo_{ss} which is able to bridging or arresting cracks.

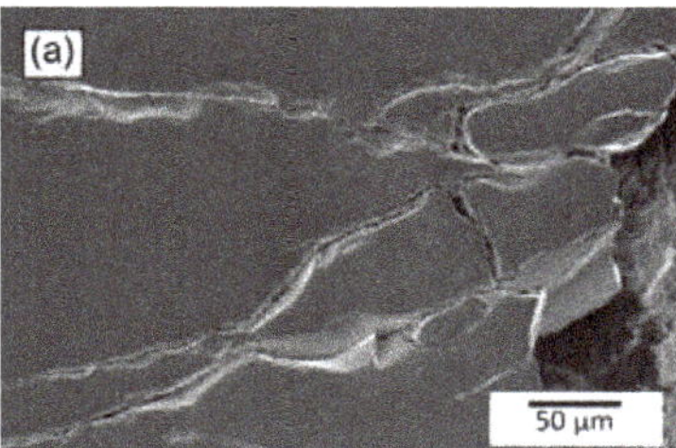 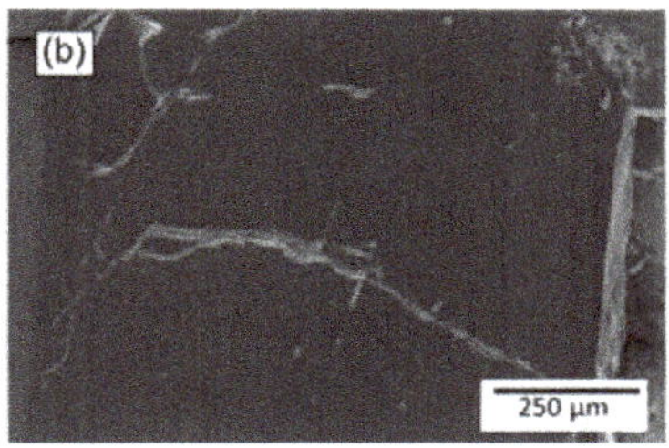 

Figure 8. Specimens after creep tests: (**a**) and (**b**) cracks in the specimens, (**c**) damaged sample.

Conclusively, Mo$_3$Si is characterized by a lower creep rates than Mo$_{ss}$ but a higher compared to Mo$_5$SiB$_2$ (Figure 7), which should be considered for novel alloy design.

To describe the creep behavior in a wide stress range, various response functions have been proposed. Overviews are presented in [30,35,36]. In this study, the variation in the creep rate with applied stress under different temperature was fitted with three constitutive equations—(i) a Norton power law, (ii) a stress range dependent constitutive model, as described in [37], and (iii) power law with temperature-dependent creep exponent as described in [38].

3.3.2. Regression with Power Law with Constant Creep Exponent

The most common way to fit the variation of the creep rate in the stationary creep regime for metallic alloys is a power law also known as Norton's law

$$\dot{\varepsilon} = A\sigma^n, \tag{1}$$

where $\dot{\varepsilon}$ is the minimum creep rate, A and n are material constants, and σ is the applied stress. For analysis and comparison of properties the dependence of the creep rate on the applied stress is commonly plotted in a graph with double logarithmic scale also called Norton-plot, like in Figure 7a,b. For non-isothermal test conditions, the Norton law coefficient A is generally analyzed in terms of an Arrhenius equation

$$A = B \cdot \exp\left(-\frac{Q}{RT}\right) \tag{2}$$

where B is material constant, T is the test temperature, Q is the apparent activation energy for the mechanism involved in the deformation process (usually close to the activation energy for self-diffusion) and R is the gas constant.

The constants and accuracy (R-squared values, R^2), determined according to the regression with the Equation (1) of the two isothermal cases, at 1093 °C and 1300 °C, are specified in Table 1, fitting curves are plotted in Figure 7a,b. For a case of constant stress $\sigma = 300$ MPa an activation energy of $Q = 542$ kJ/mol was determined. The corresponding Arrhenius-plot with experimental data and fitted curve ($R^2 = 0.988$) is plotted in Figure 7c. The determined activation energy is close to the activation energy for integrated diffusion in Mo$_3$Si, calculated from the growth kinetics data, $Q_{int} = 502$ kJ/mol [39]. As the curves at 1093 °C and 1300 °C show different slopes, of 4.6 and 1.3, it is not possible to fit all experimental data at once with the Equation (1) using (2) for the Norton law coefficient A. Thus, other constitutive equations are discussed in Sections 3.3.3 and 3.3.4, which allow the consideration of the change in the slope.

Table 1. Results of regression with power law Equation (1).

T	n	A, $(\text{s}^{-1}\cdot\text{MPa}^n)$	R^2
1093 °C	4.6	5.08×10^{-21}	0.878
1300 °C	1.3	3.85×10^{-10}	0.722

3.3.3. Regression with Stress Range Dependent Constitutive Model

Another way to describe the stationary creep behavior was described in [37] and suggests that the minimum creep rate is the sum of the linear and the power law stress functions. In this case, the constitutive equation would be

$$\dot{\varepsilon} = \dot{\varepsilon}_0 \frac{\sigma}{\sigma_0} + \dot{\varepsilon}_0 \left(\frac{\sigma}{\sigma_0}\right)^n = \dot{\varepsilon}_0 \frac{\sigma}{\sigma_0}\left[1 + \left(\frac{\sigma}{\sigma_0}\right)^{n-1}\right] \tag{3}$$

where n, $\dot{\varepsilon}_0$ and σ_0 are material constants. For non-isothermal conditions, the material constants $\dot{\varepsilon}_0$ and σ_0 must be replaced by Arrhenius type temperature functions, thus

$$\dot{\varepsilon}_0 = a \cdot \exp\left(-\frac{Q_a}{RT}\right), \ \sigma_0 = b \cdot \exp\left(-\frac{Q_b}{RT}\right) \tag{4}$$

where a and b are material constants, Q_a and Q_b are activation energies. The values of $\dot{\varepsilon}_0$ and σ_0 represent the point on the Norton plot, in which the extended curves of linear and power law creep would intersect. The parameter n represents the slope of the power law range.

The constants and accuracy (R-squared values, R^2) determined according to regression with stress range dependent constitutive model are specified in Table 2 and the fitting results are plotted on Figure 9. Despite the ability to describe the typical transition from linear to power law creep, not enough data are available in the power law range in order to determine the constants accurately. Thereby, the results fit with a good accuracy for $n = 7$ as well as for $n = 13$ (R-squared values are 0.981 and 0.998, respectively). Although the R-squared value for $n = 13$ is slightly higher, it cannot be concluded that it describes the properties better, because this value is determined by the several points at high stress levels and the further trend of the curve is not defined without data at higher stresses. Thus, a more detailed analysis of the properties at higher stresses is recommended for more accurate analysis taking into account the experimental deviations.

Table 2. Results of regression with stress range dependent constitutive model (3) using (4).

n	a, $(\text{s}^{-1}\cdot\text{MPa})$	b, $(\text{s}^{-1}\cdot\text{MPa}^n)$	Q_a, (kJ/mol)	Q_b, (kJ/mol)	R^2
7	2.68×10^{13}	2129	590	19	0.981
13	1.77×10^{13}	2364	577	17	0.998

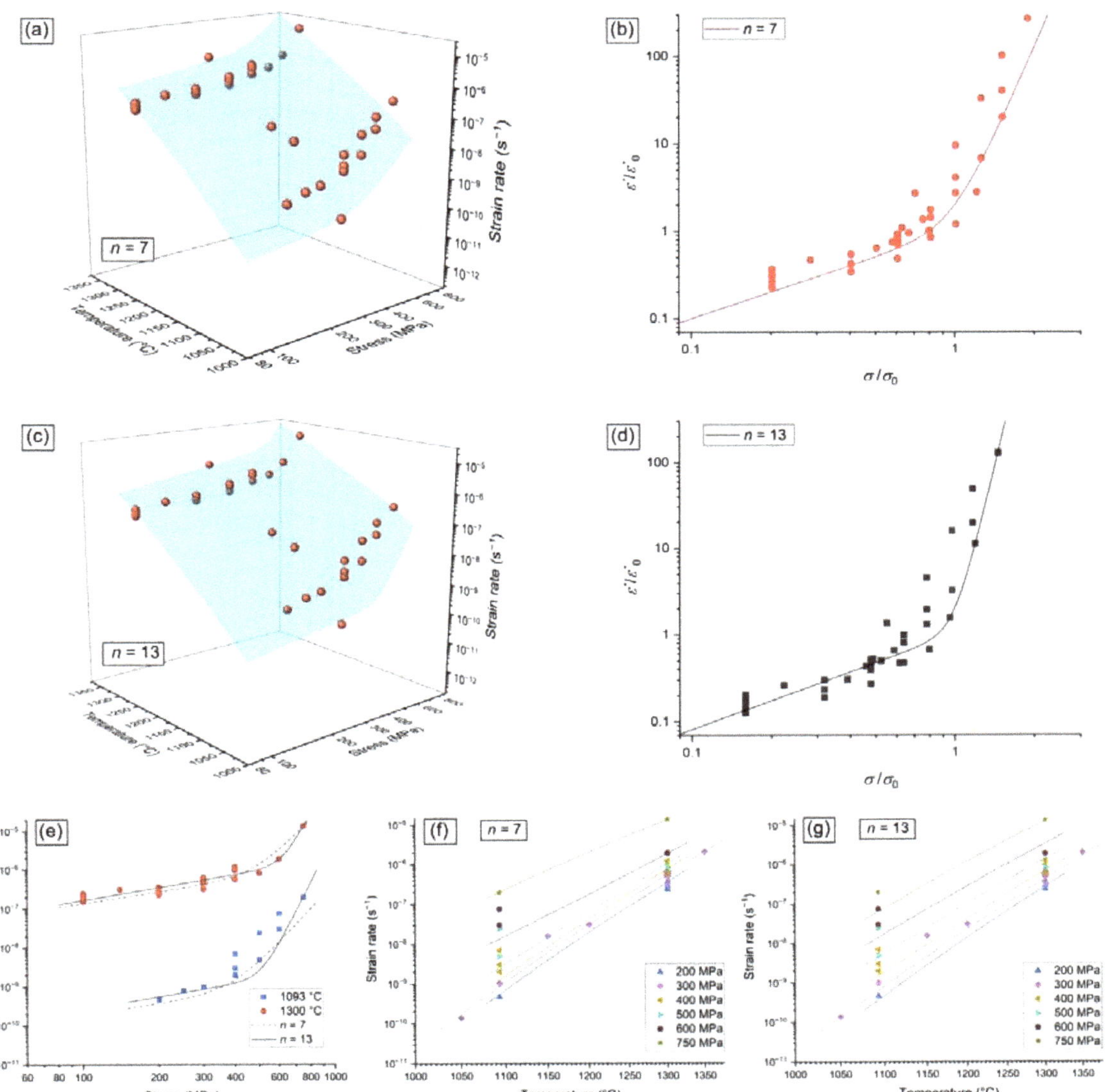

Figure 9. Fitting the experimental results with stress range dependent constitutive model: (**a**) and (**b**) $n = 7$; (**c**) and (**d**) $n = 13$; (**e**) comparison $n = 7$ and $n = 13$; (**f**) and (**g**) Arrhenius plot for $n = 7$ and $n = 13$, respectively.

3.3.4. Regression with Power Law Using Temperature-Dependent Creep Exponent

Another model to describe the stationary creep properties was proposed in [38]. It suggests that for non-isothermal conditions the creep exponent n in Equation (1) varies directly as a function of the absolute temperature and can be replaced by the temperature-dependent function:

$$n = c + \frac{d}{T} \tag{5}$$

where c and d are material constants. This correlation implies that all isothermal curves representing the variation of the strain rate versus the applied load intersect at the same point, named the pivot point $(\dot{\varepsilon}_p, \sigma_p)$. The pivot point depends on the material but is interrelated for the same class of materials. Figure 10d represents this correlation for different oxide ceramics and spinel. This relation as well as linearity between n and $1/T$ was verified in [38] with an incredibly good accuracy using experimental data for 27 materials.

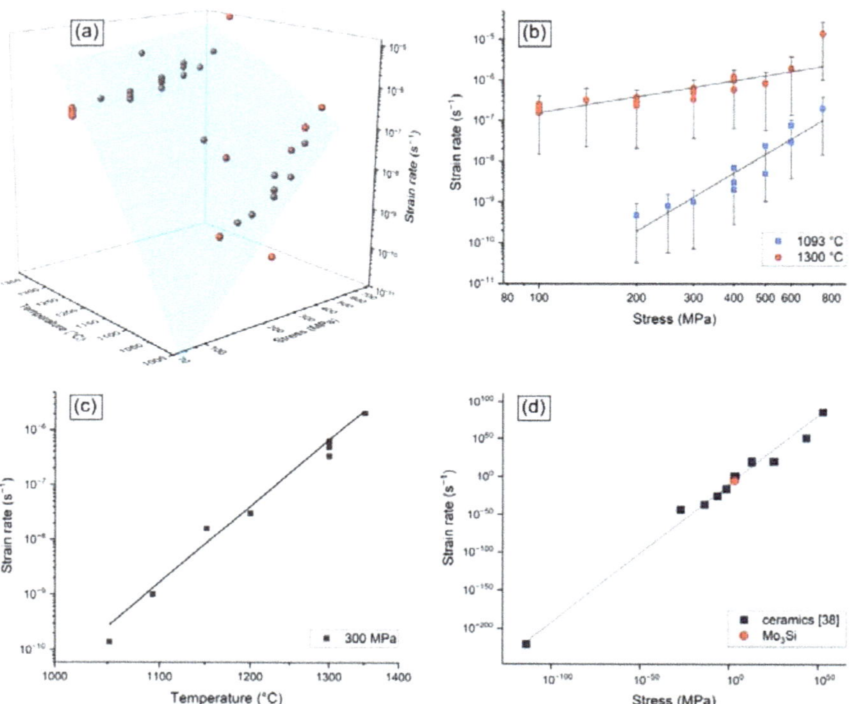

Figure 10. Fitting the experimental results with power law with temperature-dependent creep exponent: (**a**) 3D plot; (**b**) Norton plot; (**c**) Arrhenius plot; (**d**) determined pivot point in the plot showing the relation of creep rate vs. stress in pivot points for different ceramic materials [38].

The constants determined according to regression with power law using temperature-dependent creep exponent (in Equations (1), (2), and (5)) are $B = 3.45 \times 10^{64}\ \mathrm{s}^{-1} \times \mathrm{MPa}^n$, $c = -21$, $d = 35{,}653$ K, $Q = 2227$ kJ/mol with accuracy (R-squared values) $R^2 = 0.956$. The fitting curves are plotted in Figure 10a–c. The fitting curves are remarkably similar to Section 3.3.2 as the model can be transformed to the same equation for isothermal cases. The determined pivot point is $\dot{\varepsilon}_p = 6.36 \times 10^{-6}\ \mathrm{s}^{-1}$ and $\sigma_p = 1840$ MPa and generally lies on the characteristic curve of pivot points for ceramic materials, described in [38], as shown in Figure 10d.

For each stress level, only 1–6 experiments were carried out. Thus, the accurate determination of standard deviation value was not possible. Therefore, 30% of the mean value was used as error bars in the Figure 10b in order to describe the significance of more data for accurate statistical analysis in more detail. More data points would allow a more accurate regression analysis and describe the properties of the alloys more precisely.

4. Discussion on Creep Mechanism

Generally, creep occurs as a result of diffusion and dislocation motion. At low stresses dislocation density is quite low. Thus, the portion of creep deformation carried by dislocations becomes insignificant. The deformation caused by diffusion can be described with linear dependence of strain rate on stress (power law with stress exponent equal to 1), whereby the dislocation creep is commonly described with power law. Thereby, as soon as dislocation creep mechanism is significant, the creep rate increases much faster as diffusion creep rate and accordingly with higher stresses the value of dislocation creep is not significant anymore [40].

Both diffusion and dislocation motion can cause the deformation along the grain boundaries as well as through the grain bulk. With increasing stress and overall creep rate, the proportion of grain boundary deformation becomes smaller. In the case of sufficiently high stresses, the grain boundary slip is insignificant with regard to the overall deformation [40].

In general, speculations about the dominant creep mechanism can be made based on the value of creep stress exponent [41]. However, the precise conclusions about creep mechanism can be made only after investigation of dislocation motion under creep deformation in a wide range of stresses and temperatures. In this study, three different constitutive models were used for regression of experimental results. The further discussion of calculated stress exponents with different constitutive models would be divided in two cases

- Regressions with (i) power law with constant creep exponent and with (ii) power law with temperature-dependent creep exponent
- regression with stress range dependent constitutive model.

4.1. Regressions with (i) Power Law with Constant Creep Exponent in Section 3.3.2 and with (ii) Power Law with Temperature-Dependent Creep Exponent in Section 3.3.4

In Section 3.3.2, each isothermal case was considered separately and the stress exponent for each isotherm was calculated. As the exponents at 1093 °C and 1300 °C are significantly different, they were combined in Section 3.3.4 with the function of temperature (Equation (5) using additional data achieved at different temperatures at 300 MPa. This enables the possibility to extrapolate the results for other temperatures in a small range. Nevertheless, the solution of this function for stress exponent at 1093 °C and 1300 °C provides similar values as in Section 3.3.2. Thus, the discussion of the possible creep mechanisms would be the same. For better lucidity, the values of the stress exponent from Section 3.3.4, in which each isothermal case is considered separately, are used in further discussion.

At 1300 °C, the stress exponent of 1.3 (Table 1) was determined. As mentioned above, stress exponent with value of 1 (linear function) is characteristic for diffusion creep, which occurs at lower stresses. The calculated value is slightly higher, which is caused by several points at higher stresses, leading to the speculation that there could be a transition in mechanism in the faster creep rate regime. This assumption was considered in the regression with stress range dependent law.

At higher temperatures, the diffusion passes through the volume of crystal (bulk diffusion, Nabarro–Herring creep). At lower temperatures, when bulk diffusion is slow, grain-boundary diffusion takes over (Coble creep). In our case Coble creep is more likely, as 1300 °C is ~68% of melting temperature of Mo_3Si (ratio T/T_m in K). Note, that the last point is an assumption, which should be considered carefully before its verification by an investigation of the dislocation's motion under creep deformation.

At 1093 °C, the stress exponent of 4.6 was determined, which corresponds to dislocation creep controlled by dislocation climb [41]. This conclusion is based on the equivalence of the activation energies of creep and self-diffusion. In addition, factors influencing self-diffusion, such as phase transformation, superimposed hydrostatic pressure, etc., similarly affect creep rates. In our case dislocation core diffusion is more likely, than dislocation bulk diffusion, as 1093 °C is ~59% of melting temperature of Mo_3Si (ratio T/T_m in K). Also here, the last assumption should be discussed carefully with regard to the necessary quantitative verification of dislocation motion.

Higher creep rate at 1093 °C as at 1300 °C can be explained by higher transition stress from linear to power law creep, which will be further explained by discussing the results of regression with stress range dependent law.

4.2. Regression with Stress Range Dependent Constitutive Model in Section 3.3.3

In this model, it was assumed that creep proceeds linearly (with creep stress exponent 1) at lower stresses and with power law creep exponent at least >1 (for most materials it is >3) at higher stresses. The extended lines of linear and power law creep would intersect in the transition point in the Norton plot with strain rate $\dot{\varepsilon}_0$ and stress σ_0, described with Equation (4).

As described in the previous discussion of the results of other regression models, the linear regime at lower stresses corresponds to diffusion creep. The most likely diffusion mechanism at investigated temperatures is Coble creep, but this assumption should be taken with great caution. Note, that at higher temperatures the stress range of linear creep is wider, as transition occurs at higher stresses and strains, what follows from the Equation (4) with Arrhenius type temperature functions for $\dot{\varepsilon}_0$ and σ_0. The variation of $\dot{\varepsilon}_0$ and σ_0 with temperature is shown in Figure 11.

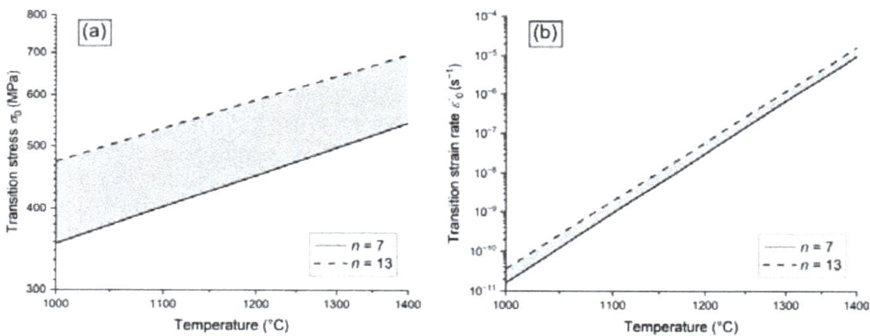

Figure 11. Variation of transition stress σ_0 (**a**) and transition strain rate $\dot{\varepsilon}_0$ (**b**) with temperature.

As described in Section 3.3.3, the wide range of constants in the power law range would provide a good fit with experiment, as not enough data are available in the power law range. Two options, both with good accuracy, $n = 7$ and $n = 13$, were presented. The mechanism of steady-state creep deformation with stress exponents > 7 is not clearly resolved [41]. The most likely mechanism is dislocation climb facilitated by short path diffusion of vacancies through the large number of dislocation cores generated at high applied stresses [41]. Controlling mechanism for such stress exponent was also supposed to be a vacancy diffusion caused by vacancy supersaturation by Sherby and Burke [42], breakup of subgrain walls by Pharr [43] and cross slip or cutting of forest dislocations by Poirier [44].

5. Conclusions

Microstructure and microhardness of three near-stoichiometric compositions of Mo₃Si (Mo-23Si, Mo-24Si, and Mo-25Si, at %) produced by arc melting were compared before and after 100 h annealing at 1600 °C. It was confirmed that Mo-23Si composition consists only of Mo₃Si phase, which remains unaffected by the subsequent annealing. Both compositions, Mo-25Si and Mo-24Si, comprise small amount of Mo₅Si₃ along the grain boundaries after processing, which completely dissolved in Mo-24Si and partially in Mo-25Si during annealing. Additionally, as-cast Mo-25Si contains small regions of Mo₃Si–Mo₅Si₃ eutectic, which are not present anymore after annealing as well. No significant change was identified in the microhardness of Mo_3Si in these specimens, even when comparing different compositions or conditions (before and after annealing). The microhardness was determined to be *HV 0.5* = 1336 $\pm$ 87, which corresponds to the data published in [4]. Thus, the composition Mo-23Si was chosen for further creep investigation as it showed phase stability with the highest Mo_3Si content. Additionally, this value was compared to the microhardness determined for the Mo_3Si phase in a two-phase zone-melted alloy Mo-15Si. Conclusively, the properties of the Mo_3Si phase in the alloy and in single-phase specimen are similar.

For the investigation of the creep properties of Mo_3Si, compression creep tests at 1093 °C and 1300 °C at varying stresses as well as at 300 MPa in the temperature range between 1050 and 1350 °C were performed. The results confirm the concept of the properties of the phases in Mo-Si-B alloys as described in the introduction and in [1–3]. It is found that Mo_3Si possesses a higher creep response than Mo_{ss} and Mo-Si-B alloys, but a lower creep response than single-phase Mo_5SiB_2. The results were fitted to three constitutive models: power law with constant creep exponent, stress range dependent, and power law with temperature-dependent creep exponent. The regression using these methods showed good agreement with the experimental data in the studied stress range. However, additional experimental research is needed for extrapolation of the properties to higher stresses levels.

Author Contributions: Conceptualization, O.K., M.K., and K.N.; Methodology, O.K., M.K., and K.N.; Formal analysis, O.K.; Investigation, O.K., S.O., I.B., T.W., and N.S.; Resources, M.K. and K.N.; Data curation, O.K., S.O., T.W., and N.S.; Writing—original draft preparation, O.K.; Writing—review and editing, O.K., S.O., I.B., K.N., and M.K.; Visualization, O.K.; Supervision, M.K. and K.N.; Project administration, O.K.; Funding acquisition, O.K., M.K., and K.N. All authors have read and agreed to the published version of the manuscript.

Funding: This research was supported by Landesgraduierten-Stipendium (State Graduate Scholarship) funded by Ministry of Science and Economy of Saxony-Anhalt according to GradFG and GradFVO of 04.02.2016 and the Research Training Group 1554/2—Micro-Macro-Interactions in Structured Media and Particle Systems (financed by DFG (Deutsche Forschungsgemeinschaft/German Research Foundation). Hardness measurements were performed in the framework of the DFG project 438070774. Financial support of the Methodisch-Diagnostisches Zentrum Werkstoffprüfung (MDZWP/Methodical-Diagnostic Center for Materials Testing) e.V., Magdeburg, Germany is greatly acknowledged.

Data Availability Statement: The data presented in this study are available on request from the corresponding author.

Acknowledgments: The authors would like to thank Martin Heilmaier and Alexander Kauffmann from Karlsruhe Institute of Technology KIT, Germany for providing access to the high temperature creep tests under vacuum.

Conflicts of Interest: The authors declare no conflict of interest. The funders had no role in the design of the study; in the collection, analyses, or interpretation of data; in the writing of the manuscript, or in the decision to publish the results.

Nomenclature and Abbreviations List

Abbreviations

BSE	backscattered electrons
EDM	electrical discharge machining
EDX	energy dispersive X-Ray
SE	secondary electrons
SEM	scanning electron microscope
XRD	X-ray diffraction

Latin Letters

A	material constant
a	material constant
B	material constant
b	material constant
c	material constant
d	material constant
E	Young modulus
n	material constant
Q	activation energy for the mechanism involved in the deformation process
Q_a	activation energy of linear creep
Q_b	activation energy of power law creep
R	gas constant
R^2	R-squared values
T	test temperature
T_m	melting temperature

Greek Letters

$\dot{\varepsilon}$	minimum creep rate
$\dot{\varepsilon}_0$	material constant
ε_p	pivot point creep rate
σ	stress
σ_0	material constant
σ_p	pivot point stress

References

1. Lemberg, J.A.; Ritchie, R.O. Mo-Si-B Alloys for Ultrahigh-Temperature Structural Applications. *Adv. Mater.* **2012**, *24*, 3445–3480. [CrossRef]
2. Dimiduk, D.M.; Perepezko, J.H. Mo-Si-B Alloys: Developing a Revolutionary Turbine-Engine Material. *MRS Bull.* **2003**, *28*, 639–645. [CrossRef]
3. Mitra, R. 5-Molybdenum silicide-based composites. In *Intermetallic Matrix Composites*; Mitra, R., Ed.; Woodhead Publishing: Cambridge, UK, 2018; pp. 95–146. ISBN 978-0-85709-346-2.
4. Rosales, I.; Schneibel, J.H. Stoichiometry and Mechanical Properties of Mo3Si. *Intermetallics* **2000**, *8*, 885–889. [CrossRef]
5. Zhang, L.; Pan, K.; Lin, J. Fracture Toughness and Fracture Mechanisms in Mo5SiB2 at Ambient to Elevated Temperatures. *Intermetallics* **2013**, *38*, 49–54. [CrossRef]
6. Liu, C.T.; Schneibel, J.H.; Heatherly, L. Processing, Microstructure, and Properties of Multiphase Mo Silicide Alloys. *MRS Online Proc. Libr. OPL* **1998**, *552*. [CrossRef]
7. Schneibel, J.H.; Liu, C.T.; Easton, D.S.; Carmichael, C.A. Microstructure and Mechanical Properties of Mo–Mo3Si–Mo5SiB2 Silicides. *Mater. Sci. Eng. A* **1999**, *261*, 78–83. [CrossRef]
8. Schneibel, J.H.; Kramer, M.J.; Ünal, Ö.; Wright, R.N. Processing and Mechanical Properties of a Molybdenum Silicide with the Composition Mo–12Si–8.5B (at %). *Intermetallics* **2001**, *9*, 25–31. [CrossRef]
9. Schneibel, J.H.; Kramer, M.J.; Easton, D.S. A Mo–Si–B Intermetallic Alloy with a Continuous α-Mo Matrix. *Scr. Mater.* **2002**, *46*, 217–221. [CrossRef]
10. Choe, H.; Schneibel, J.H.; Ritchie, R.O. On the Fracture and Fatigue Properties of Mo-Mo3Si-Mo5SiB2 Refractory Intermetallic Alloys at Ambient to Elevated Temperatures (25 °C to 1300 °C). *Metall. Mater. Trans. A* **2003**, *34*, 225–239. [CrossRef]
11. Kruzic, J.J.; Schneibel, J.H.; Ritchie, R.O. Fracture and Fatigue Resistance of Mo–Si–B Alloys for Ultrahigh-Temperature Structural Applications. *Scr. Mater.* **2004**, *50*, 459–464. [CrossRef]
12. Schneibel, J.H.; Tortorelli, P.F.; Ritchie, R.O.; Kruzic, J.J. Optimization of Mo-Si-B Intermetallic Alloys. *Metall. Mater. Trans. A* **2005**, *36*, 525–531. [CrossRef]
13. Mitra, R.; Srivastava, A.K.; Eswara Prasad, N.; Kumari, S. Microstructure and Mechanical Behaviour of Reaction Hot Pressed Multiphase Mo–Si–B and Mo–Si–B–Al Intermetallic Alloys. *Intermetallics* **2006**, *14*, 1461–1471. [CrossRef]
14. Choe, H.; Chen, D.; Schneibel, J.H.; Ritchie, R.O. Ambient to High Temperature Fracture Toughness and Fatigue-Crack Propagation Behavior in a Mo–12Si–8.5B (at %) Intermetallic. *Intermetallics* **2001**, *9*, 319–329. [CrossRef]
15. Bewlay, B.P.; Jackson, M.R.; Gigliotti, M.F.X. Niobium Silicide High Temperature In Situ Composites. In *Intermetallic Compounds—Principles and Practice*; John Wiley & Sons, Ltd.: London, UK, 2002; pp. 541–560. ISBN 978-0-470-84585-1.
16. Schneibel, J.H.; Lin, H.T. Creep Properties of Molybdenum Silicide Intermetallics Containing Boron. *Mater. High Temp.* **2002**, *19*, 25–28. [CrossRef]
17. Hasemann, G.; Kaplunenko, D.; Bogomol, I.; Krüger, M. Near-Eutectic Ternary Mo-Si-B Alloys: Microstructures and Creep Properties. *JOM* **2016**, *68*, 2847–2853. [CrossRef]
18. Jain, P.; Kumar, K.S. Tensile Creep of Mo–Si–B Alloys. *Acta Mater.* **2010**, *58*, 2124–2142. [CrossRef]

19. Bewlay, B.P.; Jackson, M.R.; Subramanian, P.R.; Zhao, J.-C. A Review of Very-High-Temperature Nb-Silicide-Based Composites. *Metall. Mater. Trans. A* **2003**, *34*, 2043–2052. [CrossRef]
20. Bewlay, B.P.; Jackson, M.R.; Zhao, J.-C.; Subramanian, P.R.; Mendiratta, M.G.; Lewandowski, J.J. Ultrahigh-Temperature Nb-Silicide-Based Composites. *MRS Bull.* **2003**, *28*, 646–653. [CrossRef]
21. Schneibel, J.H. Beyond Nickel-Base Superalloys. In *Processing and Fabrication of Advanced Materials XIII, Proceedings of the Conference Organized by National University of Singapore [NUS], Singapore Institute of Manufacturing Technology, Co-Sponsored by American Society for Materials International (ASM Int.) (The Materials Information Society, USA), Singapore, 6–8 December 2004*; Gupta, M., Srivatsan, T.S., Lim, C.Y.H., Varin, R.A., Eds.; Stallion Press: Singapore, 2005; Volume 2, pp. 563–574. ISBN 981-05-3000-5.
22. Kauss, O.; Tsybenko, H.; Naumenko, K.; Hütter, S.; Krüger, M. Structural Analysis of Gas Turbine Blades Made of Mo-Si-B under Transient Thermo-Mechanical Loads. *Comput. Mater. Sci.* **2019**, *165*, 129–136. [CrossRef]
23. Kauss, O.; Naumenko, K.; Hasemann, G.; Krüger, M. Structural Analysis of Gas Turbine Blades Made of Mo-Si-B Under Stationary Thermo-Mechanical Loads. In *Advances in Mechanics of High-Temperature Materials*; Naumenko, K.D., Krüger, M., Eds.; Advanced Structured Materials; Springer: Cham, Switzerland, 2020; Volume 117, pp. 79–91. ISBN 978-3-030-23868-1.
24. Krüger, M.; Kauss, O.; Naumenko, K.; Burmeister, C.; Wessel, E.; Schmelzer, J. The Potential of Mechanical Alloying to Improve the Strength and Ductility of Mo–9Si–8B–1Zr Alloys—Experiments and Simulation. *Intermetallics* **2019**, *113*, 106558. [CrossRef]
25. Gokhale, A.B.; Abbaschian, G.J. The Mo-Si (Molybdenum-Silicon) System. *J. Phase Equilibria* **1991**, *12*, 493–498. [CrossRef]
26. Gnesin, I.; Gnesin, B. Composition of the Mo-Mo3Si Alloys Obtained via Various Methods. *Int. J. Refract. Met. Hard Mater.* **2020**, *88*, 105188. [CrossRef]
27. Bolbut, V.; Bogomol, I.; Loboda, P.; Krüger, M. Microstructure and Mechanical Properties of a Directionally Solidified Mo-12Hf-24B Alloy. *J. Alloys Compd.* **2018**, *735*, 2324–2330. [CrossRef]
28. Bolbut, V.; Bogomol, I.; Bauer, C.; Krüger, M. Gerichtet erstarrte Mo-Zr-B-Legierungen. *Mater. Werkst.* **2017**, *48*, 1113–1124. [CrossRef]
29. Ha, S.-H.; Yoshimi, K.; Maruyama, K.; Tu, R.; Goto, T. Compositional Regions of Single Phases at 1800 °C in Mo-Rich Mo–Si–B Ternary System. *Mater. Sci. Eng. A* **2012**, *552*, 179–188. [CrossRef]
30. Frost, H.J.; Ashby, M.F. *Deformation-Mechanism Maps: The Plasticity and Creep of Metals and Ceramics*; Pergamon Press: Oxford, UK, 1982; ISBN 978-0-08-029338-7.
31. Hasemann, G.; Bogomol, I.; Schliephake, D.; Loboda, P.I.; Krüger, M. Microstructure and Creep Properties of a Near-Eutectic Directionally Solidified Multiphase Mo–Si–B Alloy. *Intermetallics* **2014**, *48*, 28–33. [CrossRef]
32. Hayashi, T.; Ito, K.; Ihara, K.; Fujikura, M.; Yamaguchi, M. Creep of Single Crystalline and Polycrystalline T2 Phase in the Mo–Si–B System. *Intermetallics* **2004**, *12*, 699–704. [CrossRef]
33. Hasemann, G. Microstructure and Properties of Near-Eutectic Mo-Si-B Alloys for High Temperature Applications. Ph.D. Thesis, Otto-von-Guericke-Universität, Magdeburg, Germany, 2017.
34. Schneibel, J.H. High Temperature Strength of Mo–Mo3Si–Mo5SiB2 Molybdenum Silicides. *Intermetallics* **2003**, *11*, 625–632. [CrossRef]
35. Naumenko, K.; Altenbach, H. Foundations of Engineering Mechanics. In *Modeling of Creep for Structural Analysis*; Springer: Berlin/Heidelberg, Germany, 2007; ISBN 978-3-540-70834-6.
36. Penny, R.K.; Marriott, D.L. *Design for Creep*, 2nd ed.; Springer: Dordrecht, The Netherlands, 1995; ISBN 978-0-412-59040-5.
37. Naumenko, K.; Altenbach, H.; Gorash, Y. Creep Analysis with a Stress Range Dependent Constitutive Model. *Arch. Appl. Mech.* **2009**, *79*, 619–630. [CrossRef]
38. Iost, A. The Correlation between the Power-Law Coefficients in Creep: The Temperature Dependence. *J. Mater. Sci.* **1998**, *33*, 3201–3206. [CrossRef]
39. Prasad, S.; Paul, A. Growth Mechanism of Phases by Interdiffusion and Atomic Mechanism of Diffusion in the Molybdenum–Silicon System. *Intermetallics* **2011**, *19*, 1191–1200. [CrossRef]
40. Maier, H.J.; Niendorf, T.; Bürgel, R. *Handbuch Hochtemperatur-Werkstofftechnik: Grundlagen, Werkstoffbeanspruchungen, Hochtemperaturlegierungen Und-Beschichtungen*; Springer Vieweg: Berlin, Germany; Springer Fachmedien Wiesbaden GmbH: Wiesbaden, Germany, 2019; ISBN 978-3-658-25314-1.
41. Murty, K.L.; Gollapudi, S.; Ramaswamy, K.; Mathew, M.D.; Charit, I. Creep deformation of materials in light water reactors (LWRs). In *Materials Ageing and Degradation in Light Water Reactors*; Elsevier: Paris, France, 2013; pp. 81–148. ISBN 978-0-85709-239-7.
42. Sherby, O.D.; Burke, P.M. Mechanical Behavior of Crystalline Solids at Elevated Temperature. *Prog. Mater. Sci.* **1968**, *13*, 323–390. [CrossRef]
43. Pharr, G.M. Some Observations on the Relation between Dislocation Substructure and Power Law Breakdown in Creep. *Scr. Metall.* **1981**, *15*, 713–717. [CrossRef]
44. Poirier, J.P. Is Power-Law Creep Diffusion-Controlled? *Acta Metall.* **1978**, *26*, 629–637. [CrossRef]

Article

Hot Deformation Behavior and Processing Map of High-Strength Nickel Brass

Qiang Liang [1,2,*]**, Xin Liu** [3]**, Ping Li** [2] **and Xianming Zhang** [1]

[1] Engineering Research Center for Waste Oil Recovery Technology and Equipment of Ministry of Education, Chongqing Technology and Business University, Chongqing 400067, China; zxm215@126.com
[2] College of Mechanical Engineering, Chongqing Technology and Business University, Chongqing 400067, China; lpcq@ctbu.edu.cn
[3] College of Material Science and Engineering, Chongqing University, Chongqing 400044, China; 20133530@cqu.edu.cn
* Correspondence: liangqianglx@hotmail.com; Tel.: +86-136-583-561-46

Received: 10 May 2020; Accepted: 10 June 2020; Published: 12 June 2020

Abstract: The flow behavior of a new kind of high-strength nickel brass used as automobile synchronizer rings was investigated by hot compression tests with a Gleeble-3500 isothermal simulator at strain rates ranging from 0.01 to 10 s^{-1} and a wide deformation temperature range of 873–1073K at intervals of 50 K. The experimental results show that flow stress increases with increasing strain rate and decreasing deformation temperature, and discontinuous yielding appeared in the flow stress curves at higher strain rates. A modified Arrhenius constitutive model considering the compensation of strain was established to describe the flow behavior of this alloy. A processing map was also constructed with strain of 0.3, 0.6, and 0.9 based on the obtained experimental flow stress–strain data. In addition, the optical microstructure evolution and its connection with the processing map of compressed specimens are discussed. The predominant deformation mechanism of Cu-Ni-Al brass is dynamic recovery when the deformation temperature is lower than 973 K and dynamic recrystallization when the deformation temperature is higher than 973 K according to optical observation. The processing map provides the optimal hot working temperature and strain rate, which is beneficial in choosing technical parameters for this high-strength alloy.

Keywords: nickel brass; hot deformation; constitutive model; processing map; workability

---

1. Introduction

Hot deformation is a complex plastic deformation processing method and can be affected by many factors, such as strain rate, deformation and microstructure. In the automobile industry, especially in the manufacturing of components, high wear resistance is essential. High-strength brass is often chosen as a component material because of its good strength, toughness, and corrosion resistance and remarkable wear resistance. High-strength brass contains an $\alpha + \beta$ or β phase which may contain some conditional elements, such as nickel, aluminum, iron, and silicon. These additions enhance the strength property due to solute strengthening, precipitation strengthening, and grain refinement [1,2]. In the case of nickel brass, used to fabricate synchronizer ring gears, it has a narrow deformation temperature range in which it deforms and a fragile temperature interval, which may lead to defects in the final component after deformation. Thus, a comprehensive study on hot compression behavior and workability is required to successfully obtain qualified products without any deformation defects.

A constitutive equation has been widely employed to describe the flow behavior of all kinds of metals, and has also been applied to the finite element method for simulating deformation processes. As a result, an accurate constitutive model is important for feasible numerical simulation. Many researchers have established constitutive models for high-strength brass [3–6], but there have been few reports on nickel brass. The Arrhenius model and modified Arrhenius model [7,8] have been extensively adopted in metals and alloys for their excellent accuracy. Therefore, the modified Arrhenius model considering the compensation of strain is employed to establish the constitutive model of Cu-Ni-Al brass. Though the constitutive model can describe flow behavior and perform finite element simulation during hot deformation, it lacks microstructure features to show its workability, which means that nickel brass lacks sufficient workability information when subjected to hot deformation processes. At present, the hot processing map is widely used to measure workability and has become a useful method for optimizing the hot deformation parameters for metals to obtain the safe and advisable regions for hot deformation processes. As yet, the processing map has been widely employed for various metals, such as magnesium alloys [9–14], aluminum alloys [15,16], titanium alloys [17–21], superalloys [22–24], steel [25–28], high entropy alloys [29,30], and copper alloys [31]. However, little research has reported on the hot deformation behavior and processing map of the Cu-Ni-Al brass alloy which is used to fabricate synchronizer rings. In addition, coarsening of the dynamic recrystallization grains and inhomogeneous microstructure are universal problems in the Cu-Ni-Al nickel brass alloy after the hot deformation process, but most practical production only focuses on the final shape of the products rather than its forming mechanism.

As a result, this new nickel brass material is essential and there is very little research concentrating on it. Therefore, in this study, hot flow behavior at the strain rates of 0.01, 0.1, 1, and 10 s^{-1} and deformation temperature ranging from 873 to 1073 K at intervals of 50 K were investigated by a Gleeble-3500 simulator, and microstructure features were also observed. The constitutive model and processing map of Cu-Ni-Al nickel brass were constructed based on the experimental data to help describe flow behavior and choose suitable hot deformation parameters. Furthermore, the combination of the constitutive model, processing map, and microstructure evolution was applied to establish guidelines for optimizing parameters and techniques during hot deformation processes.

2. Materials and Methods

The material used in the tests was an extruded nickel brass alloy cylinder, offered by Luzhou Long River Mechanical Company Ltd., Luzhou, China. The chemical composition and microstructure of the material are shown in Table 1 and Figure 1, respectively. The crystal structure of this brass is body-centered cubic (bcc), and the phase is a β-phase matrix with granular hard-strengthening particles. The specimens were prepared as cylinders with a radius of 8mm and a height of 12 mm for hot compression tests. The tests were carried out on an isothermal simulator (Gleeble-3500, Dynamic Systems Inc., New York, NY, USA) with strain rates of 0.01, 0.1, 1, and 10 s^{-1} and deformation temperatures ranging from 873 to 1073 K at intervals of 50 K. The prepared specimens were heated to experimental temperatures at a heating rate of 5 K·s^{-1} and held for 3 min to ensure uniform temperature distribution. Then hot compression tests were conducted with a 60% reduction in height, meaning that the total true strain of each specimen was 0.9. A graphite lubricant was used to reduce friction between the interface of specimen ends and the experimental apparatus during the compression process. Finally, the true stress and true strain curves were obtained automatically by a computer-equipped monitor. After the tests, the specimens were quenched in room-temperature water to keep their hot forming microstructure.

The compressed specimens were cut along the compression axis by a wire electrical discharge machine (BMG-640X, Treasure Marge Precision Machinery Ltd., Suzhou, China) for subsequent microstructure observations. The sections were grinded, polished, and etched in a corrosive solution of 15 mL HNO_3 + 21 mL CH_3COOH + 14 mL distilled water for 10 s. The optical microstructure of the center sections was examined by an optical metallographic microscope (Leica DM ILM, Leica Microsystems Inc., Wetzlar, Germany).

Table 1. Chemical composition of Cu-Ni-Al nickel brass alloy (mass fraction, %)

Elements	Cu	Ni	Al	Si	Fe	Zn
Mass fraction (wt.%)	54~56	6~7.3	3~4.3	2~2.5	0.5~1	Rest

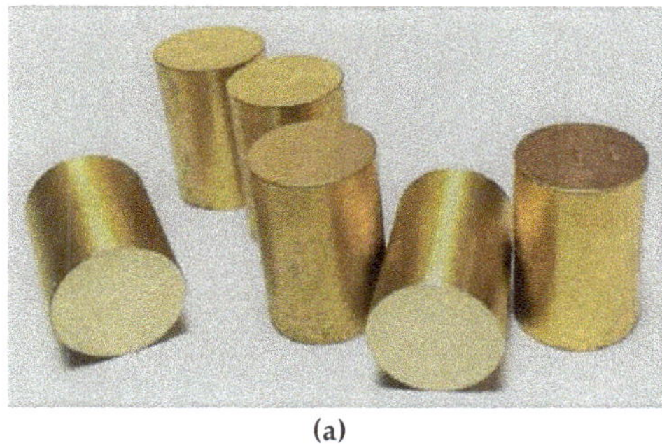

(a)

(b)

Figure 1. (**a**) Cu-Ni-Al nickel brass cylinders. (**b**) The microstructure of Cu-Ni-Al nickel brass before hot compression tests.

3. Results and Discussion

3.1. True Stress and True Strain Curves

The true stress–true strain curves of Cu-Ni-Al nickel brass obtained from compression tests are presented in Figure 2. It can be seen that the deformation temperature and strain rate have important effects on true flow stress under all deformation conditions. The curves clearly show that true stress increases when the deformation temperature is decreased from 1073 to 873 K or the strain rate is increased from 0.01 to 10 s^{-1}, which also indicates that Cu-Ni-Al nickel brass is a deformation temperature- and strain rate-sensitive material. This is because these processes reduce the movements of dislocation: the migration of grain boundaries is limited by lower deformation temperatures and there is not enough time for dynamic softening (dynamic recovery and dynamic recrystallization) at higher strain rates. On one hand, the obstructions of dislocation motion and crystal slip become easy due to the increased average kinetic energy of atoms at relatively high deformation temperatures [32]. On the other hand, higher temperatures promote the mobility at grain boundaries, which results in dislocation attenuation and the nucleation and growth of dynamically recrystallized grains [33]. In Figure 2, the curves with lower strain rate (≤ 1 s^{-1}) undergo high stress, then increase slowly or keep relatively steady along with increasing strain, which means that work hardening dominates compared with the dynamic softening mechanism, especially at lower deformation temperatures. At the same strain rate, true stress decreases with increasing deformation temperature because the atomic kinetic energy and atomic vibration amplitude of material increase at high deformation temperatures, which improves the coordination of grains and eventually leads to increasing plasticity and decreasing strength. At the same deformation temperature, true stress increases with increasing strain rate because there is not enough time for softening and work hardening is the main deformation mechanism. When the strain rate is 10 s^{-1}, as shown in Figure 2d, the true stress curves rapidly reach a peak with a value that is initially like that of the other curves, suddenly decreases visibly, then increases, and eventually stays steady, which means that work hardening and softening are dynamically balanced. This phenomenon at the small strain stage indicates discontinuous dynamic recrystallization under the conditions of a high strain rate. The value of true stress ranges from 0 to 90 MPa which indicates that this material is very soft at temperatures of 873 to 1073 K.

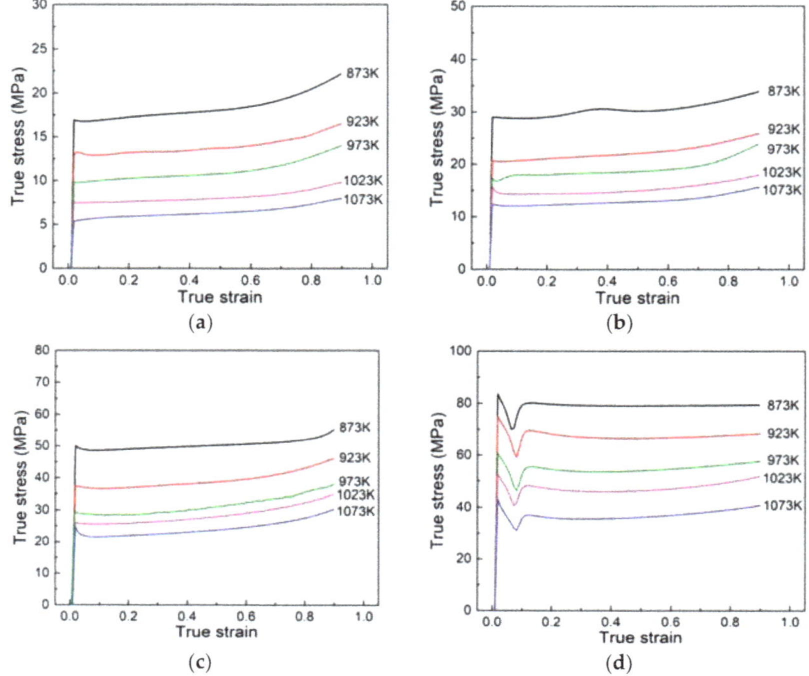

Figure 2. True stress–strain curves of Cu-Ni-Al brass alloy under different strain rates, (**a**) 0.01 s^{-1}, (**b**) 0.1 s^{-1}, (**c**) 1 s^{-1}, (**d**) 10 s^{-1}.

3.2. Arrhenius Model

To better understand the relationship between true stress and some deformation parameters during hot compression tests, such as strain rate, temperature, strain, etc., the Zener–Hollomon model (Arrhenius model) [34] was adopted. The model is usually presented as the following equations:

$$\dot{\varepsilon} = AF(\sigma)\exp(-Q/RT) \tag{1}$$

$$F(\sigma) = A_1\sigma^{n_1} \text{ for } \alpha\sigma < 0.8 \tag{2}$$

$$F(\sigma) = A_2\exp(\beta\sigma) \text{ for } \alpha\sigma > 1.2 \tag{3}$$

$$F(\sigma) = A[\sinh(\alpha\sigma)]^n \text{ for all } \alpha\sigma \tag{4}$$

Among the above equations, ε is strain rate (s^{-1}) and σ is characteristic flow stress (MPa), where peak stress is used to build a model. T is absolute temperature (K). Q is deformation activation energy (KJ·mol^{-1}). R is the gas constant (8.314 J·mol^{-1}·K^{-1}). A, A_1, A_2, n, n_1, α and β are material constants that have no relationship with temperature, where $\alpha = \beta/n_1$. Substituting the above Equations (2)–(4) into Equations (5)–(7) and taking the natural logarithm results in:

$$\ln\dot{\varepsilon} = \ln A_1 + n_1\ln\sigma - \left(\frac{Q}{RT}\right) \tag{5}$$

$$\ln\dot{\varepsilon} = \ln A_2 + \beta\sigma - \left(\frac{Q}{RT}\right) \tag{6}$$

$$\ln\dot{\varepsilon} = \ln A + n\ln(\sinh(\alpha\sigma)) - \left(\frac{Q}{RT}\right) \tag{7}$$

According to Equations (5) and (6), the values of n_1 and β can be determined to be 4.35852 and 0.13252 by calculating the average slopes of the linear fitted lines of the $\ln \dot{\varepsilon} - \ln \sigma_p$ and $\ln \dot{\varepsilon} - \sigma_p$ curves plotted on Figure 3a,b. The value of α can be calculated to be 0.030405 MPa^{-1} from the determined n_1 and β.

Equations (8)–(9) can be transformed according to Equation (7), and $\ln \dot{\varepsilon} - \ln\left(\sinh\left(\alpha\sigma_p\right)\right)$ has a linear relationship with $\ln \dot{\varepsilon}$ and 1000/T in each equation, respectively. Therefore, the deformation activation energy Q is determined by the linear fitting lines of the $\ln \dot{\varepsilon} - \ln\left(\sinh\left(\alpha\sigma_p\right)\right)$ and $\ln\left(\sinh\left(\alpha\sigma_p\right)\right)-1000T^{-1}$ curves, as shown in Figure 3c,d. The constants n and S are the slopes of these two curves, and the average values of n and S are 3.16684 and 4.846375, respectively. Q is determined by the $\ln \dot{\varepsilon} - \ln\left(\sinh\left(\alpha\sigma_p\right)\right)$ and $\ln\left(\sinh\left(\alpha\sigma_p\right)\right)-1000T^{-1}$ curves, and its calculated value is 127.6 KJ·mol^{-1}. Additionally, much research has shown that high activation energy is generally attributed to dynamic recovery (DRV) and dynamic recrystallization (DRX). Thus, the dominant mechanism here is dynamic recrystallization (DRX) due to the low activation energy Q. ln$A$-Q/RT is the intercept of the $\ln \dot{\varepsilon} - \ln\left(\sinh\left(\alpha\sigma_p\right)\right)$ curve, as shown in Figure 3c, which can be calculated from the fitting curve of $\ln \dot{\varepsilon} - \ln\left(\sinh\left(\alpha\sigma_p\right)\right)$: its average value is 1,616,711.73.

$$n = \left.\frac{\partial \ln \dot{\varepsilon}}{\partial \ln[\sinh(\alpha\sigma)]}\right|_T \tag{8}$$

$$Q = Rn \left.\frac{\partial \ln[\sinh(\alpha\sigma)]}{\partial(1/T)}\right|_{\dot{\varepsilon}} = RnS \tag{9}$$

As a result, the constitutive model based on peak value at all experimental deformation conditions was constructed below; the equation was given as:

$$\dot{\varepsilon} = 1616711.73\sinh(0.030405\sigma_p)^{3.16684} \exp(-127600.73/RT) \tag{10}$$

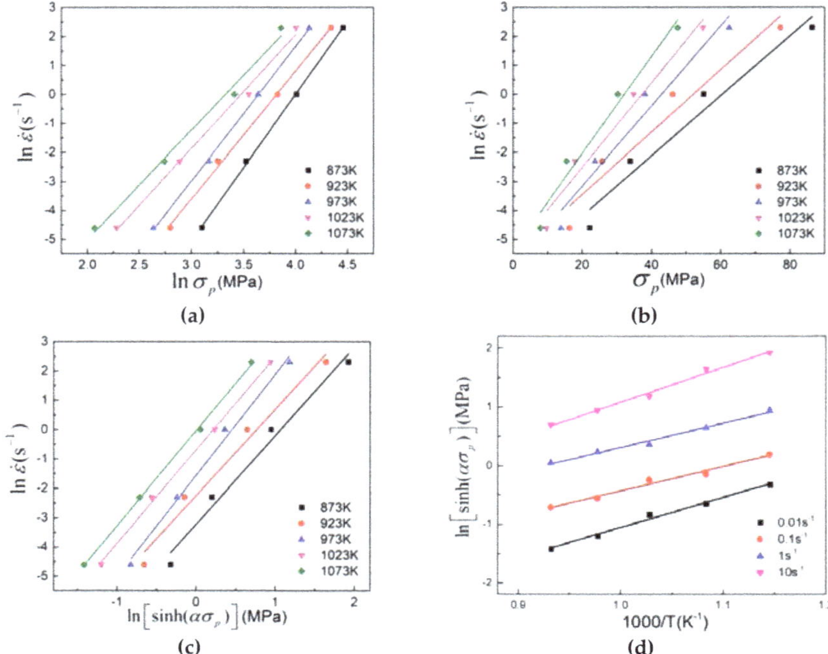

Figure 3. The relationship plots between (**a**) $\ln\dot{\varepsilon} - \ln\sigma_p$, (**b**) $\ln\dot{\varepsilon} - \sigma_p$, (**c**) $\ln\left[\sinh(\alpha\sigma_p)\right] - \ln\dot{\varepsilon}$, (**d**) $\ln\left[\sinh(\alpha\sigma_p)\right] - 1000T^{-1}$.

The Arrhenius model based on peak stress is not exhaustive for describing the hot deformation behavior of Cu-Ni-Al nickel brass under other levels of strain; therefore, Lin [35,36] modified the Arrhenius model and adopted strain as a parameter of Arrhenius model. Zener and Hollomon [34] proposed a relationship between deformation temperature and strain rate, expressed as:

$$Z = \dot{\varepsilon}\exp(Q/RT) \tag{11}$$

By connecting Equations (1)–(4) and (11), a modified Arrhenius model was constructed, considering different strain ranges from 0.05 to 0.85 with an interval of 0.05 to describe the flow behavior of the Ni-Cu-Al nickel brass alloy at other levels of strain. Material constants, such as α, n, Q and $\ln A$, are related to strain, as shown in Figure 4.

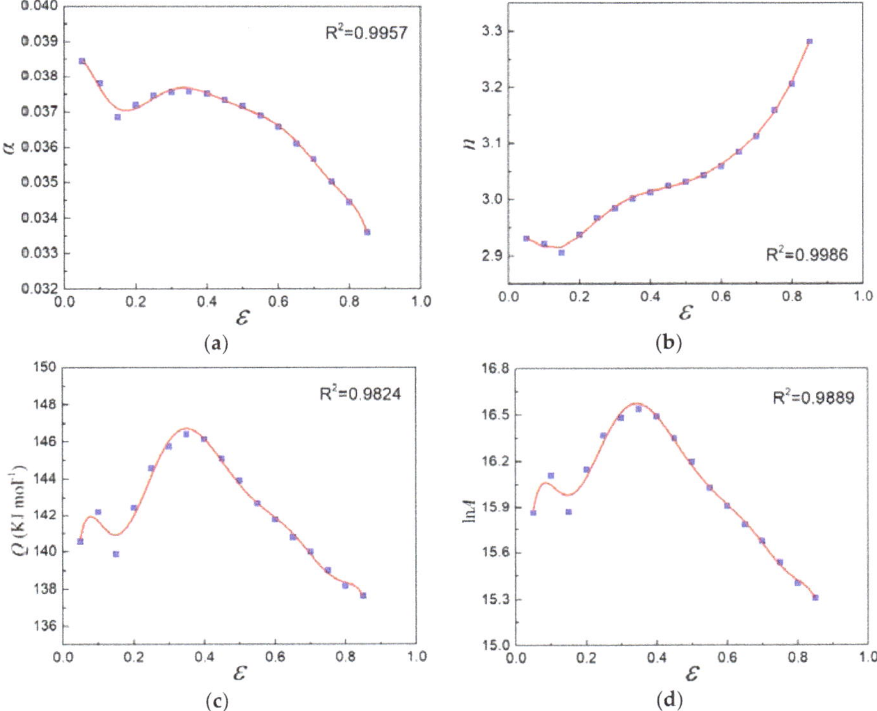

Figure 4. Eighth-order polynomial regression results between true strain and (**a**) α, (**b**) n, (**c**) Q, (**d**) $\ln A$.

The calculation results indicate that the material constants vary with strain, and that the constitutive model equation for each amount of strain is independent. Therefore, to make this Arrhenius-type constitutive model applicable to other levels of strain, the eighth-order polynomial regression method was used to obtain the relationship between strain and these material constants. The relation between these parameters and strain is given by Equation (12) and the determination coefficients are in Table 2. Finally, a modified Arrhenius-type constitutive model was constructed using an introduced Z parameter, and is given by Equation (13).

$$\begin{cases} \alpha(\varepsilon) = B_0 + B_1\varepsilon + B_2\varepsilon^2 + B_3\varepsilon^3 + B_4\varepsilon^4 + B_5\varepsilon^5 + B_6\varepsilon^6 + B_7\varepsilon^7 + B_8\varepsilon^8 \\ n(\varepsilon) = C_0 + C_1\varepsilon + C_2\varepsilon^2 + C_3\varepsilon^3 + C_4\varepsilon^4 + C_5\varepsilon^5 + C_6\varepsilon^6 + C_7\varepsilon^7 + C_8\varepsilon^8 \\ Q(\varepsilon) = D_0 + D_1\varepsilon + D_2\varepsilon^2 + D_3\varepsilon^3 + D_4\varepsilon^4 + D_5\varepsilon^5 + D_6\varepsilon^6 + D_7\varepsilon^7 + D_8\varepsilon^8 \\ \ln A(\varepsilon) = E_0 + E_1\varepsilon + E_2\varepsilon^2 + E_3\varepsilon^3 + E_4\varepsilon^4 + E_5\varepsilon^5 + E_6\varepsilon^6 + E_7\varepsilon^7 + E_8\varepsilon^8 \end{cases} \tag{12}$$

$$\sigma = \frac{1}{\alpha(\varepsilon)} \ln\left\{ (Z/A(\varepsilon))^{\frac{1}{n(\varepsilon)}} + \left[(Z/A(\varepsilon))^{\frac{2}{n(\varepsilon)}} + 1 \right]^{\frac{1}{2}} \right\} \tag{13}$$

where $Z = \dot{\varepsilon}\exp[Q(\varepsilon)/RT]$.

Table 2. The coefficients of the eighth-order polynomial functions.

α	Value	n	Value	Q	Value	$\ln A$	Value
B_0	0.03728	C_0	2.92090	D_0	123.98543	E_0	13.94705
B_1	0.07507	C_1	1.20573	D_1	669.68819	E_1	75.97630
B_2	−1.48702	C_2	−29.70380	D_2	−9346.20325	E_2	−1031.15598
B_3	11.32611	C_3	253.75623	D_3	62,537.37801	E_3	6797.12177
B_4	−43.91601	C_4	−1012.34798	D_4	−224,395.85636	E_4	−24,157.72629
B_5	95.56359	C_5	2187.15643	D_5	459,440.69083	E_5	49,073.23415
B_6	−118.39965	C_6	−2644.25515	D_6	−539,568.91491	E_6	−57,213.34877
B_7	78.04056	C_7	1687.53405	D_7	338,812.43677	E_7	35,676.65667
B_8	−21.24861	C_8	−443.12339	D_8	−88,246.43448	E_8	−9230.40473

Figure 5 shows the comparison between the experimental values and predicted values using this modified equation. The average absolute relative error (AARE), employed to quantitatively describe the accuracy of the modified constitutive model, is 6.35%, and the root mean squared error (RMSE) is 2.71 MPa, meaning that this modified constitutive model considering strain compensation has a good correlation capability.

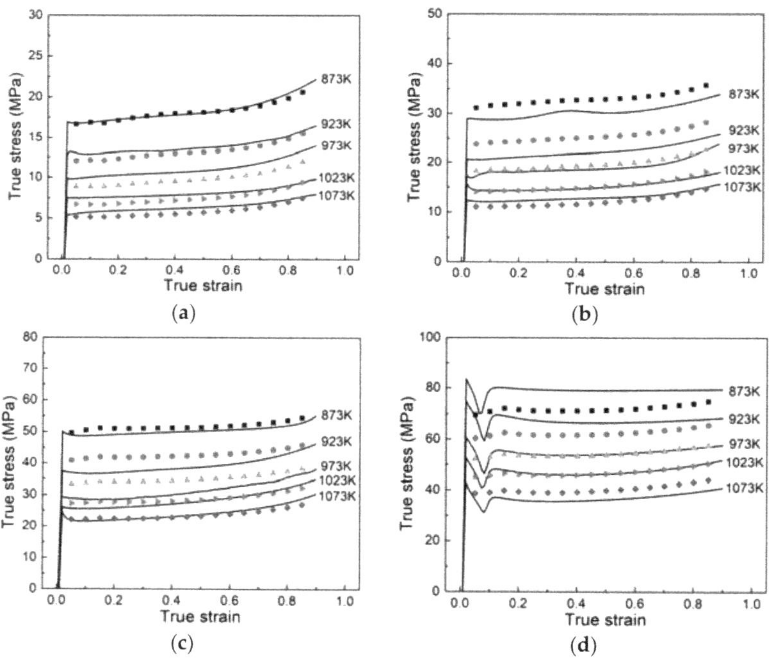

Figure 5. Comparison of experimental (black line) and predicted (dots) values under strain rates of (**a**) 0.01 s^{-1}, (**b**) 0.1 s^{-1}, (**c**) 1 s^{-1}, (**d**) 10 s^{-1}.

3.3. Processing Map

3.3.1. The Principle of the Processing Map

The processing map was constructed based on the dynamic material model (DMM) to analyze the intrinsic workability of various alloys. In DMM theory, the power dissipation (P) of materials during hot deformation contains two parts: (1) part G which results in heat generation during deformation; (2) part J which leads to microstructural changes. Each part can be represented in terms of flow stress, strain, and strain rate. In the dynamic material model, the total input power (P) of the system is given by the below Equation (14) [37,38]:

$$P = \sigma\dot{\varepsilon} = G + J = \int_0^{\dot{\varepsilon}} \sigma d\dot{\varepsilon} + \int_0^{\sigma} \dot{\varepsilon} d\sigma \tag{14}$$

where G represents dissipation by plastic work, J represents dissipation by microstructure evolution, σ is flow stress and $\dot{\varepsilon}$ is strain rate. The proportion of G and J during the hot forming process is defined by strain rate sensitivity (m), and m is given by:

$$m = \frac{\partial J}{\partial G} = \frac{\dot{\varepsilon}\partial\sigma}{\sigma\partial\dot{\varepsilon}} = \frac{\partial(\ln\sigma)}{\partial(\ln\dot{\varepsilon})} \tag{15}$$

To describe power dissipation conditions during the hot forming process quantitatively, the power dissipation coefficient η is used, and different values of η may imply different forming mechanisms. η is defined by:

$$\eta = \frac{J}{J_{\max}} = \frac{2m}{m+1} \tag{16}$$

where the ideal linear dissipation unit $m = 1$ and the maximum value of J is $J_{\max}$.

The 3D change maps of strain rate sensitivity m and power dissipation coefficient η are graphics constituted of curves with the same values under a specific strain. Generally, a higher η-value indicates good workability, but this is not absolute. Flow instabilities that also have a high η-value, such as adiabatic shear bands and small cracks, should be considered and avoided. According to the principle of maximum entropy production rate, Prasad [37,39] created a principle to determine whether the material has workability or not based on DMM theory. The principle is given by:

$$\xi(\dot{\varepsilon}) = \frac{\partial\ln\left(\frac{m}{m+1}\right)}{\partial\ln\dot{\varepsilon}} + m < 0 \tag{17}$$

where the instability parameter $\xi(\dot{\varepsilon})$ is a function of deformation temperature, strain rate, and strain, because m is related to deformation temperature and strain rate at a specific strain value. Generally, when strain is given, the first step is to find the true stress at all conditions, and then m and $\xi(\dot{\varepsilon})$ can be calculated and the instability map can be obtained by drawing the contour map of $\xi(\dot{\varepsilon})$ versus deformation temperature T and $\ln\dot{\varepsilon}$. For this principle, when the value of $\xi(\dot{\varepsilon})$ is negative, the material under the corresponding conditions will be unstable and may lead to instabilities such as flow localization and adiabatic shear bands, etc., which may cause the failure of final components. Therefore, the construction of the processing map demonstrating materials' workability is significant for deformation processes.

3.3.2. Construction of Processing Map

In order to establish the processing map, true stress at the strain levels of 0.3, 0.6, and 0.9 under all deformation conditions was obtained from the true stress–true strain curves. Flow stress increased rapidly at the initial stage, then the curves increased slightly or stayed steady with increasing strain; this phenomenon indicates that the work hardening and dynamic softening mechanisms both appeared during the hot deformation process, and that dynamic softening gradually becomes the dominant mechanism, especially in the stable flow curves. The obtained data enable the plotting of curves showing $\ln\sigma$ versus $\ln\dot{\varepsilon}$ using polynomial fitting. According to the cubic spline polynomial function in Equation (18), the cubic spline curves for $\ln\sigma$ versus $\ln\dot{\varepsilon}$ are obtained, as shown in Figure 6.

Subsequently, taking the partial derivative of Equation (18), the strain rate sensitivity coefficient m can be appropriately expressed by Equation (19).

$$\ln \sigma = a + b(\ln \dot{\varepsilon}) + c(\ln \dot{\varepsilon})^2 + d(\ln \dot{\varepsilon})^3 \tag{18}$$

$$m = \left. \frac{\partial(\ln \sigma)}{\partial(\ln \dot{\varepsilon})} \right|_{\varepsilon,T} = b + 2c(\ln \dot{\varepsilon}) + 3d(\ln \dot{\varepsilon})^2 \tag{19}$$

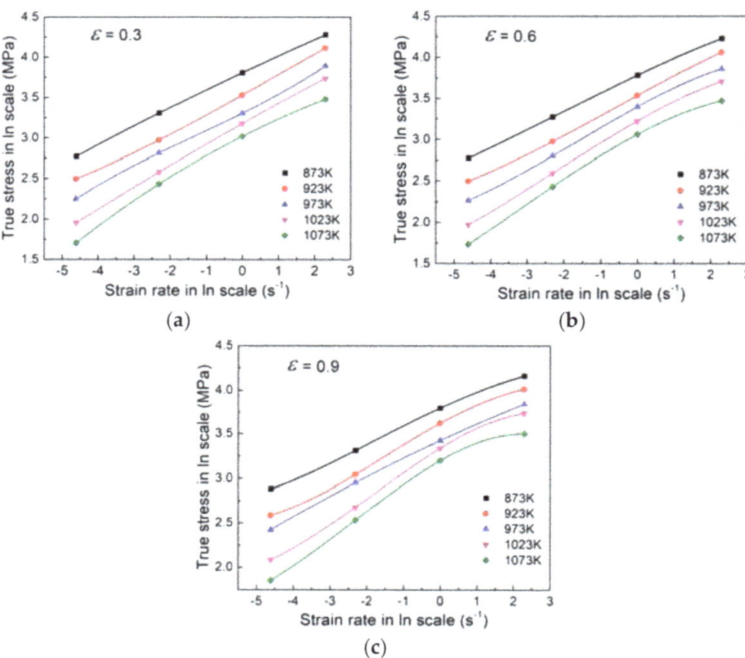

Figure 6. The relation between true stress and strain rate using ln scale, (**a**) $\varepsilon = 0.3$, (**b**) $\varepsilon = 0.6$, (**c**) $\varepsilon = 0.9$.

After making cubic spline curves of $\ln \sigma$ versus $\ln \dot{\varepsilon}$, the polynomial coefficients a, b, c and d can be obtained by fitting, and so the values of the partial derivative m under all experimental deformation conditions can be calculated using Equation (19). The fluctuation in m reflects internal microstructure evolution during the hot deformation process; it is an important index for describing superplastic deformation behaviors. High strain rate sensitivity is more likely to obtain good plasticity properties and low strain rate sensitivity may imply the formation of defects inside the material [40]. Figure 7 is the 3D change map of the strain rate sensitivity m at different strains. It can be seen that m is increased with increasing deformation temperature and decreasing strain rate, in general. This is because variation in m is related to the non-basal slip mechanism [20]. The critical stress of non-basal slip is moderate when the deformation temperature is higher; at the same time, at lower strain rates, enough time enables non-basal slip to become the dominant forming mechanism and raise the power dissipation from microstructure evolution, which leads to higher values of m at both high deformation temperatures and low strain rates. Furthermore, strain is a significant factor for m, as the value of m is increased with increasing strain. In Figure 7, the m-values of nickel brass range from 0 to 0.33; the area filled by warm colors such as red corresponds to high m-values and the area filled by cold colors such as blue corresponds to low m-values. Specifically, the m-value of the Ni-Cu-Al nickel brass alloy is approximately 0 for high strain rates with high deformation temperatures, and a maximum value of about 0.33 for low strain rates with high deformation temperatures. Furthermore, the peak *m*-value at strain of 0.3 is higher than at 0.6 and 0.9, and the *m*-value at strain of 0.3 is mostly also higher than at strain of 0.6 and 0.9 under the same deformation conditions. Moreover, the high *m*-value area at strain of 0.9 is larger than at strain of 0.3 and 0.6, and the area of high *m*-value is increased with increasing strain, which also confirms that strain has an effect on microstructure evolution.

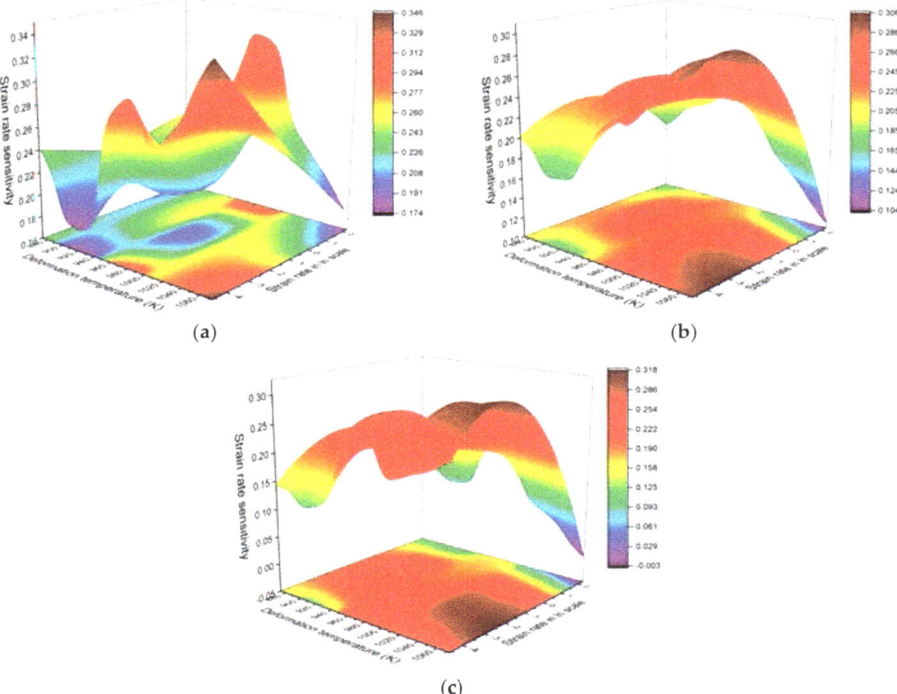

Figure 7. The 3D change maps of strain rate sensitivities at various strain levels: (**a**) strain at 0.3, (**b**) strain at 0.6, (**c**) strain at 0.9.

It is known that the power dissipation coefficient η describes the relative rate of internal entropy produced during the hot deformation process. According to the *m*-values, the power dissipation coefficient η can be calculated using Equation (16) and so the contour maps of the power dissipation coefficient η were constructed using an interpolation method based on a series of limited values of η at strain levels of 0.3, 0.6, and 0.9. As presented in Figure 8, the power dissipation coefficient η tends to increase with increasing deformation temperatures and decreasing strain rate, and the η-values range from 0.06 to 0.49. The regions colored in red, orange and yellow, associated with $\eta > 0.4$, represent the best conditions for hot deformation, which are at 0.01–1s^{-1} with deformation temperatures from 990 to 1073 K and at 0.8–10 s^{-1} with deformation temperatures from 900 to 1010 K at strain of 0.3. With increasing strain, the η-values decrease, and the best conditions for hot deformation associated with $\eta > 0.4$ are in the region of 0.02–1 s^{-1} along with deformation temperatures from 1000 to 1073 K at strain of 0.9, and the area of best hot deformation is smaller than at the strain levels of 0.3 and 0.6, which shows that the area associated with $\eta > 0.4$ shrinks with increasing strain. In addition, η changes significantly at about 950 K: η decreases from 873 to 923 K and then increases. This phenomenon may due to the brittle hot working region of brass. Simultaneously, the power dissipation coefficient η is also related to dynamic recovery and dynamic recrystallization, but there is only a dynamic recrystallization mechanism for brass due to its low stacking fault energy.

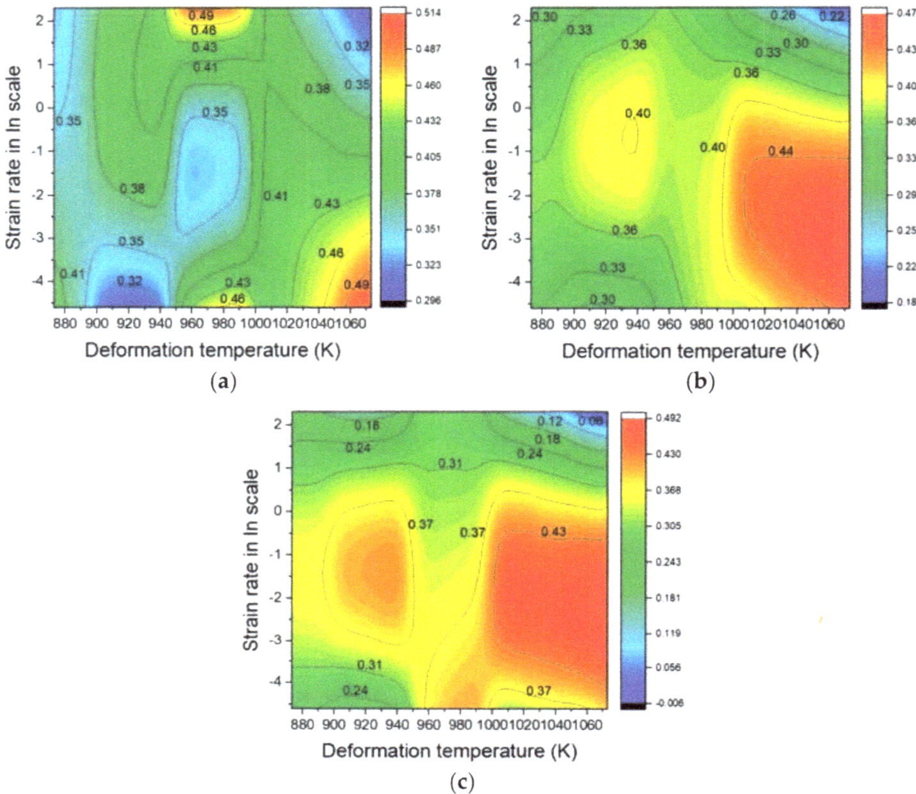

Figure 8. The contour maps of power dissipation η at various strain levels: (**a**) strain at 0.3, (**b**) strain at 0.6, (**c**) strain at 0.9.

The map of the instability parameter $\xi(\dot{\varepsilon})$ calculated by Equation (17) is plotted in Figure 9. The colored region (black and gray) shows that the billets deformed under the corresponding deformation conditions are unstable according to the criteria. It can be shown that the values of $\xi(\dot{\varepsilon})$ at strain of 0.3 are all positive which means that the material can deform without any defects at strain of 0.3. With increasing strain, negative values of $\xi(\dot{\varepsilon})$ begin to appear in the region of high strain rate with high temperatures. At strain of 0.9, the colored area with $\xi(\dot{\varepsilon}) < 0$ becomes larger than at strain of 0.3 and 0.6. Therefore, the region of instability is larger with increasing strain; this also explains why some complex components need several processes to be shaped as designed, especially for those components that are too difficult to be filled by only one process. As a result, the appearance of this region of instability indicates that this region is prone to the occurrence of defects, so hot deformation in this region should be avoided.

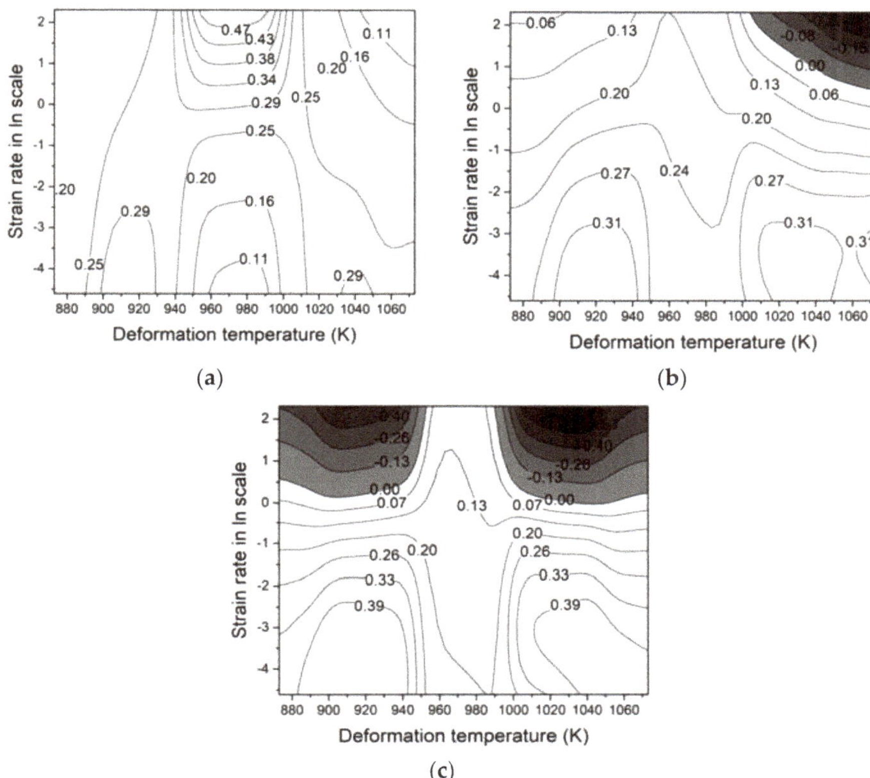

Figure 9. The contour maps of the instability region at various strain levels: (**a**) strain at 0.3, (**b**) strain at 0.6, (**c**) strain at 0.9.

The processing map was adopted to analyze microstructure features. The processing map consists of strain rate sensitivity m (blue line) and power dissipation coefficient η (red line), and the workability principle has been constructed as shown in Figure 10 at the strain levels of 0.3, 0.6, and 0.9, which was also used to confirm the microstructure characteristics after deformation compared with the theoretical calculations. The grey region of the map means that the corresponding deformation condition may lead to defects, and the other region of the map means that the material under the corresponding deformation temperature and strain rate is safe to deform without defects. There are regions framed by a green box, which means that they are not instability regions. Figure 10a shows that all deformation conditions are stable for deforming; with increasing strain, the instability region is expanded. The safe region at strain of 0.9 is DOM#1 while the deformation temperature is 873–1073 K with strain of 0.01–0.85 s^{-1}, and DOM#2 while the deformation temperature is 955–983 K with strain of 0.85–10 s^{-1}. The power dissipation coefficient and strain rate sensitivity are also important elements to consider and they are discussed below.

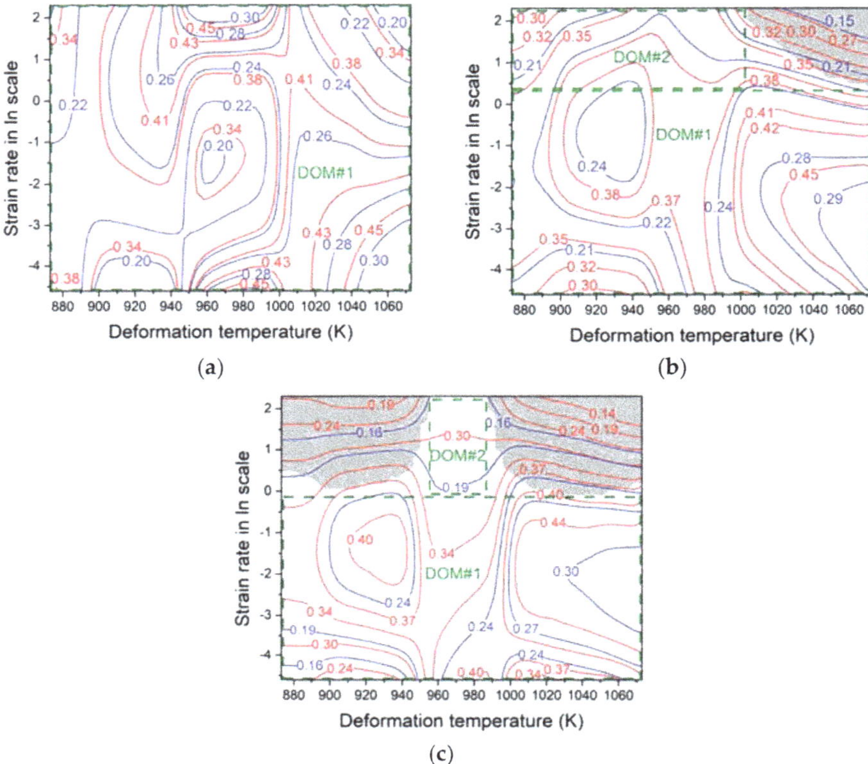

Figure 10. The processing maps consisting of strain rate sensitivity, power dissipation coefficient, and instability region (**a**) strain at 0.3, (**b**) strain at 0.6, (**c**) strain at 0.9.

3.4. Microstructure Observations

Microstructure is significantly affected by strain rate sensitivity and the power dissipation coefficient, which means that it has a connection with strain, strain rate, and deformation temperature. Therefore, the microstructure of compressive specimens was observed by an optical microscope (OM) to evaluate and identify the processing map constructed above. The OM images of deformed specimens are shown in Figures 11 and 12: the gray hard particles are distributed evenly on the Ni-Cu-Al brass β-phase base. Figure 11a–d shows the microstructure after deformation in the safe working region, while Figure 12a–d shows the microstructure after deformation in the instability region. Figure 11a,b,d shows the strain rate of 0.1 s^{-1}. In Figure 11a, the corresponding m and η are 0.22 and 0.34, respectively, and the unchanged microstructure at deformation temperature of 873 K compared with the original microstructure indicates that dynamic recovery is the dominant forming mechanism during hot deformation. In Figure 11b, at the deformation temperature of 973 K, the corresponding m and η are 0.20 and 0.36, the microstructure has slightly changed compared with the original sample's microstructure because of heating and deformation, and some small recrystallization grains have appeared around hard particles. With increasing strain rate, these recrystallization grains appear more and the grain boundaries are more manifest than at the strain rate of 0.1 s^{-1}, as shown in Figure 11c. In Figure 11d, at the deformation temperature of 1073 K, the microstructure is different compared with the original sample due to heating at high temperature and deformation, the grain boundary becomes vague, and dislocation cells are clear to distinguish, which also means that it is possible that the deformation mechanism is dynamic recrystallization (DRX) because more energy has been used in the microstructure evolution, with a higher power dissipation coefficient (η = 0.47).

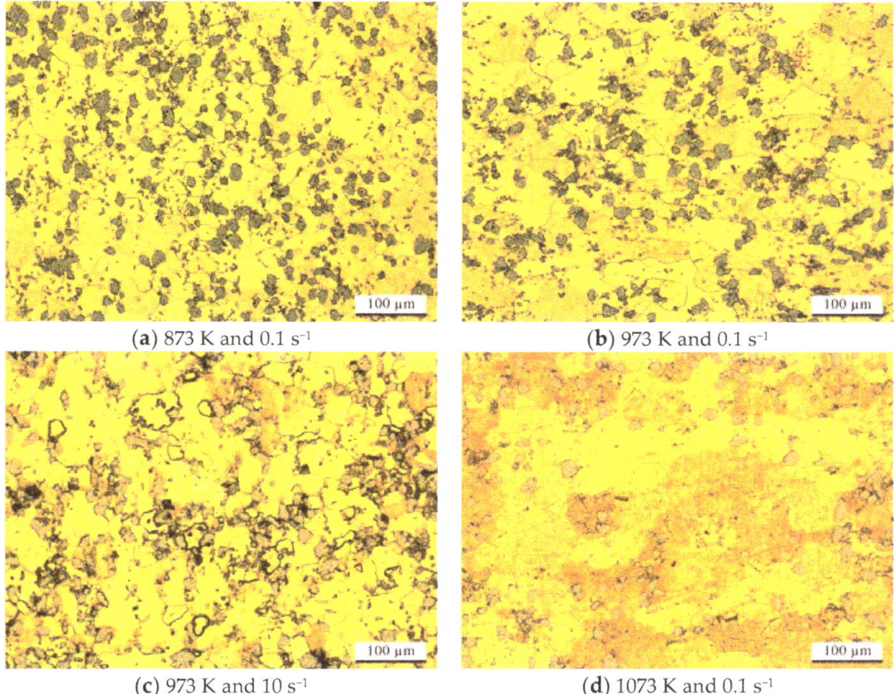

(**a**) 873 K and 0.1 s⁻¹   (**b**) 973 K and 0.1 s⁻¹

(**c**) 973 K and 10 s⁻¹   (**d**) 1073 K and 0.1 s⁻¹

Figure 11. OM images of Cu-Ni-Al brass at safe region after deformation. Abbreviations: OM, optical microscope.

In Figure 12, the corresponding deformation conditions in the optical microstructure pictures are possibly unstable for deformation and some instability microdefects have appeared, such as mixed crystal and flow localization. Figure 12a,c,d shows the microstructure at the strain rate of 10 s⁻¹ and different deformation temperatures. In Figure 12a, at the deformation temperature of 873 K, it is visible that flow localization has occurred with a power dissipation coefficient η of 0.19, and hard particles have assembled due to deformation. In Figure 12c,d, with increasing temperature, the microstructure changes a lot, the grains grow with heating, the corresponding power dissipation η is 0.19 and 0.13, respectively, and higher strain rates lead to flow localization due to the short deformation time. In Figure 11b, mixed crystal appears because of the high temperature and high strain rate. It can be concluded that, at the conditions of low m-value (<0.16) and low η-value (<0.24), the risk of the occurrence of micro deformation defects is high and these two values can be employed as a criterion for choosing proper working technique parameters.

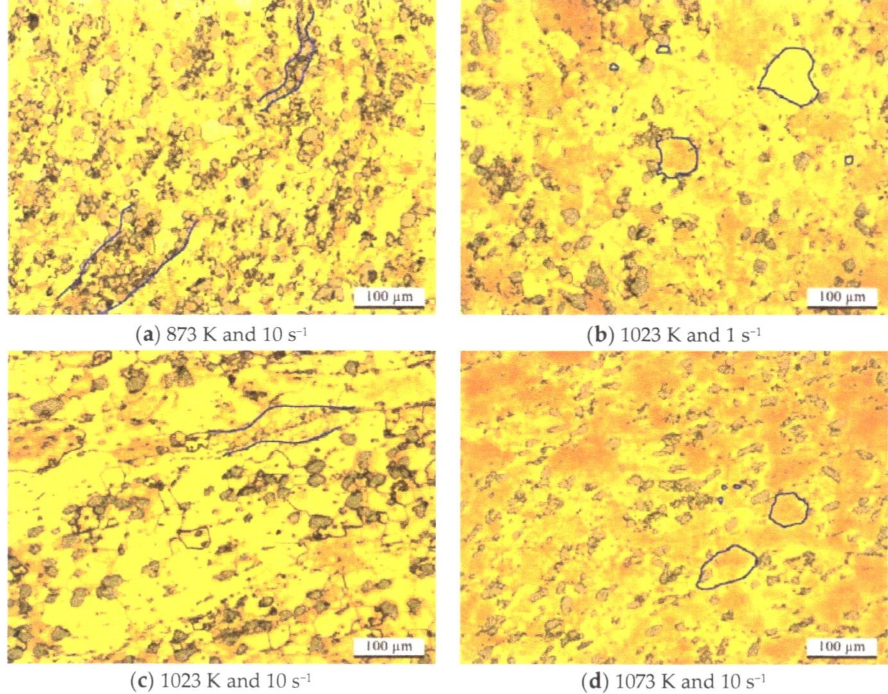

(**a**) 873 K and 10 s⁻¹   (**b**) 1023 K and 1 s⁻¹

(**c**) 1023 K and 10 s⁻¹   (**d**) 1073 K and 10 s⁻¹

Figure 12. OM images of Cu-Ni-Al brass in the instability region after deformation.

4. Conclusions

A series of compression tests were conducted on a Gleeble-3500 isothermal simulator, and true stress–true strain curves were obtained at temperatures ranging from 873–1073 K and strain rates of $0.01, 0.1, 1,$ and 10 s^{-1}. The conclusions of this paper are as follows:

(1) The true stress of the Cu-Ni-Al brass alloy increases with decreasing deformation temperature and increasing strain rate.

(2) The Arrhenius constitutive model of Cu-Ni-Al brass considering compensation for strain has been established based on experimental data; the material constants α, n, Q and $\ln A$ in the constitutive equation are functions of strain. In addition, the predictive evaluation index AARE is 6.35%, and this can be used to describe the flow behavior of Cu-Ni-Al brass effectively.

(3) Processing maps were constructed to identify which deformation conditions are safe for deformation. The results show that the safe region for deformation at the strain rate of 0.9 is deformation temperatures of 873–1073K with strain of 0.01–0.85 s^{-1}, and deformation temperatures of 955–983 K with strain of 0.85–10 s^{-1}.

(4) The microstructure observations show that the main softening mechanisms are dynamic recovery and dynamic recrystallization. The phenomenon of dynamic recrystallization with newly refined grains occurs when the deformation temperature is higher than 973 K. Finally, the main instability microdefects are mixed crystals and flow localization, which should be avoided in physical production.

Author Contributions: Data collection and model construction, Q.L. and X.L.; funding acquisition, Q.L., X.Z. and P.L.; writing—original draft, X.L. and Q.L.; writing—review and editing, Q.L., X.L., X.Z., and P.L. All authors have read and agreed to the published version of the manuscript.

Funding: This work was funded by the Chongqing Basic Science and Advanced Technology Research Program (grant number cstc2017jcyjAX0175), Chongqing Municipal Education Commission Foundation (grant numbers KJQN201900836, KJZD-M201900802 and KJZD-K201800801), and Chongqing Technology and Business University (grant number 1752009).

Acknowledgments: The authors sincerely appreciate the support of material used for experiments from the Luzhou LongRiver Mechanical Company Ltd., Luzhou, China.

Conflicts of Interest: The authors declare no conflict of interest.

Nomenclature

Symbol/Acronym	Full Name	Symbol/Acronym	Full Name
OM	Optical microscope	σ	Flow stress (MPa)
DMM	Dynamic material model	$\dot{\varepsilon}$	Strain rate (s^{-1})
β	Brass beta-phase (bcc)	ε	True strain
R	Gas constant (8.314 J.mol^{-1}.K^{-1})	$A, A_1, A_2, n, n_1, \alpha, \beta, S$	Material constants
T	Deformation temperature (K)	Q	Deformation activation energy (KJ·mol^{-1})
P	Power dissipation	G	Heat generation
J	Microstructural changes	$\xi(\dot{\varepsilon})$	Instability parameter
m	Strain rate sensitivity index	η	Power dissipation index

References

1. Mindivan, H.; Çimenoğlu, H.; Kayali, E.S. Microstructures and wear properties of brass synchroniser rings. *Wear* **2003**, *254*, 532–537. [CrossRef]
2. Shufeng, L.; Hisashi, I.; Katsuyoshi, K. Microstructure, Phase Transformation, Precipitation Behavior and Mechanical Properties of P/M Cu40Zn–1.0 wt.% Ti Brass Alloy via Spark Plasma Sintering and Hot Extrusion. *J. Mater. Sci. Technol.* **2013**, *11*, 16–22.
3. Padmavardhani, D.; Prasad, Y.V.R.K. Characterization of hot deformation behavior of brasses using processing maps: Part II. β Brass and α-β brass. *Metall. Trans. A* **1991**, *22*, 2993–3001. [CrossRef]
4. Padmavardhani, D.; Prasad, Y.V.R.K. Effect of zinc content on the processing map for hot working of α brass. *Mater. Sci. Eng., A* **1992**, *157*, 43–51. [CrossRef]
5. Pernis, R.; Kasala, J.; Bořuta, J. High temperature plastic deformation of CuZn30 brass-Calculation of the activation energy. *Kovove Mater.* **2010**, *48*, 41–46. [CrossRef]
6. Spigarelli, S.; Mehtedi, M.E.; Cabibbo, M.; Gabrielli, F.; Ciccarelli, D. High temperature processing of brass: Constitutive analysis of hot working of Cu–Zn alloys. *Mater. Sci. Eng., A* **2014**, *615*, 331–339. [CrossRef]
7. Mosleh, A.; Mikhaylovskaya, A.; Kotov, A.; Kwame, J.; Aksenov, S. Superplasticity of Ti-6Al-4V Titanium Alloy: Microstructure Evolution and Constitutive Modelling. *Materials* **2019**, *12*, 1756. [CrossRef] [PubMed]
8. Mosleh, A.; Mikhaylovskaya, A.; Kotov, A.; Pourcelot, T.; Aksenov, S.; Kwame, J.; Portnoy, V. Modelling of the Superplastic Deformation of the Near-α Titanium Alloy (Ti-2.5Al-1.8Mn) Using Arrhenius-Type Constitutive Model and Artificial Neural Network. *Metals* **2017**, *7*, 568. [CrossRef]
9. Zhi, C.; Ma, L.; Jia, W.; Huo, X.; Fan, Q.; Huang, Z.; Le, Q.; Lin, J. Dependence of deformation behaviors on temperature for twin-roll casted AZ31 alloy by processing maps. *J. Mater. Res. Technol.* **2019**, *8*, 5217–5232. [CrossRef]
10. Srinivasan, N.; Prasad, Y.V.R.K.; Rama Rao, P. Hot deformation behaviour of Mg–3Al alloy—A study using processing map. *Mater. Sci. Eng., A* **2008**, *476*, 146–156. [CrossRef]

11. Ding, X.; Zhao, F.; Shuang, Y.; Ma, L.; Chu, Z.; Zhao, C. Characterization of hot deformation behavior of as-extruded AZ31 alloy through kinetic analysis and processing maps. *J. Mater. Process. Technol.* **2020**, *276*, 116325. [CrossRef]

12. Cheng, W.; Bai, Y.; Ma, S.; Wang, L.; Wang, H.; Yu, H. Hot deformation behavior and workability characteristic of a fine-grained Mg-8Sn-2Zn-2Al alloy with processing map. *J. Mater. Sci. Technol.* **2019**, *35*, 1198–1209. [CrossRef]

13. Chen, X.; Liao, Q.; Niu, Y.; Jia, Y.; Le, Q.; Ning, S.; Hu, C.; Hu, K.; Yu, F. Comparison study of hot deformation behavior and processing map of AZ80 magnesium alloy casted with and without ultrasonic vibration. *J. Alloys Compd.* **2019**, *803*, 585–596. [CrossRef]

14. Guo, Y.; Xuanyuan, Y.; Lia, C.; Sen, Y. Characterization of Hot Deformation Behavior and Processing Maps of Mg-3Sn-2Al-1Zn-5Li Magnesium Alloy. *Metals* **2019**, *9*, 1262. [CrossRef]

15. Liu, Y.; Geng, C.; Lin, Q.; Xiao, Y.; Xu, J.; Kang, W. Study on hot deformation behavior and intrinsic workability of 6063 aluminum alloys using 3D processing map. *J. Alloys Compd.* **2017**, *713*, 212–221. [CrossRef]

16. Wang, H.; Wang, C.; Mo, Y.; Wang, H.; Xu, J. Hot deformation and processing maps of Al–Zn–Mg–Cu alloy under coupling-stirring casting. *J. Mater. Res. Technol.* **2019**, *8*, 1224–1234. [CrossRef]

17. Lin, X.; Dong, F.; Zhang, Y.; Yuan, X.; Huang, H.; Zheng, B.; Wang, L.; Wang, X.; Luo, L.; Su, Y.; et al. Hot-deformation behaviour and hot-processing map of melt-hydrogenated Ti6Al4V/(TiB + TiC). *Int. J. Hydrogen Energy* **2019**, *44*, 8641–8649. [CrossRef]

18. Liu, Q.; Hui, S.; Tong, K.; Yu, Y.; Ye, W.; Song, S.-Y. Investigation of high temperature behavior and processing map of Ti-6Al-4V-0.11Ru titanium alloy. *J. Alloys Compd.* **2019**, *787*, 527–536. [CrossRef]

19. Long, S.; Xia, Y.-F.; Wang, P.; Zhou, Y.-T.; Gong-ye, F.-J.; Zhou, J.; Zhang, J.-S.; Cui, M.-L. Constitutive modelling, dynamic globularization behavior and processing map for Ti-6Cr-5Mo-5V-4Al alloy during hot deformation. *J. Alloys Compd.* **2019**, *796*, 65–76. [CrossRef]

20. Ma, L.; Wan, M.; Li, W.; Shao, J.; Bai, X. Constitutive modeling and processing map for hot deformation of Ti–15Mo–3Al-2.7Nb-0.2Si. *J. Alloys Compd.* **2019**, *808*, 151759. [CrossRef]

21. Wu, F.; Xu, W.; Jin, X.; Zhong, X.; Wan, X.; Shan, D.; Guo, B. Study on Hot Deformation Behavior and Microstructure Evolution of Ti55 High-Temperature Titanium Alloy. *Metals* **2017**, *7*, 319.

22. Srinivasa, N.; Prasad, Y.V.R.K. Hot working characteristics of nimonic 75, 80A and 90 superalloys: A comparison using processing maps. *J. Mater. Process. Technol.* **1995**, *51*, 171–192. [CrossRef]

23. Wu, Y.; Liu, Y.; Li, C.; Xia, X.; Huang, Y.; Li, H.; Wang, H. Deformation behavior and processing maps of Ni3Al-based superalloy during isothermal hot compression. *J. Alloys Compd.* **2017**, *712*, 687–695. [CrossRef]

24. Zhang, B.; Liu, X.; Yang, H.; Ning, Y. The Deformation Behavior, Microstructural Mechanism, and Process Optimization of PM/Wrought Dual Superalloys for Manufacturing the Dual-Property Turbine Disc. *Metals* **2019**, *9*, 1127. [CrossRef]

25. Cai, Z.; Ji, H.; Pei, W.; Tang, X.; Huang, X.; Liu, J. Hot workability, constitutive model and processing map of 3Cr23Ni8Mn3N heat resistant steel. *Vacuum* **2019**, *165*, 324–336. [CrossRef]

26. Li, N.; Zhao, C.; Jiang, Z.; Zhang, H. Flow behavior and processing maps of high-strength low-alloy steel during hot compression. *Mater. Charact.* **2019**, *153*, 224–233. [CrossRef]

27. Łukaszek-Sołek, A.; Krawczyk, J.; Śleboda, T.; Grelowski, J. Optimization of the hot forging parameters for 4340 steel by processing maps. *J.Mater. Res. Technol.* **2019**, *8*, 3281–3290. [CrossRef]

28. Fu, X.-Y.; Bai, P.-C.; Yang, J.-C. Hot Deformation Characteristics of 18Cr-5Ni-4Cu-N Stainless Steel Using Constitutive Equation and Processing Map. *Metals* **2020**, *10*, 82. [CrossRef]

29. Jeong, H.T.; Park, H.K.; Park, K.; Na, T.W.; Kim, W.J. High-temperature deformation mechanisms and processing maps of equiatomic CoCrFeMnNi high-entropy alloy. *Mater. Sci. Eng. A* **2019**, *756*, 528–537. [CrossRef]

30. Kim, W.J.; Jeong, H.T.; Park, H.K.; Park, K.; Na, T.W.; Choi, E. The effect of Al to high-temperature deformation mechanisms and processing maps of Al0.5CoCrFeMnNi high entropy alloy. *J. Alloys Compd.* **2019**, *802*, 152–165. [CrossRef]

31. Zhang, Y.; Liu, P.; Tian, B.; Liu, Y.; Li, R.; Xu, Q. Hot deformation behavior and processing map of Cu-Ni-Si-P alloy. *Trans. Nonferrous Met. Soc. China* **2013**, *23*, 2341–2347. [CrossRef]

32. Lin, Y.C.; Nong, F.-Q.; Chen, X.-M.; Chen, D.-D.; Chen, M.-S. Microstructural evolution and constitutive models to predict hot deformation behaviors of a nickel-based superalloy. *Vacuum* **2017**, *137*, 104–114. [CrossRef]

33. Quan, G.; Tong, Y.; Luo, G.; Zhou, J. A characterization for the flow behavior of 42CrMo steel. *Comput. Mater. Sci.* **2010**, *50*, 167–171. [CrossRef]

34. Zener, C.; Hollomon, J.H. Effect of Strain Rate Upon Plastic Flow of Steel. *J. Appl. Phys.* **1944**, *15*, 22–32. [CrossRef]

35. Lin, Y.C.; Chen, M.-S.; Zhang, J. Modeling of flow stress of 42CrMo steel under hot compression. *Mater. Sci. Eng. A* **2009**, *499*, 88–92. [CrossRef]

36. Lin, Y.C.; Xia, Y.-C.; Chen, X.-M.; Chen, M.-S. Constitutive descriptions for hot compressed 2124-T851 aluminum alloy over a wide range of temperature and strain rate. *Comput. Mater. Sci.* **2010**, *50*, 227–233. [CrossRef]

37. Prasad, Y.V.R.K.; Gegel, H.L.; Doraivelu, S.M.; Malas, J.C.; Morgan, J.T.; Lark, K.A.; Barker, D.R. Modeling of dynamic material behavior in hot deformation: Forging of Ti-6242. *Metall. Trans. A* **1984**, *15*, 1883–1892. [CrossRef]

38. Prasad, Y.V.R.K.; Rao, K.P. Processing maps and rate controlling mechanisms of hot deformation of electrolytic tough pitch copper in the temperature range 300–950 °C. *Mater. Sci. Eng. A* **2005**, *391*, 141–150. [CrossRef]

39. Murty, S.V.S.N.; Rao, B.N. On the development of instability criteria during hotworking with reference to IN 718. *Mater. Sci. Eng. A* **1998**, *254*, 76–82. [CrossRef]

40. Long, S.; Xia, Y.-F.; Hu, J.-C.; Zhang, J.-S.; Zhou, J.; Zhang, P.; Cui, M.-L. Hot deformation behavior and microstructure evolution of Ti-6Cr-5Mo-5V-4Al alloy during hot compression. *Vacuum* **2019**, *160*, 171–180. [CrossRef]

 *metals*

Article

New Developments in Understanding Harper–Dorn, Five-Power Law Creep and Power-Law Breakdown

Michael E. Kassner

Department of Chemical Engineering and Materials Science, University of Southern California, Los Angeles, CA 90089, USA; kassner@usc.edu

Received: 1 September 2020; Accepted: 23 September 2020; Published: 25 September 2020

Abstract: This paper discusses recent developments in creep, over a wide range of temperature, that may change our understanding of creep. The five-power law creep exponent (3.5–7) has never been explained in fundamental terms. The best the scientific community has done is to develop a natural three power-law creep equation that falls short of rationalizing the higher stress exponents that are typically five. This inability has persisted for many decades. Computational work examining the stress-dependence of the climb rate of edge dislocations may rationalize the phenomenological creep equations. Harper–Dorn creep, "discovered" over 60 years ago, has been immersed in controversy. Some investigators have insisted that a stress exponent of one is reasonable. Others believe that the observation of a stress exponent of one is a consequence of dislocation network frustration. Others believe the stress exponent is artificial due to the inclusion of restoration mechanisms, such as dynamic recrystallization or grain growth that is not of any consequence in the five power-law regime. Also, the experiments in the Harper–Dorn regime, which accumulate strain very slowly (sometimes over a year), may not have attained a true steady state. New theories suggest that the absence or presence of Harper–Dorn may be a consequence of the initial dislocation density. Novel experimental work suggests that power-law breakdown may be a consequence of a supersaturation of vacancies which increase self-diffusion.

Keywords: creep; Harper-Dorn; power-law-breakdown

---

1. Introduction

This paper will review recent work that appears to allow a better understanding of the basis of elevated temperature creep in single phase ceramics, minerals, metals, and class M (pure metal behavior) alloys. This paper does not directly address multiphase materials such as superalloys and dispersion-strengthened alloys. Elevated temperature creep can be described by an examination of Figure 1 for high purity aluminum. This data was compiled by Blum [1] and represents trends in steady-state creep behavior that are widely accepted. This plot describes steady-state creep where hardening processes are believed to be exclusively balanced by dynamic recovery. Figure 1 shows three regions of steady-state creep behavior ranging from:

(1) Temperatures near the melting point with a slope of 1 (stress exponent of 1), (HD).
(2) Temperatures above approximately $0.6T_m$ where the constant slopes for crystalline materials vary from 3.5–7 (5-PL).
(3) Below about $0.6T_m$, the stress exponent is no longer constant and power-law breakdown is observed (PLB) This is sometime referred to as intermediate temperature creep that extends to roughly $0.3T_m$.

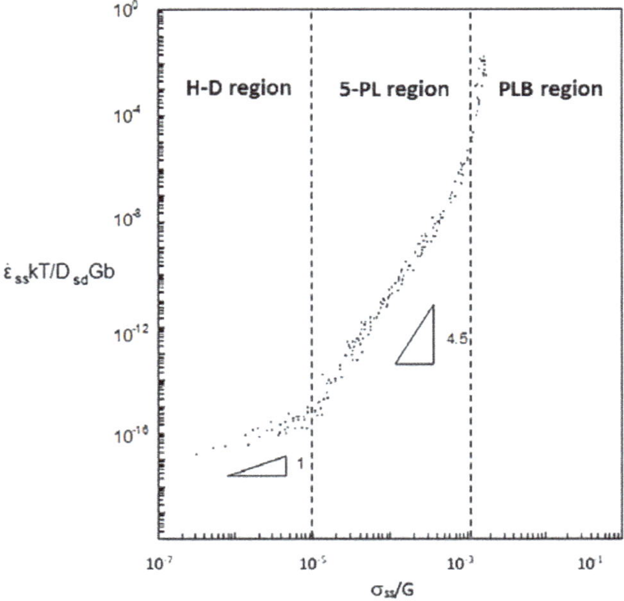

Figure 1. The compensated steady-state strain-rate versus the modulus-compensated steady-state stress for 99.999 pure Al, based on [1].

The point here is that each of these regions have been poorly understood in the past. More recent work appears to offer clarification of the creep mechanisms for each of these regions. Each section will be described separately.

2. General Creep Plasticity Considerations and Five Power-Law Creep

Before each section is described, there are some general features that Harper–Dorn and five power-law creep (and perhaps PLB) have in common. First, there are three microstructural dislocation features that are often evident in the three creep regimes. These are the Frank network dislocation density, subgrains that may or may not be present, and the misorientation across boundaries (related to the spacing of the dislocations in the subgrain boundaries). Work by the author [2–5] showed that the rate-controlling process for plasticity in these regimes involved the Frank networks. Creep in both regimes can be described without consideration of the presence of subgrains. Several other investigators, e.g., Evans and Knowles [6], Kumar et al. [7], Northwood et al. [8], Ardell [9], Lagneborg et al. [10], Burton [11], and many others also considered Frank networks to be the microstructural feature associated with the rate-controlling process for elevated-temperature creep. The author further described a generalized equation based on the network model of Evans and Knowles [6]. This will turn out to be relevant to both the Harper–Dorn regime and the five power-law regime and possibly the power-law breakdown (PLB). The author wrote a detailed derivation of the network model [7]. The derivation is lengthy and only the highlights will be presented here:

First the climb velocity, v_c, must be calculated and the original expression comes from Weertman [12]. v_c is determined from the vacancy concentration gradient influenced by the fact that the formation energy for a vacancy is altered by the climb (applied) stress.

$$v_c = 2\pi D_L \cdot [(\Omega\sigma/kT)] \cdot \ln(R_o/b) \tag{1}$$

where Ω is the atomic volume, D_L is the lattice self-diffusion coefficient, and R_o is the diffusion distance. The activation energy for both five-power-law creep (5PL) and low stress creep (so-called Harper–Dorn (HD) creep) is that for lattice self-diffusion suggesting that in both regimes, edge dislocation climb is rate-controlling (in PLB, the activation energy appears to decrease from that of the 5-PL and HD regimes [13,14]). Note in particular that the stress exponent = 1 in Equation (1).

Weertman suggests that the approximate average dislocation velocity is:

$$v' \sim v_c x_g / x_c \tag{2}$$

where x_c is the climb distance and x_g is the glide distance. Also,

$$\dot{\varepsilon}_{ss} = \rho_m b v' \tag{3}$$

where ρ_m is the mobile dislocation density. It has been phenomenologically observed that frequently in the five-power regime,

$$\rho = (\sigma_{ss}/Gb)^p \tag{4}$$

where ρ is the total dislocation density. This is not the Taylor equation although it is frequently claimed such since many believe that p is about 2.0. Actually, the proper strengthening equation is:

$$\sigma_y\big|_{\dot{\varepsilon},T} = \sigma_0' + \alpha MGb(\rho)^{\frac{1}{2}} \tag{5}$$

where σ_0' is the strengthening effect of solutes/impurities, Peierls stress, etc., and α is a constant and varies from 0.2 to 0.4 among metals. It should be pointed out that, even in high purity Al (99.999% pure), σ_0' can be a significant (e.g., over half) contributor to elevated temperature strength. That is, the strength at a fixed temperature and strain rate is the sum of an athermal (dislocation hardening) and the thermally activated term σ_0'.

Weertman suggested that the total dislocation density is equal to the mobile dislocation density (probably a bit unrealistic) which leads to:

$$\dot{\varepsilon}_{ss} = K \cdot D_{sd}/bG\Omega/kT\left(x_g/x_c\right) \cdot (\sigma/G)^3 \tag{6}$$

which is often referred to as the "natural three power law" equation. The problem is that the stress exponents in the "five-power law regime" are 3.5–7 and typically five. The exponent for high purity aluminum in Figure 1 is about 4.5. The creep plasticity community has not been able to derive an equation for five-power law creep and this has been a deficiency for a very long time. There have been mathematical gymnastics performed to derive a higher stress exponent (including Weertman [15]), but these have not been embraced by the community, and the natural three power law creep equations have persisted [16].The author also showed that in both Cu and Al, long-range internal stresses are not evident and the only the applied stress is relevant [17].

3. The So-Called Harper–Dorn Regime

Blum's plot in Figure 1 also suggests that at very low stresses (often in association with high temperatures) a new creep regime has been reported (over 60 years ago) that was termed Harper–Dorn (HD) creep [18]:

$$\dot{\varepsilon}_{ss} = A_{HD}\left(\frac{D_{sd}Gb}{kT}\right)\left(\frac{\sigma}{G}\right)^1 \tag{7}$$

where A_{HD} is a constant and the other terms have their usual meanings. Many investigators observed that unlike the higher stress five-power law regime, the steady-state dislocation density is constant in the so-called Harper–Dorn regime. As pointed out in a review by the author [19], the power of one has been observed in a variety of materials including Al [5–7,20], Ti [21], Fe [22], Co [23], Zr [24], and Sn [20].

It has also been suggested to be observed in ceramics such as CaO [25], UO_2, [26] MgO [27,28], NaCl [29,30], and also ice [31]. In the case of Al and Pb [20], a failure to observe Harper–Dorn occurred, and five power-law was observed instead at stresses associated Harper–Dorn observations. The early low-stress experiments, primarily of metals, indicated that Harper–Dorn includes:

1. Activation energy about equal to lattice self-diffusion,
2. Grain-size independence with grain boundary shearing,
3. Steady-state stress exponent of one,
4. Dislocation density that appears independent of stress, and
5. A primary creep stage.

Items 2 and 5 were considered important as these preclude Nabarro-Herring diffusional creep [3] (which may not exist). Nabarro-Herring does not have a primary creep stage and must have a grain size dependence. There have been several theories to justify Equation (7) [32–36] and all, basically, suggest that, since ρ is constant, only dislocation climb affects the stress exponent with a value of one (Equation (1)). Interestingly, more recent work by the author and others [37–39] indicates that at least under some conditions, the stress exponent is larger than one, as in Figure 2. The author's work suggest that this is a consequence of a non-constant dislocation density. In fact, for Al single crystals deformed in the so-called Harper–Dorn regime, the exponent "p" is roughly two in Equation (4) giving rise to a stress exponent of about three. It is possible that, in some cases at high (Harper–Dorn) temperatures and/or low stresses, the network dislocation density is not constant as widely presumed.

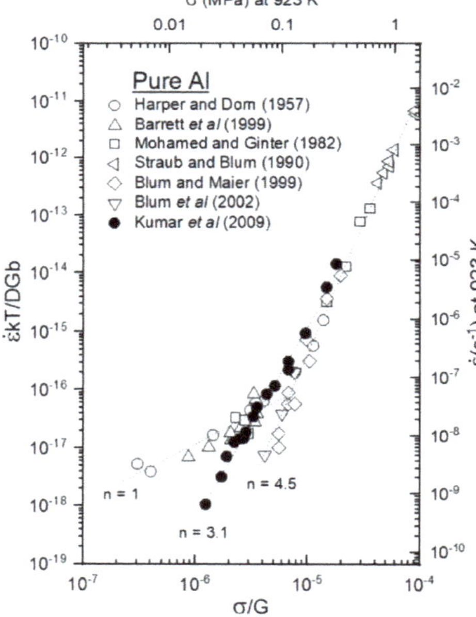

Figure 2. A comparative presentation of the main studies on pure Al at very low stresses and at temperatures close of $0.99T_m$. G is the shear modulus and b is the Burgers vector. D is the lattice self-diffusion coefficient.

The author speculated [39,40] that variation in the stress-exponent behavior in the Harper–Dorn regime might be explained by the value of the starting dislocation density in the material (just before the application of a stress) and the observation that even after annealing single crystals for a year, the dislocation density remains a constant; it reaches a "frustration" value (ρ_f). There is no quantitatively way to predict this value at this time. Further annealing cannot decrease below the value of ρ_f even after annealing at very high temperatures for over a year in the three materials examined, LiF, NaCl and Al, all single crystals.

The frustration dislocation density is fairly low and initial values of ρ just before applying the stress at elevated may be higher than the frustration value as illustrated in Figure 3. But also, as the author discovered [41] for Al, the initial dislocation density can be lower than ρ_f. Nes and Nost [42] discovered that very slow cooling of Al from the melt can produce dislocation densities that are especially low, significantly below the frustration dislocation density. In contrast, Figure 4 shows that the initial dislocation density is relatively high for LiF. Nes and Nost [42] also found a stress-dependency of the dislocation density in the Harper–Dorn regime. Dislocation density measurements using X-ray topography showed that ρ was dependent on stress by $\sigma^{1.3}$. Thus, in two cases [41,42], it appears that if the starting dislocation density is below ρ_f, then Harper–Dorn creep may not be observed and that (regular or normal) five-power law creep is observed instead. Blum et al. [38], also observed five-power law creep in Al, but did not determine the initial dislocation density. Additional experiments were performed by McNee and coworkers [40] on 99.999 pure Al. These investigators, consistent with Blum et al. and the author's work [38,41], observed the extension of five-power-law creep into Harper–Dorn regime.

Most recently, Singh et al. [39] were the first to discover that low stress/high-temperature creep in LiF is consistent with Harper–Dorn creep (i.e., Equation (7)). Work on Pb by Mohammed et al. [20] also found that this metal does not evince Harper–Dorn creep. Dislocation densities were not measured, see Figure 3. Plots the dislocation density of Al deformed to steady state both within the five-power law and the Harper–Dorn regime.

Mohammed et al. [43] reported that the microstructure of the 99.9995 Al includes wavy grain boundaries, an inhomogeneous dislocation density distribution as determined by etch pits, small new grains forming at the specimen surface and large dislocation density gradients across grain boundaries. Well-defined subgrains were not observed. However, they found that the microstructure of the deformed 99.99 Al consists of a well-defined array of subgrains and exhibits five power law behavior. These observations led Mohammed et al. to conclude that the restoration mechanism taking place during so-called "Harper–Dorn creep" includes discontinuous dynamic recrystallization (DRX) in addition to dynamic recovery, at least in Al. This restoration mechanism could give rise to the periodic accelerations observed in the creep curve. It was suggested that it is difficult to accurately determine the stress exponent due to the appearance of periodic accelerations in the creep curves. However, Mohammed et al. [43] claimed that $n = 1$ exponents are only obtained if creep curves up to small strains (1–2%) are analyzed, as was typically the case in the past. Mohamed et al. estimated stress exponents of about 2.5 at larger strains for high purity polycrystalline DRX specimens of Al. Also, Singh et al. [25] in 99.999% pure LiF single crystals observed a power-1.5 behavior [14] in the HD regime. The exponent was independent of strain and the dislocation density is constant for a fixed stress for a period of over one year.

It is important to utilize single crystals in HD experiments as grain boundary migration can be an independent restoration mechanism [44]. Care must also be taken to ensure that discontinuous dynamic recrystallization is not occurring in Al as DRX may occur even at ambient temperature albeit at relatively large strains. [45,46]. It is possible that Mohammed et al. observed grain growth as an additional restoration mechanism rather than DRX [44]. Kumar et al. [41] never observed DRX nor subgrains in Al single crystals deformed in the HD regime and observed a stress exponent slightly over 3. Power-one is legitimate in some cases even in the absence of restoration mechanisms that are not significant in five power-law creep where steady-state is a balance between dynamic recovery and hardening. It should be pointed out that McNee et al. [47] also tested Al of both 99.999 and 99.99%purity and did not observe Harper–Dorn behavior with either purity. It should be mentioned that Blum et al. [38] did not confirm Harper–Dorn creep in 99.99% pure oligo-crystal aluminum. Rather they observed normal five-power law creep extend to very low stresses within the HD regime [38]. The starting dislocation density was not measured. According to Blum et al., one explanation for others not observing stress exponents of one (Harper–Dorn creep) is that there were insufficient strains to reach steady-state.

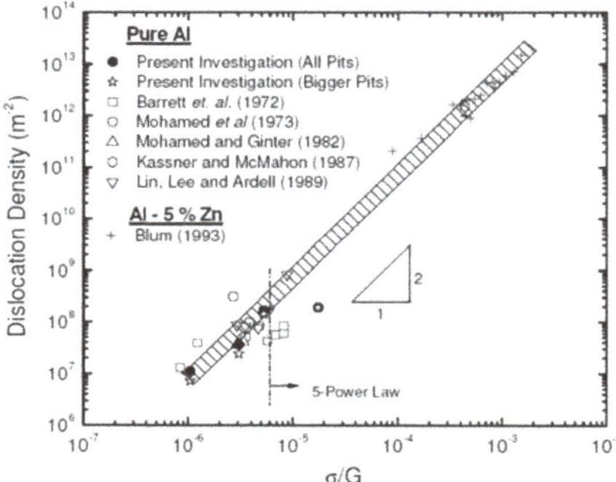

Figure 3. The stress dependence of the dislocation density in Al is measured by etch-pits at lower stresses and transmission electron microscopy (TEM) at higher stresses. The dislocation density does not always remain constant in the low-stress (HD) regime. The dislocation density increases with stress and it is consistent with the projected values based on 5-power-law trends.

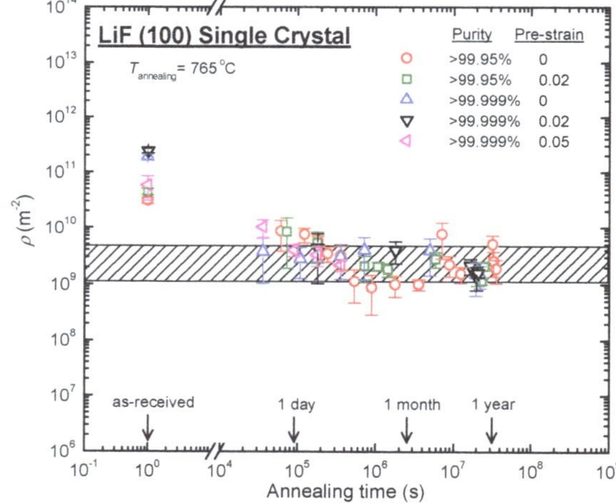

Figure 4. Variation of the network dislocation density, as measured using etch-pit method, as a function of static annealing time.

Figure 5 demonstrates that whenever the starting dislocation density is at or greater than ρ_f, then substantial hardening may not be observed as the dislocation density reduces to a frustration value (ρ_f) and a low (e.g., 1–2) stress exponent is observed. The as-grown material will recover some of the dislocation density before the application of the stress once the as-received material equilibrates to the test temperature. However, if the dislocation density is low then hardening can occur and (normal) five-power law behavior is observed. The basis for dislocation network frustration is unclear. Certainly, further decreases from ρ_f is thermodynamically favored. Ardell [9] proposed that coarsening of the network requires that Frank's rule is satisfied at the nodes. The sum of the Burgers vectors must always sum to zero. Ardell suggested that frustration occurs when further coarsening is impossible when Frank's rule can no longer be satisfied.

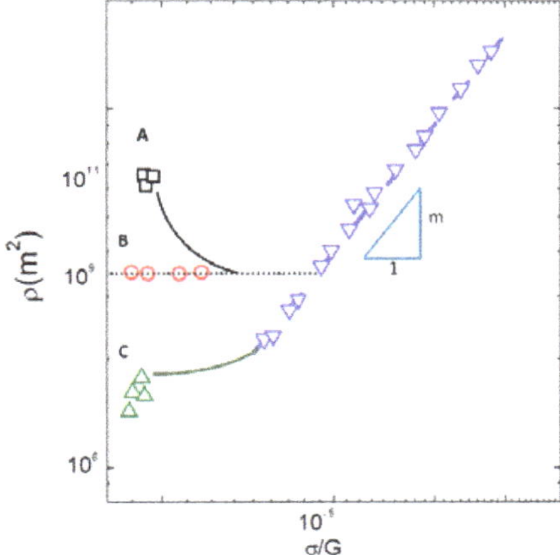

Figure 5. Schematic representation for the observed dislocation density behavior in the low-stress regime based on the initial dislocation densities. In "A", the initial dislocation density is relatively high; it is higher than the frustration dislocation density marked by the dotted line at "B". If the dislocation density starts at "A" (which would be the dislocation density on the application of the stress in the creep test and perhaps lower than the "as received" dislocation density) it would decrease to "B" and subsequently remain constant. This would give rise the Harper-Dorn creep with a stress exponent of one. With a starting dislocation density of "C", on the other hand, the dislocation density is relatively low (less than the frustration density for higher initial dislocation density values such as "A") on the application of the stress, then the dislocation density will continually increase with increasing stress and five-power-law behavior may be observed. The exponent "m" is the steady-state strain rate sensitivity which is about five.

The LiF data is illustrated in Figure 6. It can be noted that the plot suggests a stress exponent of about 1.6, as mentioned earlier, in the HD regime (which is greater than the value of one in Equation (7)). Interestingly, other materials such as Cu and CaO [25] deformed in the HD regime also have values closer to 2 than 1. Recent work by Mompiou et al. and Kabir et al. [48,49] suggested that the climb velocity dependence on the stress has a stress exponent greater than 1 based on computational modeling of the climb process. Thus, there may be inaccuracy in Equation (1). The stress exponent of 1 in Equation (7) is a consequence of the effect of the applied stress on the dislocation climb-rate based on the analytic projection by Weertman. The deviation of the stress exponent may imply that the stress exponent for five power-law creep and Harper–Dorn creep (constant dislocation density with changes in stress) could be both higher than 3 (Figure 1) and 1, respectively. It should be mentioned that Harper and Dorn did not use the applied stress in obtaining a stress exponent of one for polycrystalline Al. Rather they subtracted a surface tension stress from the applied stress. The extrapolation of the stress to zero strain-rate results in a positive stress. Another investigation did not observe this. Had Harper and Dorn not subtracted this stress, a stress exponent of three would have been observed. Overall, it appears that Harper–Dorn (stress exponents of one) is not commonly observed in Al as is widely believed.

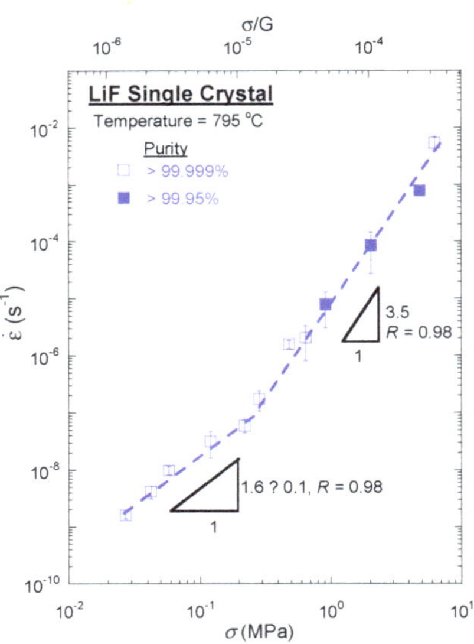

Figure 6. Variation of the steady-state strain rate as function of the true stress in LiF.

4. Power law Breakdown

There have been various explanations for PLB [50–54]: (1) It has been proposed that there is a change in the rate-controlling mechanism of steady-state plastic flow from dislocation climb-control in 5-PL to glide-control in PLB. However, this proposition is sometimes based on the presence of internal stress which does not appear to be reasonable as the long-range internal stresses (usually suggested to give rise to high local stresses at dislocation heterogeneities) appears to be low in both the PL and PLB regimes [17]. (2) Only recently has vacancy supersaturation from plasticity been experimentally verified in the PLB regime. The levels of supersaturation appear sufficient to create higher creep-rates, and thus PLB. The excess vacancies would decrease the activation energy for diffusion, as observed [13,14]. (3) The changes in the diffusion coefficient from D_{sd} to D_p (dislocation pipe diffusion coefficient) with large strain plasticity within the PLB regime may independently contribute to the observation of PLB [14]. Another point must be made in that references [55,56] experimentally verify vacancy supersaturation at ambient temperature. Other, somewhat higher temperatures, within PLB were not checked. Of course, higher temperature X-ray diffraction experiments are more difficult than those at ambient temperature. Perhaps future experiments could be performed at other temperatures within PLB to fully verify the coincidence between the onset of PLB and vacancy supersaturation.

5. Conclusions

Power-law, power-law-breakdown, and the so-called Harper–Dorn creep have been re-examined in terms of some of the latest developments within these creep regimes. It appears that within the Harper–Dorn regime, stress exponents that are higher than the expected value of one can be observed depending on the initial dislocation density. Very low starting dislocation densities may yield stress exponents closer to those observed in normal five power law creep. The five power-law creep exponent may be finally explained by incorporating more realistic dislocation climb-rate stress-exponents that may exceed a value of one according to recent computational work. Recent X-ray work assessing the vacancy concentration in plastically deformed materials suggests that power-law breakdown may be explained by vacancy supersaturation based on moving jogs leading to enhanced creep rates.

Funding: This research received no external funding.

Acknowledgments: The author wishes to thank the Choong Hoon Cho Chair at the University of Southern California for the support in writing this article.

Conflicts of Interest: The author declares no conflict of interest.

References

1. Straub, S.; Blum, W. Does the 'Natural' Third Power Law of Steady-state Hold for Pure Aluminum? *Scr. Metall. Mater.* **1990**, *24*, 837–1842. [CrossRef]
2. Kassner, M.E. A Case for Taylor Hardening During Primary and Steady-State Creep in Aluminum and Type 304 Stainless Steel. *J. Mater. Sci.* **1990**, *25*, 1997–2003. [CrossRef]
3. Kassner, M. *Fundamentals of Creep in Metals and Alloys*; Elsevier: Amsterdam, The Netherlands, 2015; pp. 1–338.
4. Kassner, M.E. Taylor hardening in five-power-law creep of metals and Class M alloys. *Acta Mater.* **2004**, *52*, 1–9. [CrossRef]
5. Kassner, M.E.; Miller, A.K.; Sherby, O.D. The Separate Roles of Forest Dislocations and Subgrains in the Isotropic Hardening of Type 304 Stainless Steel. *Metall. Trans.* **1982**, *13*, 1977–1986. [CrossRef]
6. Evans, H.; Knowles, G. A model of creep in pure materials. *Acta Metall.* **1977**, *25*, 963–975. [CrossRef]
7. Kumar, P.; Kassner, M.E. Theory for very low stress ("Harper-Dorn") creep. *Scr. Mater.* **2009**, *60*, 60–63. [CrossRef]
8. Ostrom, P.; Lagneborg, R. A Dislocation Link Length Model for Creep. *Res. Mech.* **1980**, *1*, 59–79.
9. Ardell, A.J. Harper–Dorn creep–The dislocation network theory revisited. *Scr. Metall.* **2013**, *69*, 541–544. [CrossRef]
10. Shi, L.; Northwood, D. Strain-hardening and recovery during the creep of pure polycrystalline magnesium. *Acta Metall. Mater.* **1994**, *42*, 871–877. [CrossRef]
11. Burton, B. The Low Stress Creep of Aluminum Near the Melting Point: Oxidation and Substructural Changes. *Phil. Mag.* **1972**, *25*, 645–659. [CrossRef]
12. Weertman, J. Theory for Steady-state Creep Based on Dislocation Climb. *J. App. Phys.* **1955**, *26*, 1213–1217. [CrossRef]
13. Kassner, M.E. The rate dependence and microstructure of high-purity silver deformed to large strains between 0.16 and 0.30T m. *Metall. Mater. Trans. A* **1989**, *20*, 2001–2010. [CrossRef]
14. Lüthy, H.; Miller, A.; Sherby, O. The stress and temperature dependence of steady-state flow at intermediate temperatures for pure polycrystalline aluminum. *Acta Metall.* **1980**, *28*, 169–178. [CrossRef]
15. Weertman, J. High Temperature Creep Produced by Dislocation Motion. In *Rate Processes in Plastic Deformation of Materials*; Li, J.C.M., Mukherjee, A.K., Eds.; American Society for Metals—Metals Park: Novelty, OH, USA, 1975; pp. 315–336.
16. Biberger, M.; Blum, W. On the natural law of steady state creep. *Scr. Metall.* **1989**, *23*, 1419–1424. [CrossRef]
17. Kassner, M.E.; Geantil, P.; Levine, L. Long range internal stresses in single-phase crystalline materials. *Int. J. Plast.* **2013**, *45*, 44–60. [CrossRef]
18. Harper, J.; Dorn, J.E. Viscous Creep of Aluminum Near Its Melting Temperature. Theory of Steady-State Creep Based on Dislocation Climb. *Acta Metall.* **1957**, *5*, 654–665. [CrossRef]
19. Kassner, M.E.; Kumar, P.; Blum, W. Harper-Dorn Creep. *Int. J. Plast.* **2007**, *23*, 980–1000. [CrossRef]
20. Mohamed, F.A.; Murty, K.L.; Morris, J.W. Harper-dorn creep in al, pb, and sn. *Metall. Mater. Trans. A* **1973**, *4*, 935–940. [CrossRef]
21. Malakondaiah, G.; Rao, P.R. Creep of Alpha-Titanium at low stresses. *Acta Metall.* **1981**, *29*, 1263–1275. [CrossRef]
22. Fiala, J.; Novotny, J.; Cadek, J. Coble and Harper-Dorn Creep in Iron at Homologous Temperatures T/Tm of 0.40-0.54. *Mater. Sci. Eng. A* **1983**, *60*, 195–206. [CrossRef]
23. Malakondaiah, G.; Rao, P.R. Viscous creep of β-Co. *Mater. Sci. Eng.* **1982**, *52*, 207–221. [CrossRef]
24. Novotny, J.; Fiala, J.; Čadek, J. Harper-Dorn creep in alpha-zirconium. *Acta Metall.* **1985**, *33*, 905–911. [CrossRef]
25. Dixon-Stubbs, P.J.; Wilshire, B. Deformation processes during creep of single and polycrystalline CaO. *Philos. Mag. A* **1982**, *45*, 519–529. [CrossRef]

26. Ruano, O.A.; Wolfenstine, J.; Wadsworth, J.; Sherby, O. Harper-Dorn and power law creep in uranium dioxide. *Acta Metall. Mater.* **1991**, *39*, 661–668. [CrossRef]
27. Ruano, O.A.; Wolfenstine, J.; Wadsworth, J.; Sherby, O.D. Harper-Dorn and Power-Law Creep in Single-Crystalline Magnesium Oxide. *J. Am. Ceram. Soc.* **1992**, *75*, 1737–1741. [CrossRef]
28. Ramesh, K.S.; Yasuda, E.; Kimura, S. Negative creep and recovery during high-temperature creep of MgO single crystals at low stresses. *J. Mater. Sci.* **1986**, *21*, 3147–3152. [CrossRef]
29. Banerdt, W.; Sammis, C. Low stress high temperature creep in single crystal NaCl. *Phys. Earth Planet. Inter.* **1985**, *41*, 108–124. [CrossRef]
30. Wolfenstine, J.; Ruano, O.A.; Wadsworth, J.; Sherby, O. Harper-dorn creep in single crystalline NaCl. *Scr. Metall. Mater.* **1991**, *25*, 2065–2070. [CrossRef]
31. Wang, J.N. Harper-Dorn creep in single crystals of lead, rutile and ice. *Philos. Mag. Lett.* **1994**, *70*, 81–85. [CrossRef]
32. Barrett, C.; Nix, W. A model for steady state creep based on the motion of jogged screw dislocations. *Acta Metall.* **1965**, *13*, 1247–1258. [CrossRef]
33. Friedel, J. *Dislocations*; Pergamon Press: Oxford, MS, USA, 1964.
34. Yavari, P.; Miller, D.A.; Langdon, T.G. An Investigation of Harper-Dorn Creep–I. Mechanical and Microstructural Characteristics. *Acta Metall.* **1982**, *30*, 871–879.
35. Mohamed, F.A.; Ginter, T.J. On the nature and origin of Harper-Dorn creep. *Acta Metall.* **1982**, *30*, 1869–1881. [CrossRef]
36. Ardell, A.; Przystupa, M. Dislocation link-length statistics and elevated temperature deformation of crystals. *Mech. Mater.* **1984**, *3*, 319–332. [CrossRef]
37. Kassner, M.E. Recent Developments in Understanding the Creep of Aluminum. *Mater. Phys. Mech.* **2018**, *40*, 1–6. [CrossRef]
38. Blum, W.; Maier, W. Harper-Dorn Creep—A Myth? *Phys. Stat. Sol. A* **1999**, *171*, 467–474.
39. Singh, S.P.; Kumar, P.; Kassner, M.E. The Low-Stress and High-Temperature Creep in LiF Single Crystals: An Explanation for the So-called Harper-Dorn Creep. *Materialia* **2020**, *13*, 100864. [CrossRef]
40. Singh, S.; Huang, X.-R.; Kumar, P.; Kassner, M.E. Short communication: Assessment of the dislocation density using X-ray topography in Al single crystals annealed for long times near the melting temperature. *Mater. Today Commun.* **2019**, *21*, 100613. [CrossRef]
41. Kumar, P.; Kassner, M.E.; Blum, W.; Eisenlohr, P.; Langdon, T.G. New observations on high-temperature creep at very low stresses. *Mater. Sci. Eng. A* **2009**, *510*, 20–24. [CrossRef]
42. Nes, E.; Nost, B. Dislocation Densities in Slowly Cooled Aluminum Single Crystals. *Phil. Mag.* **1966**, *124*, 855–865.
43. Mohammed, F.A. The Role of Impurities during Creep and Superplasticity at Very Low Stresses. *Metall. Mater. Trans. A* **2002**, *33*, 261–278.
44. McQueen, H.; Blum, W.; Straub, S.; Kassner, M. Dynamic grain growth a restoration mechanism in 99.999 Al. *Scr. Metall. Mater.* **1993**, *28*, 1299–1304. [CrossRef]
45. Kassner, M.; Pollard, J.; Evangelista, E.; Cerri, E. Restoration mechanisms in large-strain deformation of high purity aluminum at ambient temperature and the determination of the existence of a Steady State. *Acta Metall. Mater.* **1994**, *42*, 3223–3230. [CrossRef]
46. Kassner, M.; McQueen, H.; Pollard, J.; Evangelista, E.; Cerri, E. Restoration mechanisms in large-strain deformation of high purity aluminum at ambient temperature. *Scr. Metall. Mater.* **1994**, *31*, 1331–1336. [CrossRef]
47. McNee, K.R.; Jones, H.; Greenwood, G.W. Low Stress Creep of Aluminum at Temperatures near to Tm. In *Creep and Fracture of Engineering Materials and Structures: Proceedings of the 9th International Conference*; Parker, J.D., Ed.; Inst. Metals: London, UK, 2001; p. 3.
48. Mompiou, F.; Caillard, D. On the stress exponent of dislocation climb velocity. *Mater. Sci. Eng. A* **2008**, *483*, 143–147. [CrossRef]
49. Kabir, M.; Lau, T.T.; Rodney, D.; Yip, S.; Van Vliet, K.J. Predicting Dislocation Climb and Creep from Explicit Atomistic Details. *Phys. Rev. Lett.* **2010**, *105*, 095501. [CrossRef] [PubMed]
50. Kassner, M.E.; Ermagan, R. Power Law Breakdown in the Creep of Metals and Alloys. *Metals* **2019**, *9*, 1345. [CrossRef]

51. Wu, M.Y.; Sherby, O.D. Unification of Harper-Dorn and power law creep through consideration of internal stress. *Acta Metall.* **1984**, *32*, 1561–1572. [CrossRef]
52. Arieli, A.; Mukherjee, A.K. *Creep and Fracture of Engineering Materials and Structures*; Wilshire, B., Owen, D.R.J., Eds.; Pineridge Press: Swansea, UK, 1981; pp. 97–111.
53. Weertman, J.; Weertman, J.R. *Physical Metallurgy*, 3rd ed.; Cahn, R.W., Haasen, P., Eds.; North Holland: Amsterdam, The Netherlands, 1983; pp. 159–1307.
54. Nix, W.; Ilschner, B. Mechanisms Controlling Creep of Single Phase Metals and Alloys. In *Strength of Metal and Alloys*; Haasen, P., Gerold, V., Kostorz, G., Eds.; Pergamon: Aachen, Germany, 1980; pp. 1503–1530.
55. Ungár, T.; Schafler, E.; Hanák, P.; Bernstorff, S.; Zehetbauer, M. Vacancy production during plastic deformation in copper determined by in situ X-ray diffraction. *Mater. Sci. Eng. A* **2007**, *462*, 398–401. [CrossRef]
56. Čížek, J.; Janeček, M.; Vlasák, T.; Smola, B.; Melikhova, O.; Islamgaliev, R.; Dobatkin, S. The Development of Vacancies during Severe Plastic Deformation. *Mater. Trans.* **2019**, *60*, 1533–1542. [CrossRef]

MDPI

St. Alban-Anlage 66

4052 Basel

Switzerland

Tel. +41 61 683 77 34

Fax +41 61 302 89 18

www.mdpi.com

Metals Editorial Office

E-mail: metals@mdpi.com

www.mdpi.com/journal/metals